Comprehensive Asymmetric Catalysis
Supplement 1

Springer

Berlin
Heidelberg
New York
Hong Kong
London
Paris
Tokyo

Eric N. Jacobsen · Andreas Pfaltz · Hisashi Yamamoto (Eds.)

Comprehensive Asymmetric Catalysis

Supplement 1

With contributions by numerous experts

 Springer

ERIC N. JACOBSEN
Department of Chemistry and Chemical Biology
Harvard University
12 Oxford Street
MA 02138 Cambridge, USA
e-mail: jacobsen@chemistry.harvard.edu

ANDREAS PFALTZ
Department of Chemistry
University of Basel
St. Johanns-Ring 19
CH-4056 Basel, Switzerland
e-mail: andreas.pfaltz@unibas.ch

HISASHI YAMAMOTO
Department of Chemistry
University of Chicago
5735 South Ellis Avenue
Chicago, IL 60637, USA
e-mail: yamamoto@uchicago.edu

ISBN 3-540-00333-9 Springer-Verlag Berlin Heidelberg New York

Cataloging-in-Publication Data applied for
Bibliographic information published by Die Deutsche Bibliothek
Die Deutsche Bibliothek lists this publication in the Deutsche Nationalbibliografie;
detailed bibliographic data is available in the Internet at <http:/dnb.ddb.de>.

Springer-Verlag Berlin Heidelberg New York
a member of BertelsmannSpringer Science + Business Media GmbH

http:/www.springer.de

© Springer-Verlag Berlin Heidelberg 2004
Printed in Germany

Typesetting: Data conversion by medio Technologies AG, Berlin
Cover: E. Kirchner, Heidelberg

Printed on acid-free paper   62/3020xv   5 4 3 2 1 0

Authors

Huw. M. L. Davies
Department of Chemistry
University at Buffalo
The State University of New York
Buffalo, NY 14260-3000
USA
e-mail: hdavies@acsu.buffalo.edu

Keiko Hatanaka
Graduate School of Engineering
Nagoya University
Chikusa, nagoya 464-8603
JAPAN

Amir H. Hoveyda
Department of Chemistry
Merkert Chemistry Center
Boston College
Chestnut Hill, MA 02467
USA
e-mail: amir.hoveyda@bc.edu

David L. Hughes
Merck and Co., inc. Mail drop R80Y-250
Rahway, NJ 07065
USA
e-mail: dave_hughes@merck.com

Harald Gröger
Graduate School of Pharmaceutical Sciences
The University of Tokyo
7-3-1 Hongo
Bunkyo-ku, Tokyo 113-0031
JAPAN

Eric N. Jacobsen
Department of Chemistry and
Chemical Biology
Harvard University
12 Oxford St.
Cambridge, MA 02138
USA
e-mail: jacobsen@chemistry.harvard.edu

Elizabeth R. Jarvo
Department of Chemistry
Merkert Chemistry Center
Boston College
Chestnut Hill, MA 02467-3860
USA

Motomu Kanai
Graduate School of Pharmaceutical Sciences
The University of Tokyo
7-3-1 Hongo
Bunkyo-ku, Tokyo 113-0031
JAPAN

Shū Kobayashi
Graduate School of Pharmaceutical Sciences
The University of Tokyo
Hongo
Bunkyo-ku, Tokyo 113-0033
JAPAN
e-mail: skobayas@mol.f.u-tokyo.ac.jp

Shigeki Matsunaga
Graduate School of Pharmaceutical Sciences
The University of Tokyo
7-3-1 Hongo
Bunkyo-ku, Tokyo 113-0031
JAPAN

Scott J. Miller
Department of Chemistry
Merkert Chemistry Center
Boston College
Chestnut Hill, MA 02467-3860
USA
e-mail: scott.miller.1@bc.edu

Kerry E. Murphy
Department of Chemistry
Merkert Chemistry Center
Boston College
Chestnut Hill, MA 02467
USA

Ryoji Noyori
Department of Chemistry and Research
Center for Materials Science
Nagoya University
Chikusa, Nagoya 464–8602
JAPAN
e-mail: noyori@chem3.chem.nagoya-u.ac.jp

Takeshi Ohkuma
Department of Chemistry and Research
Center for Materials Science
Nagoya University
Chikusa, Nagoya 464–8602
JAPAN

Takashi Ohshima
Graduate School of Pharmaceutical Sciences
The University of Tokyo
7-3-1 Hongo
Bunkyo-ku, Tokyo 113-0031
JAPAN

Richard R. Schrock
Department of Chemistry
Massachusetts Institute of Technology
Cambridge
Massachusetts 02319
USA
e-mail: rrs@mit.edu

Masakatsu Shibasaki
Graduate School of Pharmaceutical Sciences
The University of Tokyo
7-3-1 Hongo
Bunkyo-ku, Tokyo 113-0031
JAPAN
e-mail: mshibasa@mol.f.u-tokyo.ac.jp

Takanori Shibata
Department of Chemistry
School of Science & Engineering
Waseda University
3-4-1 Okubo
Shinjuku-ku, Tokio, 169-8555
JAPAN
e-mail: tshibata@waseda.jp

Kenso Soai
Department of Applied Chemistry
Faculty of Science

Tokyo University of Science
Kagurazaka
Shinjuku-ku, Tokyo 162-8601
JAPAN
e-mail: soai@rs.kagu.tus.ac.jp

Masaharu Ueno
Graduate School of Pharmaceutical Sciences
The University of Tokyo
Hongo
Bunkyo-ku, Tokyo 113-0033
JAPAN

Petr Vachal
Department of Chemistry and Chemical
Biology
Harvard University
12 Oxford St.
Cambridge, MA 02138
USA

Erasmus M. Vogl
Graduate School of Pharmaceutical Sciences
The University of Tokyo
7-3-1 Hongo
Bunkyo-ku, Tokyo 113-0031
JAPAN

Masahiko Yamaguchi
Department of organic Chemistry
Graduate School of Pharmaceutical Sciences
Tohoku University
Aoba, Sendai 980-8578
JAPAN
e-mail: yama@mail.pharm.tohoku.ac.jp

Hisashi Yamamoto
Department of Chemistry
University of Chicago
5735 South Ellis Avenue
Chicago, IL 60637
USA
e-mail: yamamoto@uchicago.edu

Naoki Yoshikawa
Graduate School of Pharmaceutical Sciences
The University of Tokyo
7-3-1 Hongo
Bunkyo-ku, Tokyo 113-0031
JAPAN

Preface

We have been gratified to see that the *Comprehensive Asymmetric Catalysis* three volume set has been received with enthusiasm by the chemical community since its publication in 1999. As was easily anticipated, advances in asymmetric catalysis have continued at an explosive pace since then. Recognition of the impact of this field on chemistry has been evidenced both in practical terms by the application of asymmetric catalytic methods in a variety of new laboratory and industrial contexts, and quite visibly through the lofty recognition of to three of the pioneers of the field in the 2001 Nobel Prize.

In order to keep this reference work as useful and fresh as possible, our plan from the start was to provide supplementary volumes on a periodic basis. These would contain updates to chapters on topics where there is substantial recent progress, and new chapters on emerging topics. We are delighted to provide you here with the first of these supplements.

Eric N. Jacobsen, Cambridge   August 2003
Andreas Pfaltz, Basel
Hisashi Yamamoto, Chicago

Contents

The chapters whose numbers have a gray background are completely new.

Contents (Vol. I–III)

Volume I

Supplement to Chapter 6.1
Hydrogenation of Carbonyl Groups

Takeshi Ohkuma, Ryoji Noyori

Department of Chemistry and Research Center for Materials Science, Nagoya University,
Chikusa, Nagoya 464–8602, Japan
e-mail: noyori@chem3.chem.nagoya-u.ac.jp

Keywords: Asymmetric hydrogenation, Asymmetric transfer hydrogenation, BINAP, Chiral alcohols, Chiral amines, Homogeneous catalysts, Simple ketones

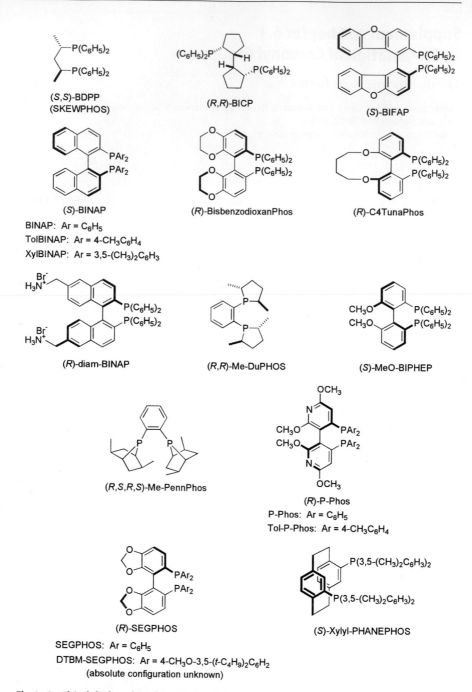

Fig. 1. C_2-Chiral diphosphine ligands (in alphabetical order)

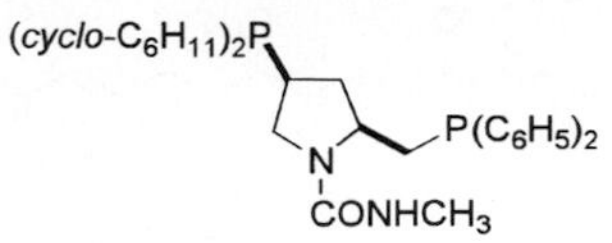

(2S,4S)-MCCPM

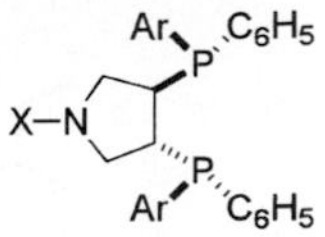

(S,R,R,R)-TMO-DEGUPHOS

Ar = 2,4,6-(CH$_3$O)$_3$C$_6$H$_2$;
X = (CH$_3$)$_3$COCO

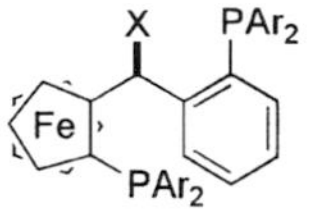

(R)-(S)-**1a**: Ar = C$_6$H$_5$; X = [(CH$_3$)$_2$CHCH$_2$]$_2$N
(R)-(S)-**1b**: Ar = 3,5-(CH$_3$)$_2$C$_6$H$_3$; X = (CH$_3$)$_2$N

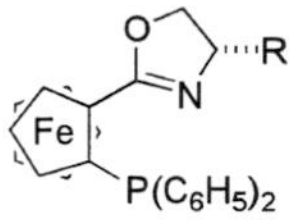

(S)-**2a**: R = (CH$_3$)$_2$CH
(S)-**2b**: R = C$_6$H$_5$

Fig. 2. Phosphine ligands without C_2 chirality

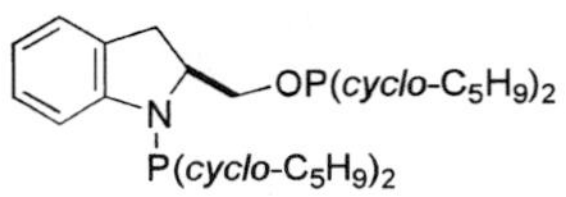

(S)-Cp,Cp-IndoNOP

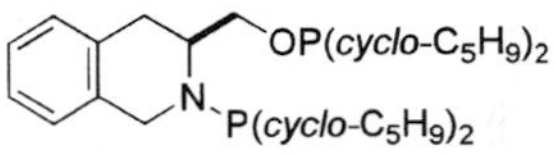

(S)-Cp,Cp-QuinoNOP

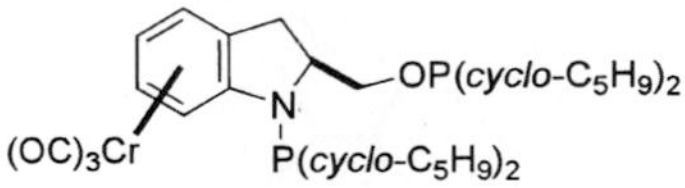

(S,2S)-Cr(CO)$_3$-Cp,Cp-IndoNOP

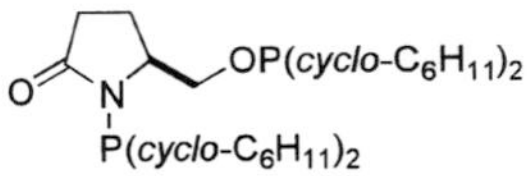

(S)-Cy,Cy-oxoProNOP

Fig. 3. Amido- or aminophosphine ligands (in alphabetical order)

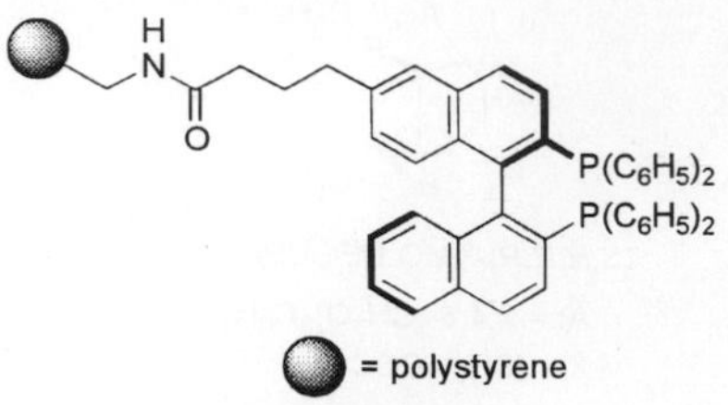

(*R*)-APBBINAP

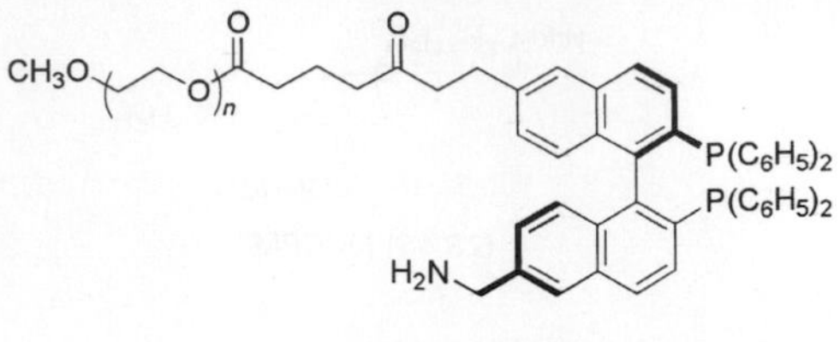

PEG-(*R*)-Am-BINAP

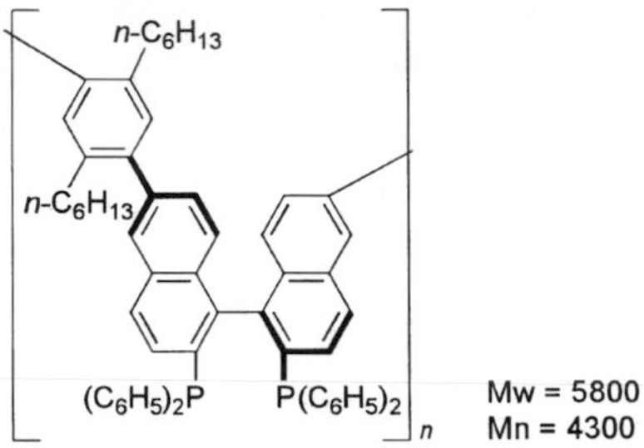

(*R*)-poly(BINAP)

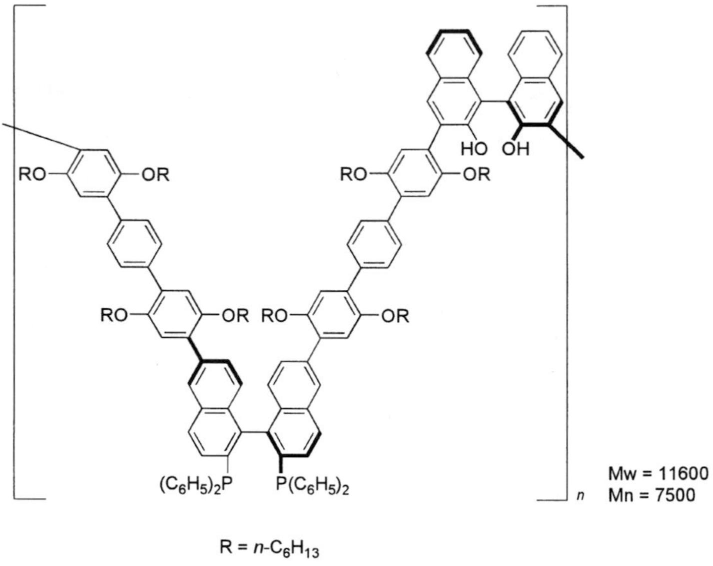

R = *n*-C_6H_{13}

(*R*,*R*)-poly(BINOL-BINAP)

Fig. 4. Immobilized BINAP ligands (in alphabetical order)

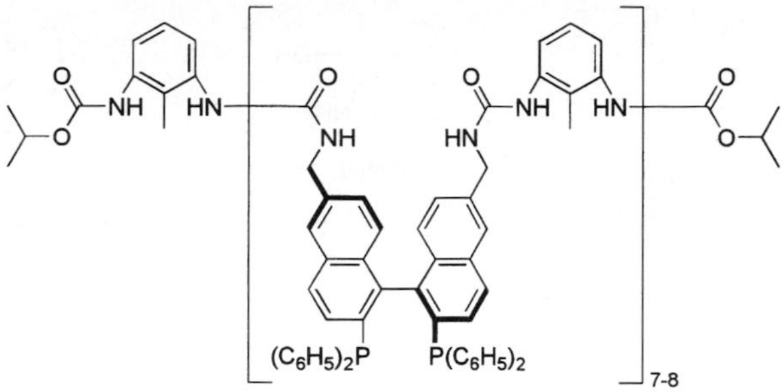

(R)-poly-NAP

Fig. 4. (continued)

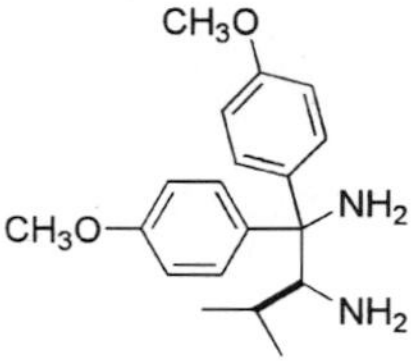

(S)-DAIPEN

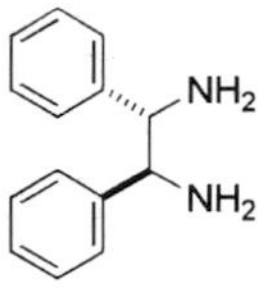

(S,S)-DPEN

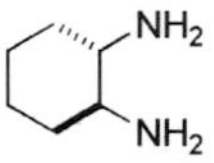

(S,S)-CYDN

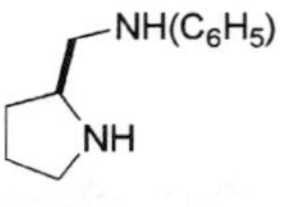

(S)-**3**

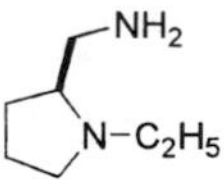

(S)-**4**

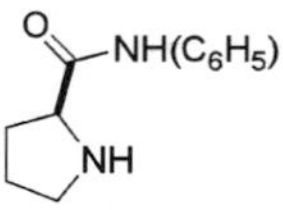

(S)-**5**

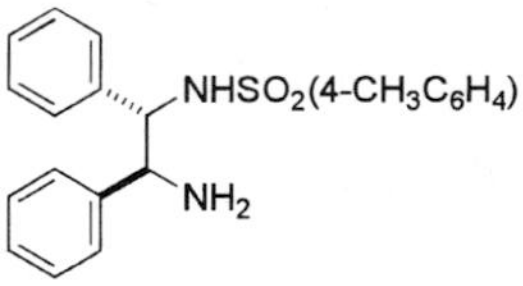

(S,S)-TsDPEN

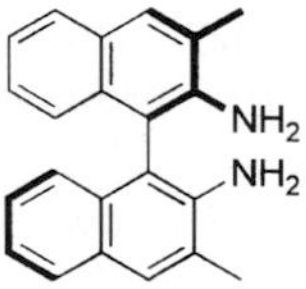

(R)-DM-DABN

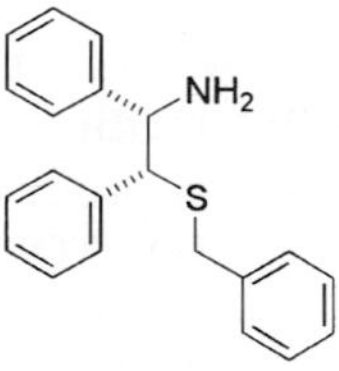

(R,S)-**6**

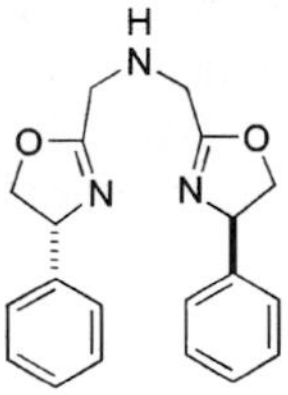

(R)-AMBOX

Fig. 5. Amine ligands

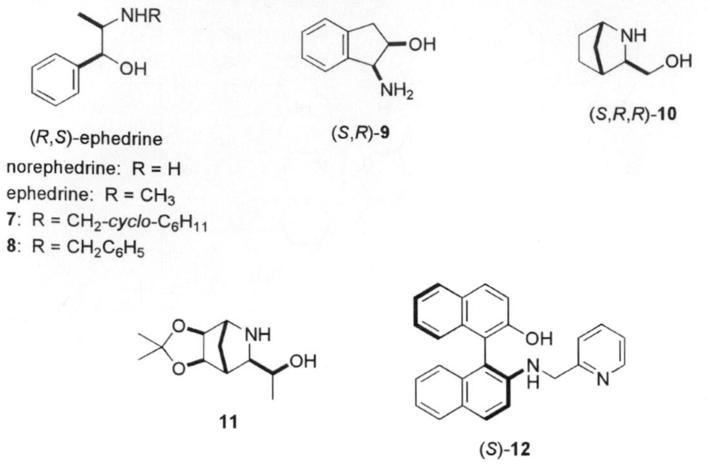

(R,S)-ephedrine

norephedrine: R = H
ephedrine: R = CH$_3$
7: R = CH$_2$-*cyclo*-C$_6$H$_{11}$
8: R = CH$_2$C$_6$H$_5$

(S,R)-**9**

(S,R,R)-**10**

11

(S)-**12**

Fig. 6. Amino alcohols

1
Hydrogenation

1.1
Simple Ketones

1.1.1
Alkyl Aryl Ketones

The discovery of Ru catalysts, which have BINAP as a chiral diphosphine-ligand and a chiral 1,2-diamine, has resulted in a notable advance in the asymmetric hydrogenation of simple ketones having no second heteroatom functionality [1–4]. The typical example is the hydrogenation of alkyl aryl ketones promoted by *trans*-RuCl$_2$(TolBINAP)(DPEN) with a strong base in 2-propanol achieving a turnover number (TON, moles of product per mole of catalyst) as high as 2,400,000 and a turnover frequency (TOF, TON sec^{-1}) of 72 [5]. However, no single asymmetric catalyst can be universal, because a structurally diverse array of substrates exists. The original (Tol)BINAP/DPEN/Ru with base catalysts exhibit sufficiently high enantioselectivity for limited ketonic substrates.

A breakthrough has been provided by the use of the (S)-XylBINAP/(S)-DAIPEN/Ru catalyst or its enantiomer [6]. The C_1-symmetric DAIPEN often gives slightly better ee than DPEN. Acetophenone was hydrogenated in the presence of *trans*-RuCl$_2$[(S)-XylBINAP][(S)-DAIPEN] and (CH$_3$)$_3$COK with a substrate/catalyst molar ratio (S/C) of 100,000 under 8 atm of H$_2$ to afford (R)-1-phenylethanol in 99% ee. As shown in Scheme 1, various alkyl aryl ketones were hydrogenated with this catalyst at a consistently high optical yield [6]. The hydrogenation conditions are tolerant of many substituents on the aro-

$$\text{Ar}\overset{\displaystyle O}{\underset{}{\Vert}}\text{R} \;+\; H_2 \xrightarrow[\;(CH_3)_2CHOH\;]{\text{catalyst}} \text{Ar}\overset{OH}{\underset{*}{\diagup}}\text{R}$$

catalyst:

trans-RuCl$_2$[(S)-xylbinap][(S)-daipen] + (CH$_3$)$_3$COK; (S,S)-**13**
trans-RuCl$_2$[(S)-xylbinap][(S,S)-dpen] + (CH$_3$)$_3$COK; (S,SS)-**14**
trans-RuHCl[(S)-binap][(S,S)-cydn] + (CH$_3$)$_2$CHOK; (S,SS)-**15**
trans-RuH(η^1-BH$_4$)[(S)-xylbinap][(S,S)-dpen]; (S,SS)-**16**
trans-RuCl$_2$[(R)-xylyl-phanephos][(S,S)-dpen] + (CH$_3$)$_3$COK; (R,SS)-**17**
[NH$_2$(C$_2$H$_5$)$_2$][{RuCl[(S)-tolbinap]}$_2$(μ-Cl)$_3$]; (S)-**18**
RuClCp*(cod)-(S)-**4** + KOH; (S)-**19**
[RhCl(cod)]$_2$-(R,S,R,S)-Me-PennPhos + 2,6-lutidine + KBr; (R,S,R,S)-**20**

Ar = 3,5-(CH$_3$)$_2$C$_6$H$_3$
trans-RuCl$_2$[(S)-xylbinap][(S)-daipen]: X = Y = Cl,
R^1 = R^2 = 4-CH$_3$OC$_6$H$_4$, R^3 = (CH$_3$)$_2$CH
trans-RuCl$_2$[(S)-xylbinap][(S,S)-dpen]: X = Y = Cl,
R^1 = H, R^2 = R^3 = C$_6$H$_5$
trans-RuH(η^1-BH$_4$)[(S)-xylbinap][(S,S)-dpen]: X = H, Y = η^1-BH$_4$,
R^1 = H, R^2 = R^3 = C$_6$H$_5$

R	Ar	Catalyst	S/C[a]	Pressure (atm)	Temp. (°C)	Yield (%)	ee (%)	Config
CH$_3$	C$_6$H$_5$	(S,S)-**13**	100,000	8	28	97	99	R
CH$_3$	C$_6$H$_5$[b]	(S,SS)-**15**	5,000	3	20	100	88	R
CH$_3$	C$_6$H$_5$	(S,SS)-**16**	100,000	8	45	100	99	R
CH$_3$	C$_6$H$_5$	(R,SS)-**17**	20,000	8	18–20	>99	99	R
CH$_3$	C$_6$H$_5$	(R,S,R,S)-**20**	100[c,d]	30	rt	97	95	S
CH$_3$	3-CH$_3$C$_6$H$_4$	(S,S)-**13**	10,000	10	28	98	100	R
CH$_3$	4-CH$_3$C$_6$H$_4$	(R,RR)-**14**	2,000	4	28	100	98	S
CH$_3$	4-n-C$_4$H$_9$C$_6$H$_4$	(R,RR)-**14**	2,000	4	28	100	98	S
CH$_3$	2,4-(CH$_3$)$_2$C$_6$H$_3$	(R,R)-**13**	2,000	4	28	99	99	S
CH$_3$	2-FC$_6$H$_4$	(S,S)-**13**	2,000	8	28	100	97	R
CH$_3$	2-FC$_6$H$_4$	(S)-**18**	1,300[c]	85	35	21	>99	–
CH$_3$	3-FC$_6$H$_4$	(R,RR)-**14**	2,000	4	28	99	98	S
CH$_3$	4-FC$_6$H$_4$	(R,R)-**13**	2,000	4	28	100	97	S
CH$_3$	2-ClC$_6$H$_4$	(R,RR)-**14**	2,000	4	28	99.5	98	S
CH$_3$	2-BrC$_6$H$_4$	(R,R)-**13**	2,000	4	28	99	96	S
CH$_3$	2-BrC$_6$H$_4$	(S)-**18**	950[c]	85	35	95	97	S
CH$_3$	3-BrC$_6$H$_4$	(R,R)-**13**	2,000	4	28	100	99.5	S

Scheme 1

R	Ar	Catalyst	S/C[a]	Pressure (atm)	Temp. (°C)	Yield (%)	ee (%)	Config
CH_3	$4\text{-}BrC_6H_4$	(S,S)-13	20,000	8	28	99.9	99.6	R
CH_3	$4\text{-}BrC_6H_4$	(S,S)-13	500	1	28	99.7	99.6	R
CH_3	$4\text{-}BrC_6H_4$	(R,SS)-17	3,000	8	18–20	>99	99	R
CH_3	$4\text{-}IC_6H_4$	(S,S)-13	2,000	8	28	99.7	99	R
CH_3	$2\text{-}CF_3C_6H_4$	(R,R)-13	2,000	4	28	99	99	S
CH_3	$3\text{-}CF_3C_6H_4$	(R,R)-13	2,000	4	28	100	99	S
CH_3	$3\text{-}CF_3C_6H_4$	(R,SS)-17	3,000	8	18–20	>99	99	R
CH_3	$4\text{-}CF_3C_6H_4$	(S,S)-13	10,000	10	28	100	99.6	R
CH_3	$2\text{-}CH_3OC_6H_4$	(R,R)-13	2,000	4	28	100	92	S
CH_3	$3\text{-}CH_3OC_6H_4$	(R,R)-13	2,000	4	28	99	99	S
CH_3	$4\text{-}CH_3OC_6H_4$	(S,S)-13	2,000	10	28	100	100	R
CH_3	$4\text{-}CH_3OC_6H_4$	(R,S,R,S)-20	100[c]	30	rt	83	94	S
CH_3	3-(R)-Glycidyl-oxyphenyl	(S,SS)-16	2,000	8	25	99	99	R,R
CH_3	$4\text{-}(C_2H_5OCO)$ C_6H_4	(S,SS)-16	4,000	8	25	100	99	R
CH_3	$4\text{-}[(CH_3)_2 CHOCO]C_6H_4$	(S,S)-13	2,000	8	28	100	99	R
CH_3	$4\text{-}NO_2C_6H_4$	(S,S)-13	2,000	8	28	100	99.8	R
CH_3	$4\text{-}NH_2C_6H_4$	(S,S)-13	2,000	8	28	100	99	R
CH_3	1-Naphthyl	(R,RR)-14	2,000	4	28	99	99	S
CH_3	2-Naphthyl	(R,RR)-14	2,000	4	28	99	98	S
C_2H_5	C_6H_5	(R,RR)-14	2,000	4	28	100	99	S
C_2H_5	C_6H_5	(S,RR)-17	3,000	5.5	18–20	>99	98	S
C_2H_5	C_6H_5	(R,S,R,S)-20	100[c]	30	rt	95	93	S
C_2H_5	$4\text{-}FC_6H_4$	(R,RR)-14	2,000	4	28	99	99	S
C_2H_5	$4\text{-}ClC_6H_4$	(S,S)-13	20,000	8	28	99.9	99	R
$(CH_3)_2CH$	C_6H_5	(R,R)-13	10,000	8	28	99.7	99	S
$Cyclo\text{-}C_3H_5$	C_6H_5	(S,S)-13	2,000	8	28	99.7	99	R
$(CH_3)_2$ $CHCH_2$	C_6H_5	(S)-19	100	10	30	98	95	R
$(CH_3)_3C$	C_6H_5	(S)-19	100	10	30	99	81	R

[a] Substrate/catalyst molar ratio. [b] Reaction with neat substrate. [c] Reaction in methanol. [d] Without addition of KBr.

Scheme 1. (continued)

matic ring, including F, Cl, Br, I, CF_3, OCH_3, $COOCH(CH_3)_2$, NO_2, and NH_2. The electronic and steric effects of substituents on enantioselectivity are relatively small. Propiophenone, isobutyrophenone, cyclopropyl phenyl ketone, and 1'- and 2'-acetonaphthone are also hydrogenated with excellent enantioselectivity. This methodology was applied to the asymmetric synthesis of a potent therapeutic agent for prostatomegary, TF-505 [7]. A *trans*-RuHCl(BINAP)(CYDN) complex with a strong base is also usable [8]. The (R)-Xylyl-PHANEPHOS and (S,S)-DPEN combined Ru catalyst in the presence of a base shows high activity and enantioselectivity [9]. Hydrogenation of acetophenone with (S,S)-BDPP/(S,S)-DPEN/Ru complex and $(CH_3)_3COK$ gave the R alcohol in 84% ee [10]. The catalyst prepared in situ from RuClCp*(COD) (Cp*=pentamethylcyclopentadienyl), chiral diamine (S)-4, and KOH promoted hydrogenation of pivalophenone to give the R alcohol in 81% ee [11]. The (R,S,R,S)-Me-PennPhos/Rh complex in the presence of 2,6-lutidine and KBr catalyzed the hydrogenation of certain aromatic ketones under 30 atm of H_2 to afford the corresponding S alcohol in up to 95% ee [12]. The addition of 2,6-lutidine and KBr increases both the reactivity and enantioselectivity. 2'-Halo-substituted acetophenones were hydrogenated with $[NH_2(C_2H_5)_2][\{RuCl[(S)-TolBINAP]\}_2(\mu-Cl)_3]$ under 85 atm of H_2 in up to >99% ee [13]. The reaction proceeds through a six-membered chelate intermediate formed by the coordination of the carbonyl oxygen atom and the halogen atom at the 2' position to the Ru center [14, 15].

$RuCl_2$(diphosphine)(1,2-diamine) complexes produce highly active species for the hydrogenation of simple ketones in the presence of a strong base. Therefore, the reaction cannot be applied to highly base-sensitive ketonic substrates. However, the newly devised *trans*-RuH(η^1-BH$_4$)[(S)-XylBINAP][(S,S)-DPEN] shows high catalytic activity without the addition of a base [16]. Hydrogenation of acetophenone using this S,SS complex with an S/C ratio of 100,000 under 8 atm of H_2 was completed in 7 h to give the R alcohol in 99% ee (Table of Scheme 1). The addition of an alkaline base to this system increased the catalytic activity, completing the hydrogenation in 45 min without a change of enantioselectivity. The base-free catalyst system successfully hydrogenated ketones having a base-sensitive functionality [16]. For example, (R)-glycidyl 3-acetylphenyl ether was hydrogenated with the S,SS catalyst to give the R,R alcohol with 99% de, leaving the base-labile epoxy ring untouched (Table of Scheme 1). Hydrogenation of ethyl 4-acetylbenzoate with the same catalyst resulted in only the ethyl (R)-4-(1-hydroxyethyl)benzoate in 99% ee without any transesterification.

Racemic 2-phenylpropiophenone, which has an enantiomerically labile α-stereogenic center under basic conditions, was hydrogenated with $RuCl_2[(S)$-XylBINAP][(S)-DAIPEN] and $(CH_3)_3COK$ to afford predominantly the 1R,2R al-

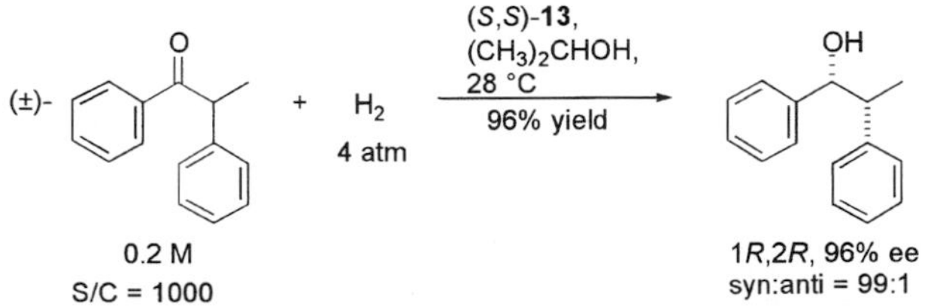

Scheme 2

cohol (96% ee, *syn:anti*=99:1) among four possible stereoisomers through dynamic kinetic resolution (Scheme 2) [2, 17]. The addition of a strong base promotes interconversion between the two enantiomers of the chiral ketone and also activates the catalyst [18]. High enantiomeric selectivity of the catalyst (*R*-ketone selectivity) and intramolecular stereo-induction (*syn* selectivity) is crucial to control two contiguous stereogenic centers.

Immobilization of homogeneous catalysts on solid materials allows their facile separation from the reaction mixture, recovery, and reuse. These catalysts will be useful for combinatorial synthesis. The BINAP/1,2-diamine/Ru catalyst was successfully immobilized by use of a polystyrene-anchored BINAP, APB-BINAP [19], as a diphosphine ligand (Scheme 3) [20]. Hydrogenation of 1'-acetonaphthone catalyzed by *trans*-RuCl$_2$[(*R*)-APBBINAP][(*R,R*)-DPEN] and (CH$_3$)$_3$COK with an S/C of 2470 in a 1:1 mixture of 2-propanol and DMF gave the *S* alcohol in 98% ee. The reactivity and enantioselectivity are comparable to those obtained under homogeneous conditions [5]. The reaction could be repeated 14 times without loss of optical yield (Table of Scheme 3). The total TON was 33,000. The hydrogenation could be performed using the same catalyst with an S/C of 12,300 [19]. Several BINAP-incorporated polymers were used

catalyst, 1:1 (CH$_3$)$_2$CHOH-DMF, 25 °C

+ H$_2$ 8 atm

S/C = 2470   total TON = 33,000

catalyst: *trans*-RuCl$_2$[(*R*)-apbbinap][(*R,R*)-dpen] + (CH$_3$)$_3$COK

Reusing run	Time (h)	Convn (%)	ee (%)
1	26	99	98
2	20	100	97
3	27	100	97
4	24	100	97
5	24	100	97
6	25	100	98
7	36	100	98
8	28	99	98
9	30	99	98
10	84	100	98
11	50	99	97
12	52	97	98
13	48	96	97
14	86	93	97

Scheme 3

for the same purpose. The (S)-Poly-NAP/(S,S)-DPEN/RuCl$_2$ complex with an S/C of 1000 in the presence of a base under 40 atm of H$_2$ catalyzed hydrogenation of 1'-acetonaphthone with 96% optical yield [21]. The reaction was repeated four times. Combination of (R)-poly(BINAP) or (R,R)-poly(BINOLBINAP) and (R,R)-DPEN also promoted the reaction with an S/C as high as 4,900 under 12 atm of H$_2$ [22, 23].

1.1.2
Diaryl Ketones

Highly enantioselective hydrogenation of 2-substituted benzophenones is achieved by the use of trans-RuCl$_2$[(S)-XylBINAP][(S)-DAIPEN] and (CH$_3$)$_3$COK with an S/C as high as 20,000 in 2-propanol (Scheme 4) [24]. Substrates having CH$_3$, CH$_3$O, F, Cl, or Br were converted to the corresponding benzhydrols in up to 99% ee. No diaryl methane derivative was observed. Chiral alcohols derived from the reaction of 2-methyl- and 2-bromo-4'-methylbenzophenones were easily converted to the antihistaminic (S)-orphenadrine and (R)-neobenodine, respectively [24]. Hydrogenation of 3- or 4-substituted benzophenones gave moderate enantioselectivity. Benzoylferrocene was hydrogenated with the (S)-TolBINAP/(S)-DAIPEN/Ru catalyst to afford the S alcohol in 95% ee.

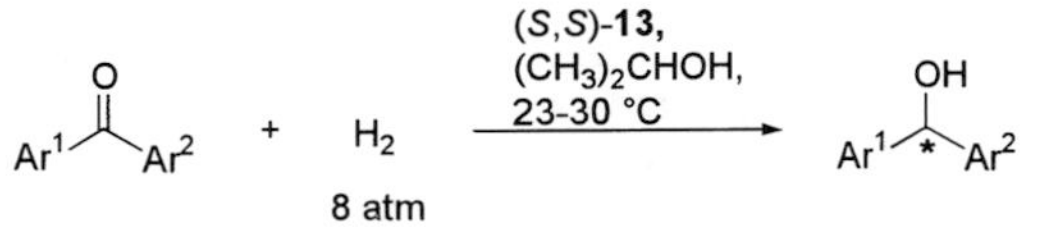

Ar^1	Ar^2	S/C^a	Yield (%)	ee (%)	Config^n
2-CH$_3$C$_6$H$_4$	C$_6$H$_5$	2,000	99	93	S
2-CH$_3$OC$_6$H$_4$	C$_6$H$_5$	2,000	100	99	S
2-FC$_6$H$_4$	C$_6$H$_5$	2,000	99	97	S
2-ClC$_6$H$_4$	C$_6$H$_5$	20,000	99	97	S
2-BrC$_6$H$_4$	C$_6$H$_5$	2,000	99	96	S
2-BrC$_6$H$_4$	4-CH$_3$C$_6$H$_4$	2,000	99	98	S
4-CH$_3$OC$_6$H$_4$	C$_6$H$_5$	2,000	95	35	R
4-CF$_3$C$_6$H$_4$	C$_6$H$_5$	2,000	99	47	S
4-CH$_3$OC$_6$H$_4$	4-CF$_3$C$_6$H$_4$	2,000	97	61	–
Ferrocenyl^b	C$_6$H$_5$	2,000	100	95	S

^{a} Substrate/catalyst molar ratio. ^{b} trans-RuCl$_2$[(S)-tolbinap][(S)-daipen] + (CH$_3$)$_3$COK was used as a catalyst.

Scheme 4

1.1.3
Heteroaromatic Ketones

A variety of ketones possessing an electron-rich or electron-deficient heteroaromatic substituent are hydrogenated with *trans*-RuCl$_2$[(*R*)-XylBINAP][(*R*)-DAIPEN] and (CH$_3$)$_3$COK to give the corresponding alcohols in consistently high ee (Scheme 5) [25]. Hydrogenation of 2-acetylfuran with the *R,R* catalyst at an S/C of 40,000 resulted in (*S*)-1-(2-furyl)ethanol in 99% ee without saturation of the furan ring. The 2- and 3-thienyl ketones were converted to the chiral alco-

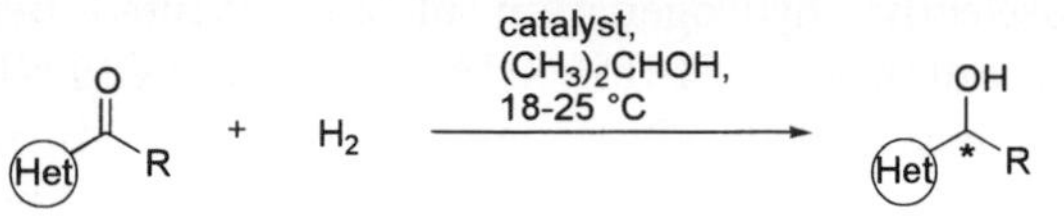

Het	R	Catalyst (S/C[a])	Pressure (atm)	Yield (%)	ee (%)	Config'n
2-Furyl	CH$_3$	(*R,R*)-**13** (40,000)	50	96	99	*S*
2-Furyl	CH$_3$	(*R,SS*)-**17** (3,000)	5.5	>99	96	*R*
2-Furyl	CH$_3$	(*R,S,R,S*)-**20** (100)	30[b]	83	96	*S*
2-Furyl	*n*-C$_5$H$_{11}$	(*R,R*)-**13** (2,000)	8	100	98	*S*
2-Furyl	CH$_2$=CH(CH$_2$)$_3$	(*R,R*)-**13** (2,000)	8	100	97	*S*
2-Thienyl	CH$_3$	(*R,R*)-**13** (5,000)	8	100	99	*S*
2-Thienyl	CH$_3$	(*S,S*)-**13** (1,000)	1	100	99	*R*
2-Thienyl	CH$_3$	RuCl$_2$[(*R,R*)-BICP](TMEDA)-(*R,R*)-DPEN+KOH (500)	4[c]	100	93	*S*
3-Thienyl	CH$_3$	(*R,R*)-**13** (5,000)	8	100	99.7	*S*
3-Thienyl	CH$_3$	(*S,RR*)-**17** (3,000)	5.5	>99	98	*S*
2-(1-Methyl)pyrrolyl	CH$_3$	(*S,S*)-**13** (1,000)	8	61	97	
2-[1-(4-Toluene-sulfonyl)]pyrrolyl	CH$_3$	(*R,R*)-**13** (1,000)[d]	8	93	98	*S*
2-Thiazolyl	CH$_3$	(*R,R*)-**13** (2,000)[e]	8	100	96	*S*
2-Pyridyl	CH$_3$	(*R,R*)-**13** (2,000)[e]	8	99.7	96	*S*
2-Pyridyl	(CH$_3$)$_2$CH	(*R,R*)-**13** (2,000)	8	100	94	*S*
3-Pyridyl	CH$_3$	(*R,R*)-**13** (5,000)	8	100	99.6	*S*
3-Pyridyl	CH$_3$	(*R,SS*)-**17** (1,500)	8	>99	99	*R*
4-Pyridyl	CH$_3$	(*R,R*)-**13** (5,000)	8	100	99.8	*S*
2,6-Diacetylpyridine		(*R,R*)-**13** (10,000)	8	99.9	100	*S,S*

[a] Substrate/catalyst molar ratio. [b] Reaction in methanol. [c] At –30 °C. [d] In 1:10 DMF-2-propanol.
[e] B[OCH(CH$_3$)$_2$]$_3$ was added. Ketone/Borate = 100.

Scheme 5

hols in >99% ee. The sulfur-containing heteroaromatic ring did not disturb the reaction. Although hydrogenation of 2-(1-methyl)pyrrolyl ketone gave the chiral alcohol in 97% ee with only 61% yield, the 1-(4-toluenesulfonyl)pyrrolyl analogue was reduced to the alcohol in 98% ee and 93% isolated yield. Hydrogenation of 2-acetylthiazol and 2-acetylpyridine was incomplete under the standard conditions. However the reaction proceeded smoothly in the presence of $B[OCH(CH_3)_2]_3$ (ketone:Ru:borate=2000:1:20) to give the corresponding chiral alcohols quantitatively [25]. The reactions of 3- and 4-acetylpyridine gave the chiral alcohols quantitatively with excellent optical purity. Double hydrogenation of 2,6-diacetylpyridine with the R,R catalyst resulted in the enantiomerically pure S,S diol without $B[OCH(CH_3)_2]_3$. The (R)-Xylyl-PHANEPHOS/ (S,S)-DPEN/Ru catalyst is also successfully applied to this reaction [9]. For instance, 3-acetylpyridine was hydrogenated to give the R alcohol quantitatively in 99% ee (Table of Scheme 5). Hydrogenation of 2-acetylthiophene with the catalyst prepared in situ from $RuCl_2[(R,R)$-BICP](TMEDA), (R,R)-DPEN, and KOH gave the S alcohol in 93% ee [26]. 2-Acetylfuran was hydrogenated with the $[RhCl(COD)]_2$, (R,S,R,S)-Me-PennPhos, 2,6-lutidine, and KBr catalyst system to afford the S alcohol in 96% ee [12].

1.1.4
Fluoro Ketones

Hydrogenation of 2,2,2-trifluoroacetophenones catalyzed by $trans$-$RuCl_2[(S)$-XylBINAP][(S)-DAIPEN] with $(CH_3)_3COK$ afforded the S trifluoro alcohols in 94–96% ee (Scheme 6) [6]. The presence of an electron-donating or electron-withdrawing group at the 4′ position had little effect on the enantioselectivity. The sense of the enantioselection was the same as that observed in the reaction of simple acetophenone.

Alkyl trifluoromethyl ketones were hydrogenated with $[Rh(OCOCF_3)\{(S)$-Cy,Cy-oxoProNOP\}]_2$ in toluene under 20 atm of H_2 to give the corresponding chiral alcohols in high ee (Scheme 7) [27]. Trifluoromethyl was efficiently differ-

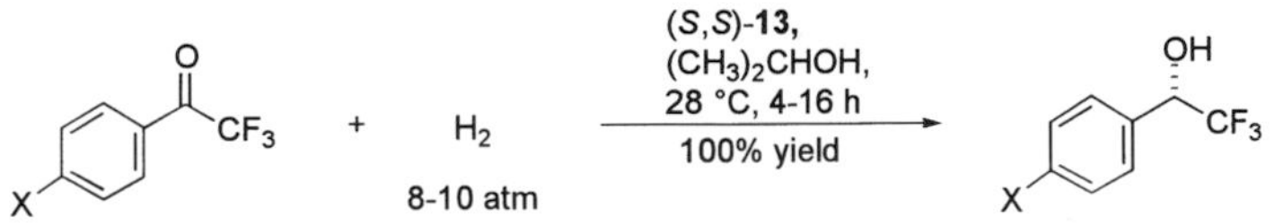

X	ee [%]
H	96
Cl	94
Br	94
CH_3O	96

Scheme 6

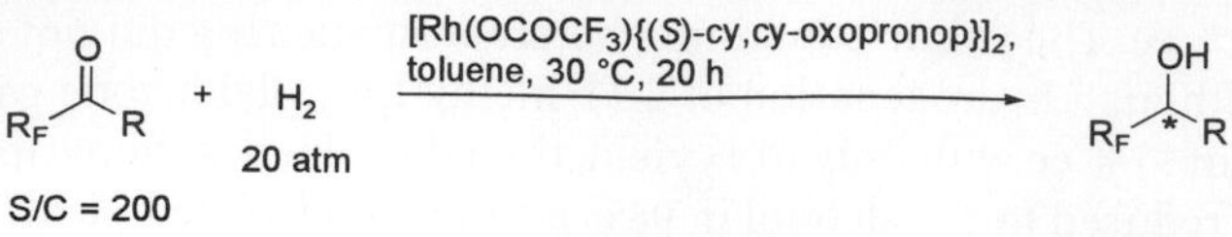

$$RF \overset{O}{\underset{}{\bigwedge}} R \;+\; H_2 \quad\xrightarrow[\text{toluene, 30 °C, 20 h}]{[Rh(OCOCF_3)\{(S)\text{-cy,cy-oxopronop}\}]_2,} \quad RF \overset{OH}{\underset{*}{\bigwedge}} R$$

20 atm

S/C = 200

R_F	R	Yield (%)	ee (%)	Config
CF_3	C_6H_5	93	73	R
CF_3	$Cyclo\text{-}C_6H_{11}$	90	97	R
CF_3	$n\text{-}C_8H_{17}$	99	97	R
CF_3	$C_6H_5CH_2OCH_2$	100	86	–
C_2F_5	$n\text{-}C_9H_{19}$	100	97	R

Scheme 7

entiated from the primary and secondary alkyl groups. *n*-Nonyl pentafluoroethyl ketone was also reduced with 97% optical yield.

1.1.5
Dialkyl Ketones

Asymmetric hydrogenation of dialkyl ketones is still a challenging target. Currently, the best catalyst for hydrogenation of *n*-alkyl methyl ketones is the (*R,S,R,S*)-Me-PennPhos/Rh complex with 2,6-lutidine and KBr [12]. 2-Hexanone was converted to (*S*)-2-hexanol in 75% ee (Table of Scheme 8). Hydrogenation of 4-methyl-2-pentanone gave the *S* alcohol in 85% ee. Cyclohexyl methyl ketone and 3-methyl-2-butanone were hydrogenated with 92% and 84% optical yield, respectively. Hydrogenation of pinacolone with the same catalysts afforded the *S* alcohol in 94% ee and 51% yield [12]. The cyclopropyl methyl and cyclohexyl methyl ketones were hydrogenated with *trans*-RuCl$_2$[(*S*)-XylBINAP][(*S*)-DAIPEN] and (CH$_3$)$_3$COK at an S/C ratio of >10,000 to afford the corresponding *R* alcohols, respectively (Table of Scheme 8) [6]. Methyl 1-methylcyclopropyl ketone was reduced with 98% optical yield. Hydrogenation of pinacolone with the (*S,R,R,R*)-TMO-DEGUPHOS/Rh catalyst gave the *S* alcohol in 84% ee [28].

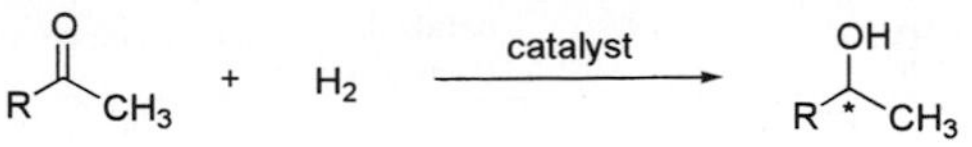

$$R\overset{O}{\underset{}{\|}}CH_3 \; + \; H_2 \xrightarrow{\text{catalyst}} R\overset{OH}{\underset{*}{|}}CH_3$$

R	Catalyst	S/C[a]	solvent	Pressure (atm)	Yield (%)	ee (%)	Config'n
n-C$_4$H$_9$	(R,S,R,S)-20	100	CH$_3$OH	30	96	75	S
(CH$_3$)$_2$CHCH$_2$	(R,S,R,S)-20	100	CH$_3$OH	30	66	85	S
(CH$_3$)$_2$CH	(R,S,R,S)-20	100	CH$_3$OH	30	99	84	S
Cyclo-C$_3$H$_5$	(S,S)-13	11,000	(CH$_3$)$_2$-CHOH	10	96	95	R
Cyclo-C$_6$H$_{11}$	(R,S,R,S)-20	100	CH$_3$OH	30	90	92	S
Cyclo-C$_6$H$_{11}$	(S,S)-13	10,000	(CH$_3$)$_2$-CHOH	8	99	85	R
(CH$_3$)$_3$C	(R,S,R,S)-20	100	CH$_3$OH	30	51	94	S
(CH$_3$)$_3$C	[Rh{(S,R,R,R)-TMO-DEGUPHOS}(COD)]BF$_4$	1,000	(CH$_3$)$_2$-CHOH	73	30	84	S
1-Methylcyclo-propyl	(S,S)-13	500	(CH$_3$)$_2$-CHOH	4	96	98	–

[a] Substrate/catalyst molar ratio.

Scheme 8

1.1.6
α,β-Unsaturated Ketones

trans-RuCl$_2$(diphosphine)(1,2-diamine) with a strong base in 2-propanol acts as an excellent catalyst for the carbonyl-selective hydrogenation of unsaturated ketones [29, 30]. Both conjugated and unconjugated enals and enones can be reduced. However, asymmetric hydrogenation of simple α,β-unsaturated ketones remains difficult due to their conformational flexibility and high sensitivity to basic conditions. Use of *trans*-RuCl$_2$[(S)-XylBINAP][(S)-DAIPEN] and K$_2$CO$_3$, a relatively weak base, as a catalyst significantly expands the scope of the substrate [6]. Benzalacetone was hydrogenated using this catalyst with an S/C of 100,000 under 80 atm of H$_2$ to afford the R allyl alcohol quantitatively in 97% ee (Table of Scheme 9). Thienyl ketone was also usable [25]. (E)-6-Methyl-2-hepten-4-one was converted with the (R)-XylBINAP/(R)-DAIPEN/Ru catalyst to the S alcohol in 90% ee [6]. This alcohol is a key building block for the synthesis of the side chain of α-tocopherol (vitamin E) [31]. More substituted, less base-sensitive substrates were hydrogenated more rapidly and conveniently by using (CH$_3$)$_3$COK as a cocatalyst. Hydrogenation of 1-acetylcycloalkenes resulted in almost perfect enantioselectivity [6]. The highly base-sensitive 3-nonene-2-one was hydrogenated with the (S)-XylBINAP/(S)-DAIPEN/Ru complex and K$_2$CO$_3$ to give the R allyl alcohol in 97% ee and in high yield under a high-dilu-

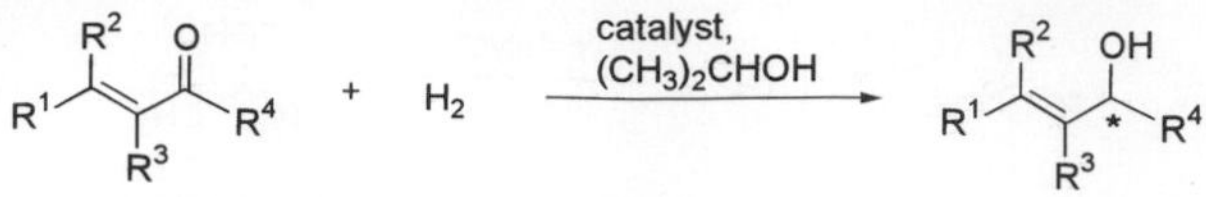

a: $R^1 = C_6H_5$; $R^2 = R^3 = H$; $R^4 = CH_3$
b: $R^1 = C_6H_5$; $R^2 = R^3 = H$; $R^4 = (CH_3)_2CH$
c: $R^1 = $ 2-thienyl; $R^2 = R^3 = H$; $R^4 = CH_3$
d: $R^1 = n\text{-}C_5H_{11}$; $R^2 = R^3 = H$; $R^4 = CH_3$
e: $R^1 = CH_3$; $R^2 = R^3 = H$; $R^4 = (CH_3)_2CHCH_2$
f: $R^1 = R^2 = R^4 = CH_3$; $R^3 = H$
g: $R^1\text{-}R^3 = (CH_2)_4$; $R^2 = H$; $R^4 = CH_3$
h: $R^1\text{-}R^3 = (CH_2)_5$; $R^2 = H$; $R^4 = CH_3$
i: $R^1\text{-}R^3 = (CH_2)_3$; $R^2 = R^4 = CH_3$
j: $R^1 = $ 2,6,6-trimethylcyclohexenyl; $R^2 = R^3 = H$; $R^4 = CH_3$ (β-ionone)

catalyst:

trans-RuCl$_2$[(S)-xylbinap][(S)-daipen] + K$_2$CO$_3$; (S,S)-**21**
trans-RuCl$_2$[(S)-xylbinap][(S,S)-dpen] + K$_2$CO$_3$; (S,SS)-**22**

Substrate	Catalyst	S/C[a]	Pressure (atm)	Yield (%)	ee (%)	Config'n
a	(S,S)-**21**	100,000	80	100	97	R
a	(S,S)-**21**	10,000	10	100	96	R
a	(R,SS)-**17**	3,000	5.5	>99	97	R
b	(S,S)-**21**	2,000	8	100	86	R
c	(R,R)-**21**	5,000	8	100	91	S
d	(S,S)-**21**	2,000	8	98	97	R
d	(S,SS)-**16**	4,000	8	95	99	R
e	(R,R)-**21**	2,000	10	100	90	S
f	(S,SS)-**22**	10,000	8	100	93	R
g	(S,S)-**13**	10,000	10	99	100	R
h	(S,S)-**13**	2,000	8	99.9	99	R
i	(S,S)-**13**	13,000	10	100	99	R
j	(S,S)-**21**	10,000	8	99	94	R

[a] Substrate/catalyst molar ratio.

Scheme 9

tion conditions (0.1 M substrate) [6]. High dilution was avoided by using *trans*-RuH(η^1-BH$_4$)[(S)-XylBINAP][(S,S)-DPEN] under base-free conditions [16]. The reaction was conducted under 2.0 M substrate concentration, resulting in the desired R alcohol in 99% ee and in 95% yield. The combination of (R)-Xylyl-PHANEPHOS and (S,S)-DPEN gave high enantioselectivity in the hydrogenation of benzalacetone [9].

1.2
Amino, Alkoxy, and Hydroxy Ketones

Pioneering studies on the asymmetric hydrogenation of α-amino ketones have been done by Kumada and Achiwa [3, 4]. Achiwa's MCCPM/Rh complex catalyzed the reaction of 2-(dimethylamino)acetophenone hydrochloride with an S/C of 100,000 under 20 atm of H_2 to give the corresponding chiral amino alcohol in 96% ee [32], although the reactions of other amino ketones showed less satisfactory rates and enantioselectivity.

trans-RuCl$_2$[(R)-XylBINAP][(R)-DAIPEN] with (CH$_3$)$_3$COK acts as an excellent catalyst for hydrogenation of α-, β-, and γ-amino ketones [33]. Hydrogenation of α-(dimethylamino)acetone with an S/C of 2,000 under 8 atm of H_2 gave the S amino alcohol in 92% ee (Table of Scheme 10), whereas 2-(dimethy amino)acetophenone was converted with the same catalyst to the R alcohol in 93% ee. The order of observed enantiodirecting ability in this hydrogenation is phenyl >(dimethylamino)methyl >methyl. Acetophenone derivatives possessing an amido group at the α position were reduced with the R,R catalyst to give the R alcohol in up to 99.8% ee [33]. The reaction was applied to the synthesis of (R)-denopamine, a β$_1$-receptor agonist used for treating congestive heart failure. Hydrogenation of β-(dimethylamino)propiophenone in the presence of a strong base is difficult due to the inherent instability of the substrate. The β-amino ketone was hydrogenated using the (S)-XylBINAP/(S)-DAIPEN/RuCl$_2$ complex with a minimum amount of (CH$_3$)$_3$COK, which resulted in the R β-amino alcohol in 97.5% ee in a 96% yield accompanied by 2% of 1-phenyl-1-propanol [33]. This problem has been solved by the use of trans-RuH(η^1-BH$_4$)[(S)-XylBINAP][(S,S)-DPEN] under base-free conditions to afford the R alcohol quantitatively in 97% ee [16]. 2-Thienyl ketone was also reduced selectively [25]. The chiral β-amino alcohols obtained are useful intermediates for the synthesis of the antidepressants (R)-fluoxetine and (S)-duloxetine [16, 25, 33]. Furthermore, a γ-amino ketone shown in Scheme 10 was hydrogenated using the (S)-XylBINAP/(S)-DAIPEN/RuCl$_2$ complex and (CH$_3$)$_3$COK with an S/C of 10,000 under 8 atm of H_2, which resulted in the R alcohol, a potent antipsychotic BMS 181100, in 99% ee [33].

Amino or amido phosphinephosphinite ligand/Rh complexes are efficient catalysts for the hydrogenation of α-amino ketones. The hydrogenation of 2-(dimethylamino)acetophenone hydrochloride catalyzed by the (S)-Cp,Cp-IndoNOP/Rh complex with an S/C of 200 under 50 atm H_2 afforded the S amino alcohol in >99% ee (Table of Scheme 10) [34]. The 3'-chloro analogue was hydrogenated with the (R)-Cy,Cy-oxoProNOP/Rh complex under 1 atm of H_2 to give the R alcohol in 96% ee. This product is a synthetic intermediate of SR 58611A, a potent atypical β-adrenoreceptor agonist [35].

Acetol was hydrogenated using [NH$_2$(C$_2$H$_5$)$_2$][{RuCl[(R)-SEGPHOS]}$_2$(μ-Cl)$_3$] with an S/C of 3,000 under 30 atm of H_2 to give (R)-1,2-propanediol in 99.5% ee (Table of Scheme 10) [36]. The enantioselectivity is higher than that obtained in the original BINAP/Ru-catalyzed reaction [14, 15]. The higher lev-

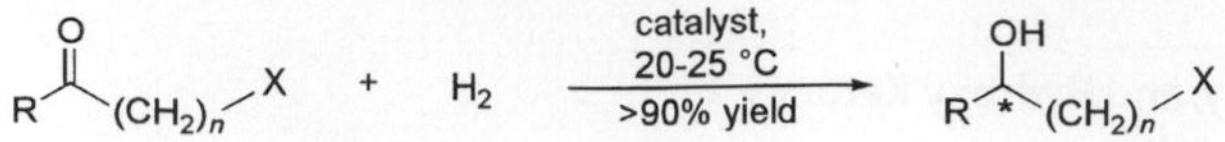

$$\underset{R}{\overset{O}{\|}}\!\!\!\diagdown\!(CH_2)_n\!\diagup\!X \ + \ H_2 \quad\xrightarrow[\text{>90\% yield}]{\substack{\text{catalyst,}\\ \text{20-25 °C}}}\quad \underset{R}{\overset{OH}{\underset{*}{|}}}\!\!\!\diagdown\!(CH_2)_n\!\diagup\!X$$

R	n	X	Catalyst (S/C[a])	Pressure (atm)	ee (%)	Config
CH_3	1	$(CH_3)_2N$	(R,R)-13 (2,000)	8	92	S
C_6H_5	1	$(CH_3)_2N$	(R,R)-13 (2,000)	8	93	R
C_6H_5	1	$Cl(CH_3)_2NH$	[Rh(OCOCF$_3$){(S)-Cp, Cp-IndoNOP}]$_2$ (200)	50	>99	S
$3\text{-}ClC_6H_4$	1	$Cl(CH_3)_2NH$	[Rh{(R)-Cy,Cy-oxo-ProNOP}(COD)]BF$_4$ (200)	1	96	R
C_6H_5	1	$C_6H_5CO(CH_3)N$	(R,R)-13 (2,000)	8	99.8	R
$4\text{-}C_6H_5CH_2OC_6H_4$	1	$C_6H_5CO[3,4\text{-}(CH_3O)_2\text{-}C_6H_3\text{-}(CH_2)_2]N$	(R,R)-13 (2,000)	8	97	R
C_6H_5	2	$(CH_3)_2N$	(S,S)-13 (10,000)[b]	8	97.5	R
C_6H_5	2	$(CH_3)_2N$	(S,SS)-16 (4,000)	8	97	R
2-Thienyl	2	$(CH_3)_2N$	(R,R)-13 (2,000)[b]	8	92	S
$4\text{-}FC_6H_4$	3	R'[c]	(S,S)-13 (10,000)	8	99	R
CH_3	1	HO	[NH$_2$(C$_2$H$_5$)$_2$][{RuCl-[(R)-SEGPHOS]}$_2$-(μ-Cl)$_3$] (3,000)	30[d]	99.5	R
C_6H_5	1	CH_3O	(R,R)-13 (2,000)	8	95	R

[a] Substrate/catalyst molar ratio. [b] *trans*-RuCl$_2$ (xylbinap)(daipen) was treated with $(CH_3)_3COK$ in $(CH_3)_2CHOH$ prior to hydrogenation. [c] $R' = F\!\!-\!\!\underset{\substack{\diagup\\ N}}{\overset{\diagdown N}{\bigcirc}}\!\!-\!N\overset{N}{\underset{}{\bigcirc}}$ [d] At 60 °C.

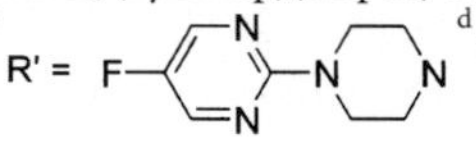

Scheme 10

$$\underset{CH_3O}{\overset{CH_3O}{\diagdown}}\!\!\!\overset{O}{\overset{\|}{\diagup}}\!\!\!\diagdown \ + \ \underset{\text{8 atm}}{H_2} \quad\xrightarrow[\substack{\text{25 °C, 8 h}\\ \text{100\% yield}}]{\substack{(R,R)\text{-}\mathbf{13},\\ (CH_3)_2CHOH,}}\quad \underset{CH_3O}{\overset{CH_3O}{\diagdown}}\!\!\!\overset{OH}{\overset{|}{\diagup}}\!\!\!\diagdown$$

S/C = 2000

98% ee

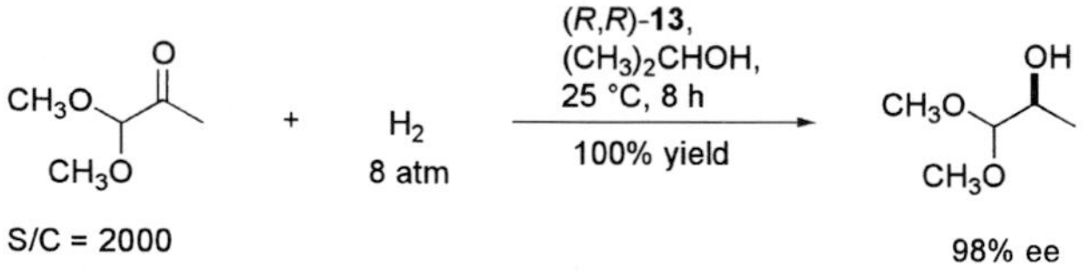

Scheme 11

el of enantioselectivity may be ascribed to the smaller dihedral angle than that of BINAP.

Hydrogenation of 2-methoxyacetophenone with (R)-XylBINAP/(R)-DAIPEN/RuCl₂ in the presence of a base gave the R alcohol in 95% ee (Table of Scheme 10) [2]. The sense of enantioselection is the same as that in hydrogenation of acetophenone. Pyruvic aldehyde dimethylacetal was hydrogenated with the R,R catalyst to afford the S alcohol in 98% ee (Scheme 11) [2]. This result shows the high enantiodirecting effect of the dimethoxymethyl group.

1.3
Dynamic and Static Kinetic Resolution of Racemic α-Substituted Cyclohexanones

α-Alkylcyclohexanones are easily racemized under basic conditions. Due to the high enantiomer discrimination ability of the (S)-BINAP/(R,R)-DPEN/Ru catalyst, racemic 2-isopropylcyclohexanone was hydrogenated in the presence of a base to give predominantly (1R,2R)-2-isopropylcyclohexanol among four possible stereoisomers [2, 18].

This transformation can be applied to 2-methoxycyclohexanone. The racemic ketone was hydrogenated with the (S)-XylBINAP/(S,S)-DPEN/Ru catalyst in the presence of a base at 5 °C and under 50 atm of H₂ to give (1R,2S)-2-methoxycyclohexanol in 99% ee (cis:trans=99.5:0.5) (Scheme 12) [37]. This alcohol can be converted to the potent antibacterial agent sanfetrinem after it is oxidized to the chiral ketone. In the same manner, racemic 2-(tert-butoxycarbonylamino)cyclo hexanone was converted with the (S)-XylBINAP/(R)-DAIPEN/Ru catalyst under basic conditions to the 1S,2R alcohol in 82% ee (cis:trans=99:1) [33].

As described above, the catalyst comprising RuCl₂ complex with a strong base is effective for the asymmetric hydrogenation through dynamic kinetic resolution. However, it is not suitable for static kinetic resolution of racemic α-substituted ketones because of the basic conditions. The newly devised trans-RuH(η¹-BH₄)[(R)-XylBINAP][(S,S)-DPEN] without any additional base allows one to

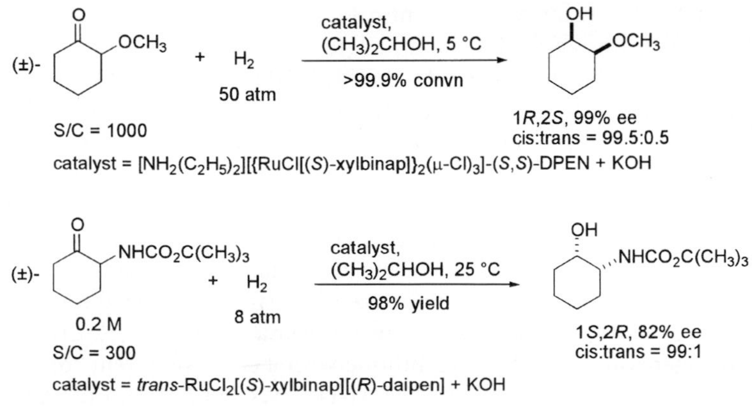

Scheme 12

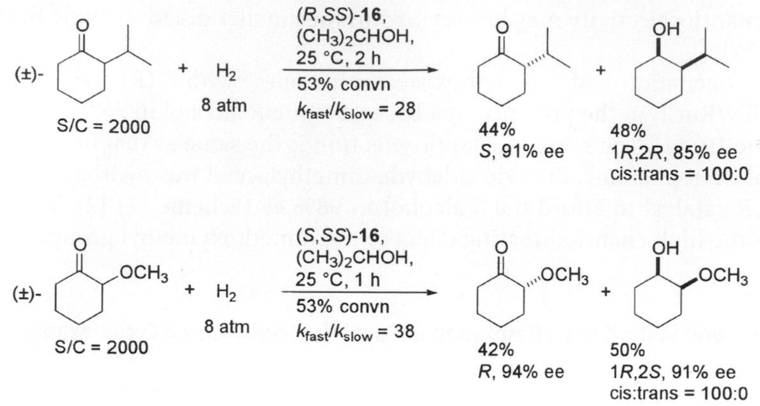

Scheme 13

obtain optically active ketones by asymmetric hydrogenation via static kinetic resolution [16]. When hydrogenation of racemic 2-isopropylcyclohexanone was stopped at 53% conversion, unreacted S ketone was recovered in 91% ee, together with the $1R,2R$ alcohol in 85% ee (*cis:trans*=100:0) (Scheme 13). The k_{fast}/k_{slow} ratio was calculated to be 28. The S,SS combined catalyst efficiently discriminated enantiomers of 2-methoxycyclohexanone by a factor of 38. At 53% conversion, the R ketone was obtained in 94% ee accompanied by the $1R,2S$ alcohol in 91% ee (*cis:trans*=100:0).

1.4
Asymmetric Activation/Deactivation

The activity and stereoselectivity of the BINAP/DPEN/Ru catalyst is highly dependent on the stereochemistry of the ligands [38, 39]. For hydrogenation of simple acyclic ketones, the combination of (R)-BINAP and (R,R)-DPEN (or S,SS) is much more reactive and enantioselective than the R,SS (or S,RR) catalyst. This system was applied to asymmetric hydrogenation using a conformationally flexible diphosphine ligand, DM-BIPHEP (Scheme 14) [40, 41]. A combination of the DM-BIPHEP/RuCl$_2$ complex with (S,S)-DPEN produced the (S)-DM-BIPHEP/(S,S)-DPEN/RuCl$_2$ complex and the R,SS combination in a 3:1 ratio. Hydrogenation of 1′-acetonaphthone with the mixed complex and base gave the R alcohol in 92% ee.

The chiral aromatic diamine, (R)-DM-DABN, selectively coordinates to the (R)-XylBINAP/RuCl$_2$ complex, providing the catalytically inactive RuCl$_2$[(R)-XylBINAP][(R)-DM-DABN] [42]. This characteristic was utilized for the asymmetric hydrogenation of ketones with a racemic XylBINAP/RuCl$_2$ complex. Thus, hydrogenation of 1′-acetonaphthone with a catalyst system consisting of the (±)-XylBINAP/RuCl$_2$ complex, (R)-DM-DABN, (S,S)-DPEN, and KOH in a 1: 0.55:0.5:2 ratio resulted in the R alcohol in 96% ee (Scheme 15).

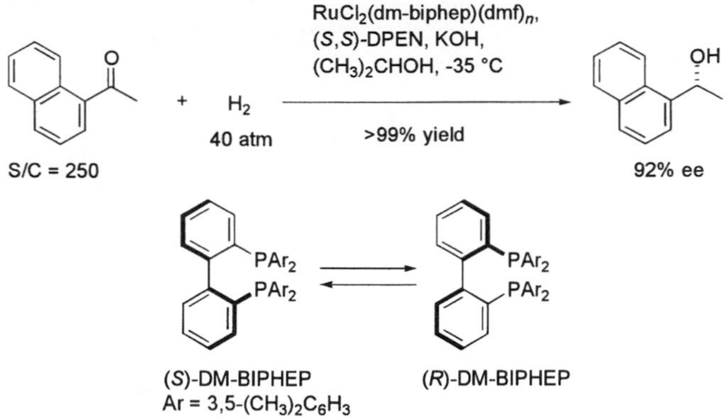

RuCl$_2$(dm-biphep)(dmf)$_n$,
(S,S)-DPEN, KOH,
(CH$_3$)$_2$CHOH, -35 °C

+   H$_2$

40 atm   >99% yield

S/C = 250   92% ee

PAr$_2$ PAr$_2$
PAr$_2$ PAr$_2$

(S)-DM-BIPHEP   (R)-DM-BIPHEP
Ar = 3,5-(CH$_3$)$_2$C$_6$H$_3$

Scheme 14

RuCl$_2$[(±)-xylbinap](dmf)$_n$,
(R)-DM-DABN,
(S,S)-DPEN, KOH,
(CH$_3$)$_2$CHOH, rt

+   H$_2$ >99% yield

8 atm

ketone:Ru:DM-DABN:DPEN = 250:1:0.55:0.5   R, 96% ee

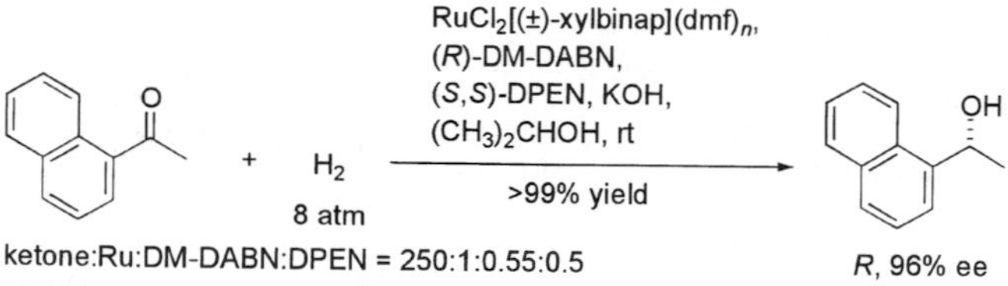

Scheme 15

1.5
Mechanistic Model of Hydrogenation of Simple Ketones

Hydrogenation of ketones has been understood as proceeding through a [2+2] reaction of a carbonyl group and metal hydride species, forming metal alkoxides. However, the phosphine/1,2-diamine/Ru-catalyzed hydrogenation of simple ketones is proposed to occur via a nonclassical metal ligand bifunctional mechanism, as shown in Scheme 16 [1, 2]. The presence of NH$_2$ groups coordinating on the metal center is crucial to achieve high catalytic activity. A RuCl$_2$ precatalyst **A** is converted to RuHX complex **B** (X=H or OR) with the aid of two equivalents of a base and a hydride source, H$_2$ and partly 2-propanol. When RuH(η^1-BH$_4$) complex is employed, the addition of a base is not necessary to produce active species [16]. The 18-electron species **B** immediately reacts with a ketone via a six-membered, pericyclic transition state **TS$_1$**, to give the didehydro species **C** with the release of an alcohol. The charge-alternating H$^{\delta-}$–Ru$^{\delta+}$–N$^{\delta-}$–H$^{\delta+}$ arrangement fits well with the C$^{\delta+}$=O$^{\delta-}$ dipole lowering the $\Delta G^{\ddagger}$. No metal alkoxide is formed. The 16-electron species **C** splits H$_2$ in a heterolytic fashion via **TS$_2$** to restore the Ru hydride **B** due to the unique Ru$^{\delta+}$–N$^{\delta-}$ dipolar bond. Alternatively, **B** may be regenerated via species **D** and **E** assisted by an alcohol and a base. Other mechanistic models have also been proposed [8, 43–45].

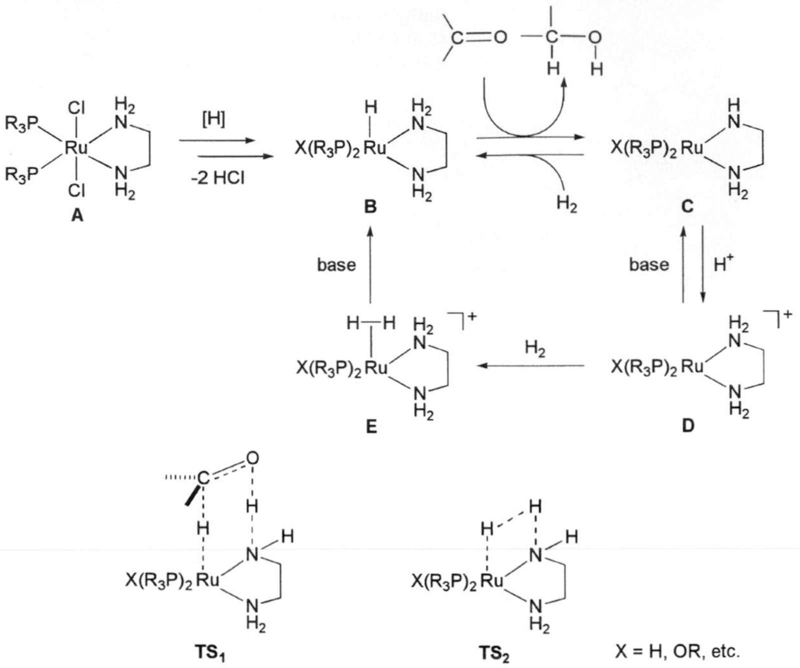

Scheme 16

1.6
Keto Esters and Amides

Asymmetric hydrogenation of α-keto esters and amides has been extensively studied with a variety of chiral Rh and Ru catalysts [3, 4, 46]. A limited number of catalysts have achieved high enantioselectivity.

A (*R*)-SEGPHOS/Ru complex effected hydrogenation of aliphatic α-keto esters (R^1=*t*-C_4H_9, $C_6H_5(CH_2)_2$) with an S/C of >1,000 to give the *R* alcohols in >95% ee (Table of Scheme 17) [36]. *N*-Benzyl benzoyl formamide was hydrogenated with the (*S,2S*)-$Cr(CO)_3$-Cp,Cp-IndoNOP or (*S*)-Cp,Cp-QuinoNOP/Rh catalyst even under 1 atm of H_2 to afford the *S* hydroxyamide in up to >99% ee [34, 47].

Highly active and enantioselective hydrogenation of ketopantolactone has been achieved by the use of [Rh(OCOCF$_3$){(*S*)-Cp,Cp-IndoNOP}]$_2$ as a catalyst (Scheme 18) [34, 47]. The reaction with an S/C of 200 under 1 atm of H_2 was completed within 10 min, giving (*R*)-pantoyl lactone in >99% ee.

Use of BINAP/Ru catalysts has achieved excellent optical yields, as high as >99%, in the hydrogenation of a variety of β-keto esters [1, 2, 14]. Following this success, methyl and ethyl 3-oxobutanoates have been used as representative substrates to check the enantioselective efficiency of new chiral ligands [3, 4, 46].

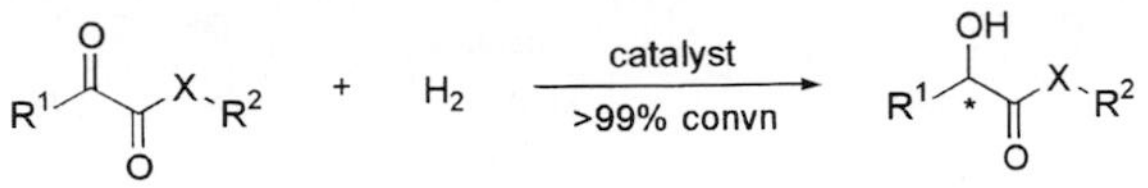

R^1	X	R^2	Catalyst (S/C^a)	Solvent	Pressure (atm)	ee (%)	Confign
t-C$_4$H$_9$	O	C$_2$H$_5$	[NH$_2$(C$_2$H$_5$)$_2$][{RuCl-[(R)-SEGPHOS]}$_2$-(μ-Cl)$_3$] (1,000)	C$_2$H$_5$OH	50	98.6	R
C$_6$H$_5$(CH$_2$)$_2$	O	C$_2$H$_5$	[NH$_2$(C$_2$H$_5$)$_2$][{RuCl-[(R)-SEGPHOS]}$_2$-(μ-Cl)$_3$] (1,500)	C$_2$H$_5$OH	50	95.7	R
C$_6$H$_5$	NH	C$_6$H$_5$CH$_2$	[RhCl{(S,2S)-Cr(CO)$_3$-Cp,Cp-IndoNOP}]$_2$ (200)	Toluene	1	97	S
C$_6$H$_5$	NH	C$_6$H$_5$CH$_2$	[RhCl{(S)-Cp,Cp-QuinoNOP}]$_2$ (200)	Toluene	50	>99	S

^{a} Substrate/catalyst molar ratio.

Scheme 17

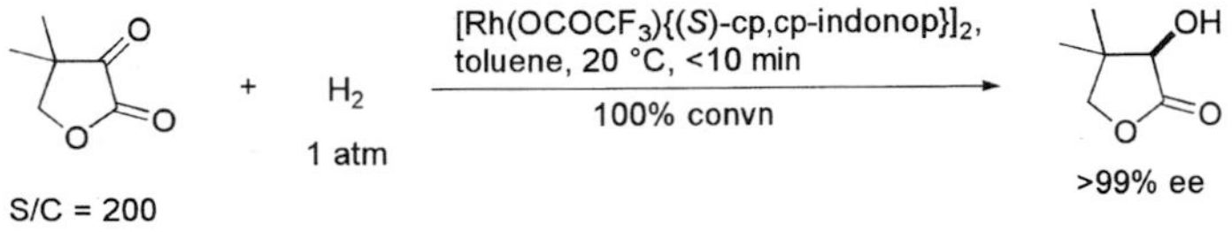

Scheme 18

Recent examples reporting an enantiomeric excess of 99% or greater are listed in the Table of Scheme 19. Some new catalyst precursors [10] and catalyst systems prepared in situ with BINAP or MeO-BIPHEP as a ligand [48, 49] have been reported. Atropisomeric diphosphines possessing heteroatoms gave high enantioselectivity [50–53]. A Ru complex with chiral 1,5-diphosphanylferrocene **1a** also acted as an efficient catalyst affording the β-hydroxyester in high ee [54].

Hydrogenation with recyclable catalysts has been reported. For example, an oligomeric (R)-poly-NAP/Ru-catalyzed hydrogenation of methyl 3-oxobutanoate with an S/C of 1,000 under 40 atm of H$_2$ gave the R alcohol quantitatively in 99% ee (Table of Scheme 19) [55]. The catalyst was reusable five times, resulting in a total TON of 4,180. A Ru complex with poly(ethylene glycol)-bound BINAP, PEG–Am-BINAP, effected the reaction with an S/C of 10,000 under 100 atm of H$_2$ [56]. Hydrogenation was performed in water with a 94% optical yield by using a Ru catalyst with the water-soluble BINAP ligand, diam-BINAP [57]. An immobilized Me-DuPhos/Rh catalyst in a poly(dimethylsiloxane) membrane was also usable [58].

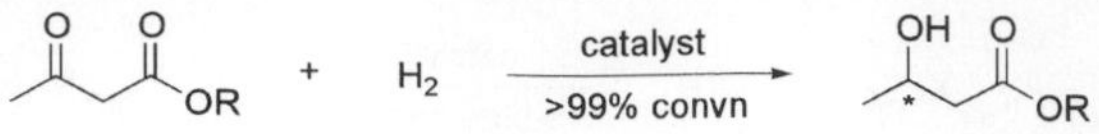

$$\text{R-C(O)-CH}_2\text{-C(O)-OR} + H_2 \xrightarrow[\text{>99\% convn}]{\text{catalyst}} \text{R-CH(OH)-CH}_2\text{-C(O)-OR}$$

R	Catalyst (S/C[a])	Solvent	Pressure (atm)	Temp. (°C)	ee (%)	Config'n
CH$_3$	trans-RuCl$_2$[(R)-BINAP]-(pyridine)$_2$+HCl (1,000)	CH$_3$OH	3.7	60	99.9	R
CH$_3$	[RuCl$_2$(COD)]$_n$+(R)-BINAP (100)	CH$_3$OH	4	rt	99	R
CH$_3$	RuCl$_3$+(S)-MeO-BIPHEP (100)	CH$_3$OH	4	50	99	S
CH$_3$	RuCl$_2$[(R)-C4TunaPhos]-(DMF)$_n$ (100)	CH$_3$OH	51	60	99.1	R
CH$_3$	RuCl$_2$[(S)-BIFAP](DMF)$_n$ (1,000)	CH$_3$OH	100	70	100	S
CH$_3$	RuCl$_2$[(R)-poly-NAP](DMF)$_n$ (1,000)	CH$_3$OH	40	50	99	R
CH$_3$	Ru[η^3-CH$_2$C(CH$_3$)CH$_2$]$_2$-[PEG-(R)-Am-BINAP]+HBr (10,000)	CH$_3$OH	100	50	99	R
C$_2$H$_5$	RuCl$_2$[(R)-bisbenzodi-oxanePhos](DMF)$_n$ (1,000)	C$_2$H$_5$OH	3.4	80–90	99.5	R
C$_2$H$_5$	RuCl$_2$[(S)-P-Phos](DMF)$_n$ (400)	C$_2$H$_5$OH	3.4	70	98.6	–
C$_2$H$_5$	Ru[η^3-CH$_2$C(CH$_3$)CH$_2$]$_2$-(COD)+(R)-(S)-1a+HBr (200)	C$_2$H$_5$OH	50	50	98.6	S

[a] Substrate/catalyst molar ratio.

Scheme 19

BINAP and MeO-BIPHEP/Ru-catalyzed hydrogenation of functionalized ketones has been applied to the synthesis of more than ten important compounds, for example, antibiotics, anti-inflammatory compounds, and anticancer agents, since 1999 [59].

It is difficult to hydrogenate benzoylacetic acid derivatives with a high optical yield. Recently, an (R)-SEGPHOS/Ru complex-catalyzed hydrogenation of the ethyl ester with an S/C of 10,000 under 30 atm of H$_2$ afforded the S alcohol in 97.6% ee (Table of Scheme 20) [36]. MeO-BIPHEP and Tol-P-Phos also performed with a high level of enantioselection [49, 60]. Hydrogenation of N-methylbenzoylacetamide with the (R)-BINAP/Ru catalyst gave the S alcohol in >99.9% ee and 50% yield [61].

α,α-Difluoro-β-keto esters were hydrogenated with (R)-BINAP/Ru and (S)-Cy,Cy-oxoProNOP/Rh complexes under 20 atm of H$_2$ to afford the R alcohols in high ee (Scheme 21) [62, 63]. Hydrogenation of ethyl 4,4,4-trifluoro-3-oxobu-

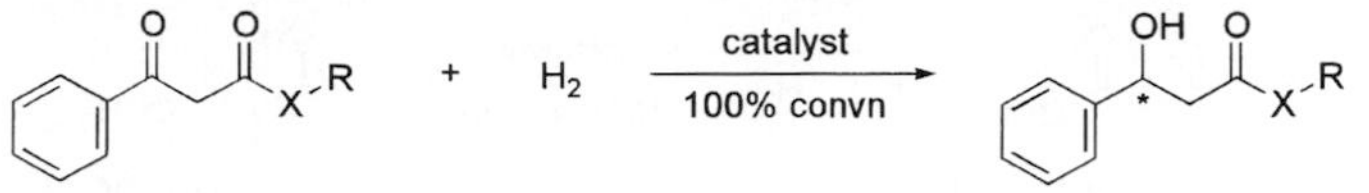

RX	Catalyst (S/C[a])	Solvent	Pressure (atm)	Temp. (°C)	ee (%)	Config'n
C_2H_5O	$RuCl_3+(S)$-MeO-BIPHEP (100)	CH_3OH	4	80	95	R
C_2H_5O	$[NH_2(C_2H_5)_2][\{RuCl-[(R)\text{-SEGPHOS}]\}_2(\mu\text{-}Cl)_3]$ (10,000)	CH_3OH	30	80	97.6	S
C_2H_5O	$RuCl_2[(S)\text{-Tol-P-Phos}]$ $(DMF)_n$ (800)	1:1 C_2H_5OH CH_2Cl_2	20	90	96.4	S
CH_3NH	$RuCl_2[(R)\text{-BINAP}](DMF)_n$ (1,800)	CH_3OH	14	100	> 99.9[b]	S

[a] Substrate/catalyst molar ratio. [b] 50% yield.

Scheme 20

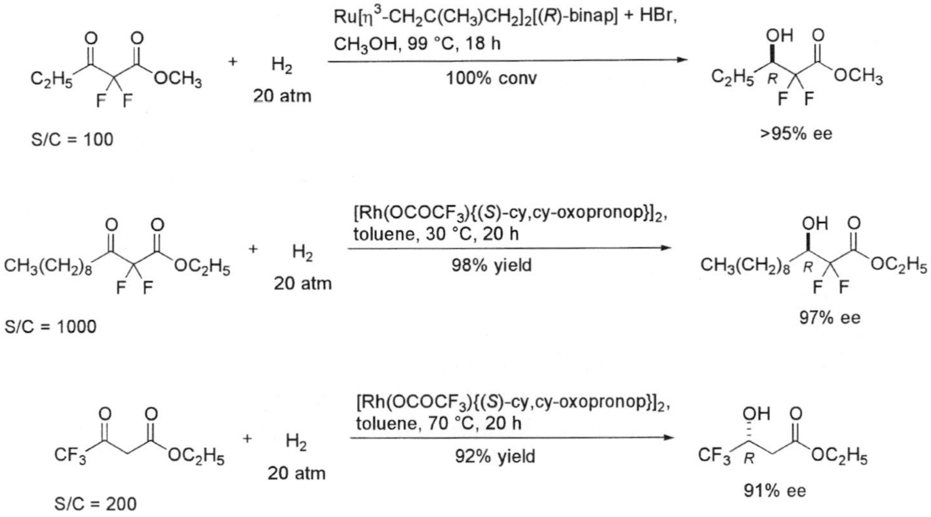

Scheme 21

tanoate catalyzed by the Cy,Cy-oxoProNOP/Rh complex gave a somewhat lower ee [63].

Highly stereoselective hydrogenation of racemic α-substituted β-keto esters via dynamic kinetic resolution [14, 17] has been reported. Hydrogenation of a racemic α-amidomethyl substrate with the (-)-DTBM-SEGPHOS/Ru catalyst resulted in the 2S,3R alcohol in 99.4% ee (syn:anti=99.3:0.7) (Scheme 22) [36]. The product was a key compound for an industrial synthesis of carbapenem antibi-

catalyst = [NH$_2$(C$_2$H$_5$)$_2$][{RuCl[(-)-dtbm-segphos]}$_2$(μ-Cl)$_3$]   syn:anti = 99.3:0.7

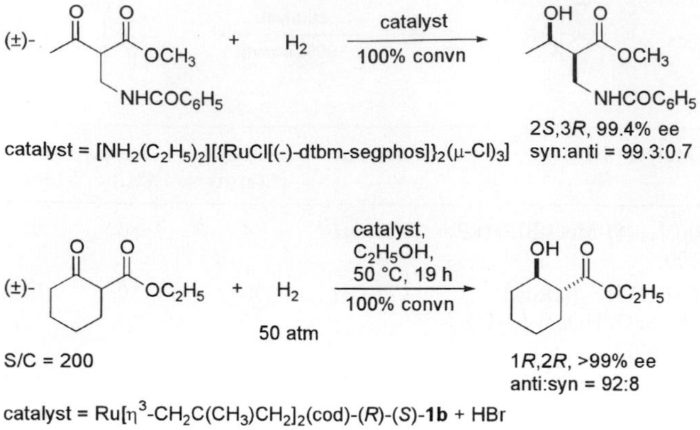

catalyst = Ru[η^3-CH$_2$C(CH$_3$)CH$_2$]$_2$(cod)-(R)-(S)-**1b** + HBr

Scheme 22

otics [14]. A Ru catalyst with a chiral ferrocenyl ligand, (R)-(S)-**1b**, catalyzed the reaction of racemic 2-ethoxycarbonylcyclohexanone to give the 1R,2R alcohol in >99% ee (*anti:syn=92:8*) [54].

1.7
Keto Sulfonates, Sulfones, and Sulfoxides

BINAP/Ru-catalyzed hydrogenation is useful for the asymmetric synthesis of β-hydroxysulfonic acids. Thus, hydrogenation of sodium β-ketosulfonates with the (R)-BINAP/Ru catalyst gave the corresponding R alcohols quantitatively in up to 97% ee (Table of Scheme 23) [64]. In a similar manner, β-ketosulfones hydrogenated with the (R)-MeO-BIPHEP/Ru catalyst gave the R alcohol in >95% ee [65]. Diastereoselective hydrogenation of (R)-β-ketosulfoxides with MeO-BIPHEP/Ru catalyst has been reported [66]. The R configuration of the substrate matched with the S catalyst, affording the S,R alcohols predominantly (Table of Scheme 24), whereas hydrogenation with the R catalyst gave a 6:94–10:90 mixture of S,R and R,R alcohols. The stereochemistry of the products was mainly controlled by the catalyst.

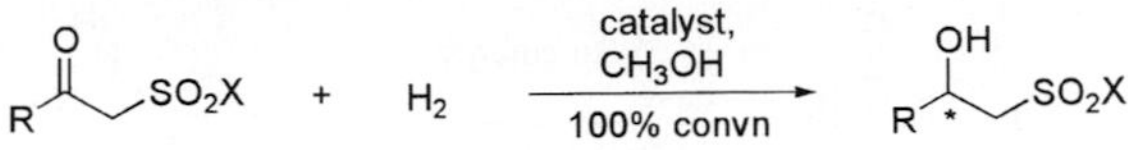

Scheme 23

R	X	Catalyst (S/C[a])	Pressure (atm)	Temp. (°C)	ee (%)	Config'n
CH$_3$	ONa	RuCl$_2$[(R)-BINAP](DMF)$_n$+HCl (200)	1	50	97	R
n-C$_{15}$H$_{31}$	ONa	RuCl$_2$[(R)-BINAP]DMF)$_n$+HCl (200)	1	50	96	R
(CH$_3$)$_2$CH	ONa	RuCl$_2$[(R)-BINAP]DMF)$_n$+HCl (200)	1	50	97	R
C$_6$H$_5$	ONa	RuCl$_2$[(R)-BINAP]DMF)$_n$+HCl (200)	1	50	96	R
CH$_3$	C$_6$H$_5$	RuBr$_2$[(R)-MeO-BIPHEP] (100)	1	65	>95	R
n-C$_5$H$_{11}$	C$_6$H$_5$	RuBr$_2$[(R)-MeO-BIPHEP] (100)	1	65	>95	R
Cyclo-C$_6$H$_{11}$	C$_6$H$_5$	RuBr$_2$[(R)-MeO-BIPHEP] (100)	1	65	>95	R
C$_6$H$_5$	C$_6$H$_5$	RuBr$_2$[(S)-MeO-BIPHEP] (100)	75	40	>95	S

[a] Substrate/catalyst molar ratio.

Scheme 23

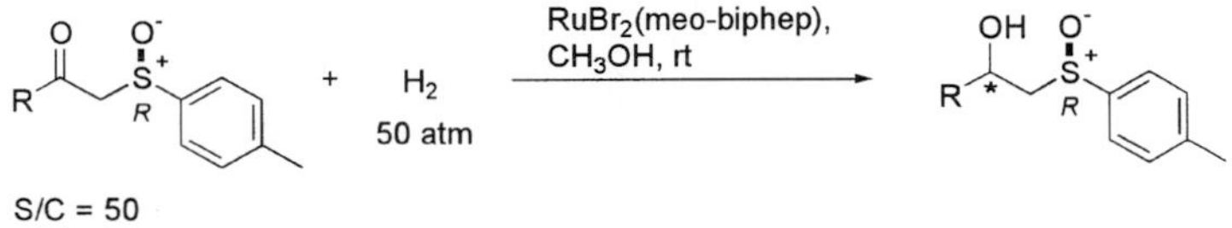

S/C = 50

R	MeO-BIPHEP	Time (h)	Yield (%)	S,R:R,R
n-C$_6$H$_{13}$	S	25	82	>99:1
n-C$_6$H$_{13}$	R	25	74	6:94
C$_6$H$_5$	S	63	70	>99:1
C$_6$H$_5$	R	63	95	10:90

Scheme 24

2
Transfer Hydrogenation

Significant progress in the transfer hydrogenation of ketones has been made by the discovery of a new type of Ru complex, RuCl[YCH(C$_6$H$_5$)CH(C$_6$H$_5$)NH$_2$](η^6-arene) (23; Y=O, 24; Y=NTs, Ts=p-toluenesulfonyl) (Scheme 25) [1, 67]. A variety of aromatic and acetylenic carbonyl compounds are reduced in an alkaline base containing 2-propanol to give the corresponding chiral alcohols in up to >99% ee [68–70]. Use of a mixture of formic acid and triethylamine as a reducing agent often leads to better results [71].

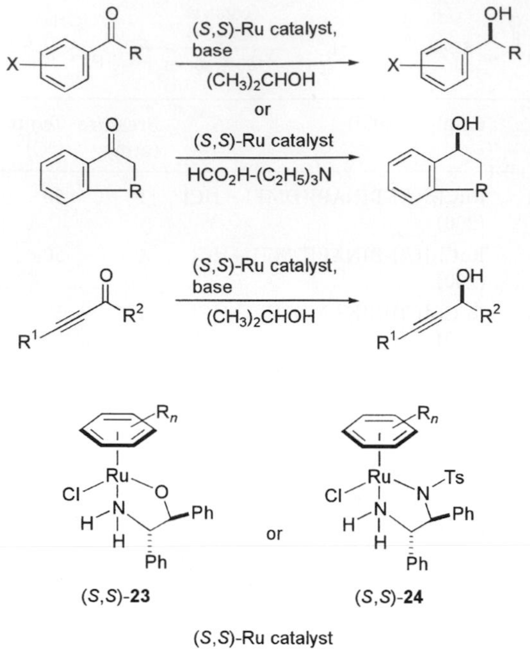

Scheme 25

2.1
Mechanistic Studies

Recently, the mechanism of this reaction has been investigated by detailed experimental [72] and theoretical methods [73–76]. As shown in Scheme 26, the reaction with 2-propanol proceeds through a coordinatively saturated 18-electron complex, $RuH[YCH(C_6H_5)CH(C_6H_5)NH_2](\eta^6\text{-arene})$. The carbonyl function of the ketonic substrate is saturated via a six-membered pericyclic transition state (TS). The 16-electron Ru-amide complex produced regenerates the 18-electron species with dehydrogenation of 2-propanol. Neither carbonyl oxygen nor alcoholic oxygen interact with the Ru center during any part of the reaction. An alkaline base is necessary merely for the generation of the 16-electron species from the 18-electron Ru chloride complex (23 or 24 in Scheme 25) by a Dcb elimination of HCl. The S,S catalyst gives the S alcohol via the favored TS_1 that is stabilized by the CH/π attractions between the η^6-arene ligand and the aromatic ring of the substrate [75]. The metal ligand bifunctional mechanism is in sharp contrast to many other systems mediated by metal complexes. Mechanistic studies on the transfer hydrogenation of ketones catalyzed by other metal complexes have also been reported [77].

This reaction has even been used for the asymmetric synthesis of useful biologically active compounds on an industrial scale [78].

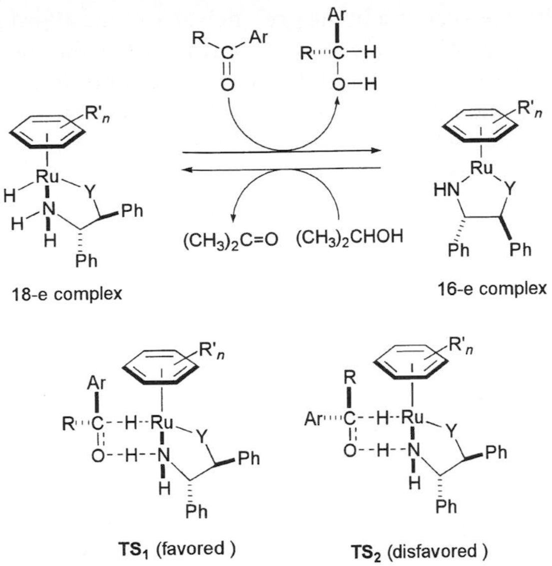

Scheme 26

2.2
Simple Ketones

Many kinds of chiral ligands have been applied to the asymmetric transfer hydrogenation of aromatic ketones [3, 4, 79, 80]. Recent examples are listed in the Table of Scheme 27. Chiral diamine 3 and amino amide 5 ligands derived from proline are effective for reaction of 2′-substituted acetophenones [81, 82]. Ru catalysts with (1R,2S)-ephedrine analogues 7 and 8 reduced acetophenone and 1′-acetonaphthone with 2-propanol in >95% optical yield [83, 84]. A series of alkyl aryl ketones were reduced using Ru catalysts with chiral 2-azanorbornyl amino alcohols 10 and 11 to give the corresponding alcohols in up to >99% ee [85, 86]. The introduction of dimethyl ketal to the ligand remarkably enhanced the catalytic activity [86]. The reaction was performed with an S/C as high as 5,000 and a TOF of 8,500 h^{-1}. A derivative of 2-amino-2′-hydroxy-1,1′-binaphthyl 12 is also effective for the reduction of acetophenone [87]. A Ru catalyst prepared from $RuCl_2[P(C_6H_5)_3]_3$ and a tridentate nitrogen-based ligand, Ph-AMBOX, in base containing 2-propanol effected the enantioselective reduction of several aromatic ketones with up to 98% ee [88]. Complexes prepared in situ from $RuCl_2[P(C_6H_5)_3]_3$ and oxazolylferrocenylphosphines 2, which exist as a diastereomixture, catalyze the transfer hydrogenation of aromatic ketones in the presence of a base in 2-propanol with high enantioselectivity [89]. The isolated complexes (S)-25 showed even higher stereoselectivity [90]. A variety of aromatic ketones were converted to the corresponding R alcohols in up to >99.9% ee. Kinetic resolution of racemic aromatic alcohols through asymmetric dehydrogenation with the S catalyst in the presence of acetone afforded the R alcohol in high ee at the appropriate conversion accompanied by the corresponding aromatic ketones and 2-propanol [90, 91]. The enantiomer differentiating ratio was as high as >368:1. RhCp*complexes **26** and **27**, which are isolobal with Ru(η^6-arene) complexes **24** (Scheme 25), are also effective for the asymmetric transfer hydrogenation of several aromatic ketones [92, 93]. An Ir catalyst with aminosulfide **6** [94] and an Os catalyst with amino alcohol **9** [95, 96] catalyzed the transfer hydrogenation of 1'-acetonaphthone with 97% and 94% optical yield, respectively.

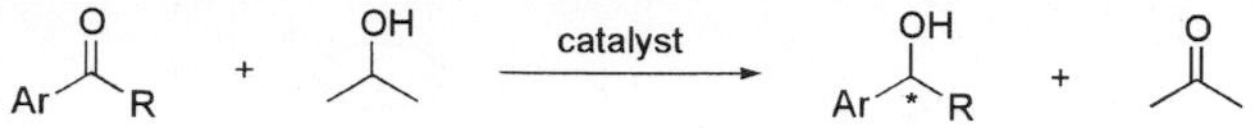

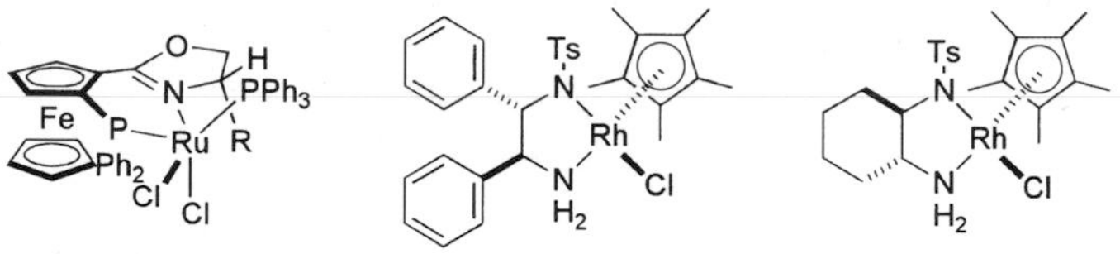

(S)-**25a**: R = (CH₃)₂CH
(S)-**25b**: R = C₆H₅   (S,S)-**26**   (R,R)-**27**

R	Ar	Catalyst (S/C[a])	Temp. (°C)	Yield (%)	ee (%)	Confign
CH₃	C₆H₅	[RuCl₂{C₆(CH₃)₆}]₂-(S,R,R)-**10**+ (CH₃)₂CHOK (200)	rt	92	95	S
CH₃	C₆H₅	[RuCl₂(p-cymene)]₂-**11**+ (CH₃)₂CHOK (5,000)	rt	96	96	S
CH₃	C₆H₅	[RuCl₂(p-cymene)]₂-(R,S)-**8**+ (CH₃)₃COK (200)	rt	91	95	R
CH₃	C₆H₅	[RuCl₂(p-cymene)]₂-(R,S)-**7**+ KOH (100)	10	91	95	R
CH₃	C₆H₅	RuCl₂[P(C₆H₅)₃]₃-(S)-**12**+ (CH₃)₃COK (100)	28	94.3	96.7	S
CH₃	C₆H₅	RuCl₂[P(C₆H₅)₃]₃-(R)-Ph-AMBOX+(CH₃)₂CHONa (100)	82	80	98	S
CH₃	C₆H₅	(S)-**25b**+(CH₃)₂CHONa (200)	rt	95	>99.7	R
CH₃	C₆H₅	(S,S)-**26**+KOH (100)	rt	80	90	S
CH₃	C₆H₅	(R,R)-**27**+(CH₃)₃COK (200)	30	85	97	R
CH₃	2-CH₃C₆H₄	RuCl₂[P(C₆H₅)₃]₃-(R)-Ph-AMBOX+(CH₃)₂CHONa (100)	82	96	98	S
CH₃	2-CH₃C₆H₄	(S)-**25a**+(CH₃)₂CHONa (200)	rt	99	>99.9	R
CH₃	4-C₂H₅C₆H₄	(R,R)-**27**+(CH₃)₃COK (200)	30	58	>99	R

Scheme 27

R	Ar	Catalyst (S/C[a])	Temp. (°C)	Yield (%)	ee (%)	Con-fign
CH$_3$	3-CH$_3$C$_6$H$_4$	(S)-**25a**+(CH$_3$)$_2$CHONa (200)	rt	98	>99.9	R
CH$_3$	2-BrC$_6$H$_4$	[RuCl$_2$(p-cymene)]$_2$-(S)-**5** (50)[b]	30	75	98.8	R
CH$_3$	2-BrC$_6$H$_4$	[RuCl$_2$(p-cymene)]$_2$-**11**+ (CH$_3$)$_2$CHOK (1000)	rt	98	95	S
CH$_3$	4-BrC$_6$H$_4$	(S)-**25a**+(CH$_3$)$_2$CHONa (200)	rt	99	>99.3	R
CH$_3$	2-CH$_3$OC$_6$H$_4$	[RuCl$_2$(p-cymene)]$_2$-(S)-**3**+(CH$_3$)$_2$CHONa (100)	rt	99	96	R
CH$_3$	3-NH$_2$C$_6$H$_4$	[RuCl$_2$(p-cymene)]$_2$-**11**+ (CH$_3$)$_2$CHOK (200)	rt	98	99	S
CH$_3$	1-Naphthyl	[RuCl$_2$(p-cymene)]$_2$-(S,R,R)-**10**+ (CH$_3$)$_2$CHOK (200)	rt	92	97	S
CH$_3$	1-Naphthyl	[RuCl$_2$(p-cymene)]$_2$-**11**+ (CH$_3$)$_2$CHOK (1000)	rt	98	>99	S
CH$_3$	1-Naphthyl	[RuCl$_2$(p-cymene)]$_2$-(R,S)-**8**+ (CH$_3$)$_3$COK (200)	rt	96	96	R
CH$_3$	1-Naphthyl	RuCl$_2$[P(C$_6$H$_5$)$_3$]$_3$-(R)-Ph-AMBOX+(CH$_3$)$_2$CHONa (100)	82	72	96	S
CH$_3$	1-Naphthyl	[IrCl(COD)]$_2$+(R,S)-**6**+ (CH$_3$)$_3$COK (200)	20	99	97	R
CH$_3$	1-Naphthyl	[OsCl$_2$(p-cymene)]$_2$+(S,R)-**9**+ (CH$_3$)$_3$COK (100)	24	80	94	S
CH$_3$	2-Furyl	(S)-**25a**+(CH$_3$)$_2$CHONa (200)	rt	66	95	R
C$_2$H$_5$	C$_6$H$_5$	RuCl$_2$[P(C$_6$H$_5$)$_3$]$_3$-(R)-Ph-AMBOX+(CH$_3$)$_2$CHONa (100)	82	77	95	S
C$_2$H$_5$	C$_6$H$_5$	(S)-**25b**+(CH$_3$)$_2$CHONa (200)	50	99	>99.7	R
n-C$_4$H$_9$	C$_6$H$_5$	[RuCl$_2$(p-cymene)]$_2$-(S,R,R)-**10**+ (CH$_3$)$_2$CHOK (200)	rt	78	95	S

[a] Substrate/catalyst molar ratio. [b] Formic acid was a hydride source.

Scheme 27. (continued)

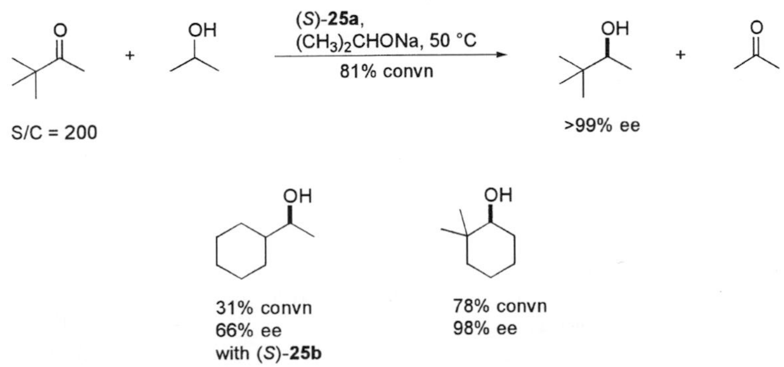

S/C = 200

(S)-**25a**, (CH$_3$)$_2$CHONa, 50 °C

81% convn

>99% ee

31% convn
66% ee
with (S)-**25b**

78% convn
98% ee

Scheme 28

Excellent enantioselectivity was achieved for the transfer hydrogenation of pinacolone by using (*S*)-**25a** as a catalyst with 2-propanol in the presence of (CH$_3$)$_2$CHONa to give the *S* alcohol in >99% ee (Scheme 28) [90]. 2,2-Dimethyl-cyclohexanone was reduced with the same catalyst with 98% optical yield. Reduction of cyclohexyl methyl ketone with (*S*)-**25b** gave the *S* alcohol in 66% ee.

Transfer hydrogenation of ketones using metal complexes with a chiral water-soluble [97, 98] and a dendritic ligand [99] was investigated for use in recycling catalysts. The reaction with immobilized catalysts has also been reported [100].

2.3
Pyridyl Ketones

The Ru-catalyzed asymmetric transfer hydrogenation was applied to heteroatom-containing ketonic substrates. Hydrogenation of 2-acetylpyridine using a catalyst consisting of [RuCl$_2$(*p*-cymene)]$_2$ and a chiral amino alcohol (*S,R*)-**8** in a base containing 2-propanol afforded the *R* alcohol quantitatively in 89% ee (Scheme 29) [101]. Several pyridyl ketones were reduced with (*S,S*)-**28** and a formic acid-triethylamine mixture to give the *S* pyridyl alcohol in up to 95% ee (Scheme 30) [102]. The chiral alcohol with a pyridyl-furyl fused ring is a synthetic intermediate of a potent anti-HIV agent, PNU-142721. Double reduction of 2,6-diacetylpyridine gave the *S,S* diol (*dl:meso*=91:9) in 99.6% ee.

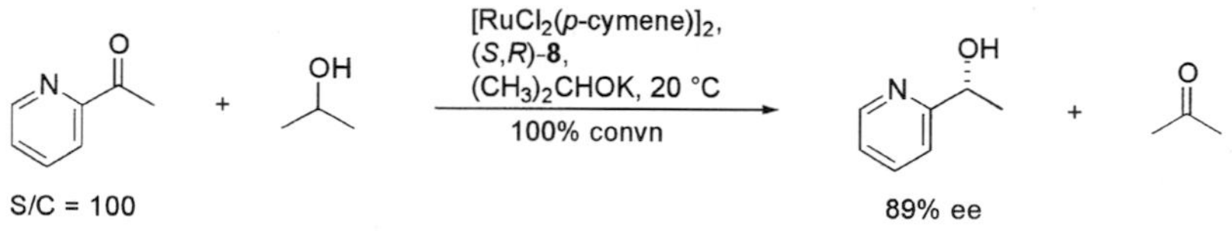

Scheme 29

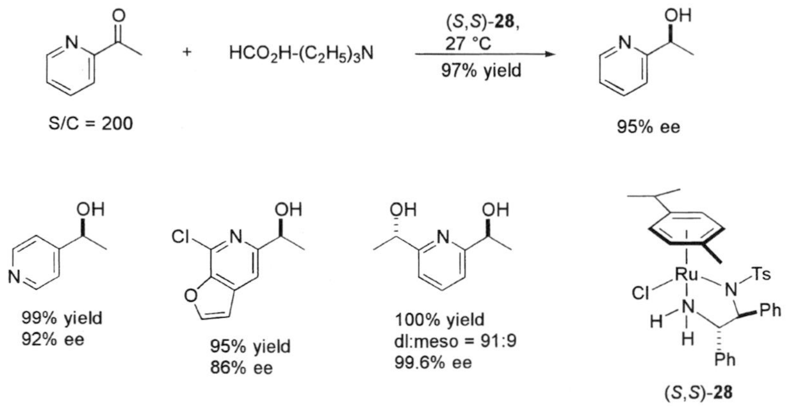

Scheme 30

2.4
α-Substituted Ketones

Transfer hydrogenation of a series of α-functionalized ketones with a chiral Ru or Rh catalyst in a formic acid-triethylamine mixture gave the corresponding chiral alcohols in high ee. 2-Cyanoacetophenone was reduced using (S,S)-28 with an S/C ratio of 1,000 to afford the S alcohol quantitatively in 98% ee (Table of Scheme 31) [103]. The sense of enantioselection was the same as that occurring with the reduction of acetophenone [71]. Reaction of 2-azido- and 2-nitroacetophenones with the same catalyst gave the R alcohols in up to 98% ee. Acetophenone containing an amide group was reduced with a catalyst consisting of [RuCl$_2$(p-cymene)]$_2$ and (R,R)-TsDPEN to give the S alcohol in 99% ee [104]. 3-t-Butoxycarbonyl(methyl)amino-1-phenoxy-2-propanone was reduced with 82% optical yield, while the absolute configuration was unknown. Reduction of 2-chloroacetophenone gave the S alcohol in 85% ee and in 68% yield [105]. 2-Hydroxy- and 2-(4-toluenesulfonyl)oxyacetophene were reduced with a [RhCl$_2$Cp*]$_2$–(R,R)-TsDPEN catalyst to give the S alcohols in moderate yields with a 94% and 93% ee, respectively [105].

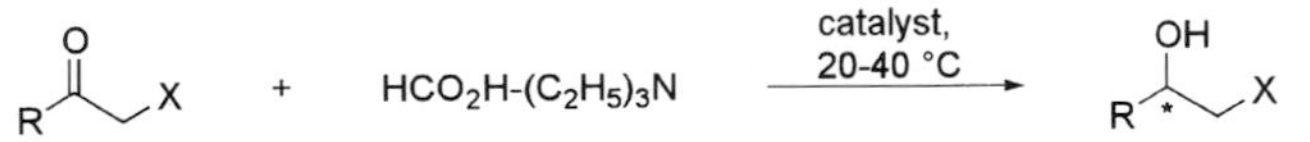

R	X	Catalyst (S/C[a])	Yield (%)	ee (%)	Confign
C$_6$H$_5$	CN	(S,S)-28 (1,000)	100	98	S
C$_6$H$_5$	t-C$_4$H$_9$OCO, (CH$_3$)N	[RuCl$_2$(p-cymene)]$_2$-(R,R)-TsDPEN (100)	82	99	S
C$_6$H$_5$OCH$_2$	t-C$_4$H$_9$OCO, (CH$_3$)N	[RuCl$_2$(p-cymene)]$_2$-(R,R)-TsDPEN (200)	50	82	–
C$_6$H$_5$	N$_2$	(S,S)-28 (100)	65	92	R
C$_6$H$_5$	NO$_2$	(S,S)-28 (200)	90	98	R
4-FC$_6$H$_4$	NO$_2$	(S,S)-28 (200)	95	96	–
C$_6$H$_5$	OH	[RhCl$_2$Cp*]$_2$-(R,R)-TsDPEN (200)	59	94	S
C$_6$H$_5$	OTs	[RhCl$_2$Cp*]$_2$-(R,R)-TsDPEN (200)	65	93	S
C$_6$H$_5$	Cl	[RuCl$_2$(p-cymene)]$_2$-(R,R)-TsDPEN (200)	68	85	S

[a] Substrate/catalyst molar ratio.

Scheme 31

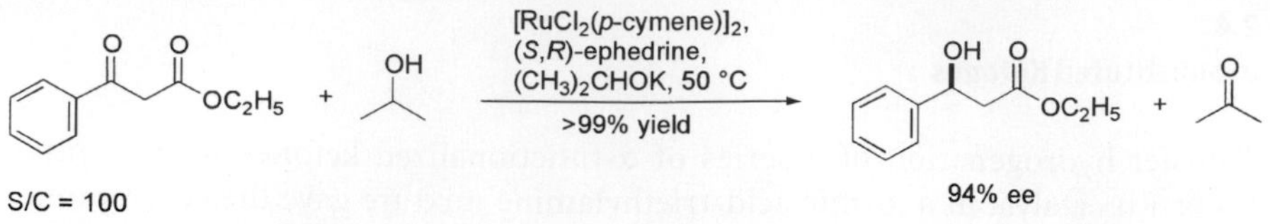

Scheme 32

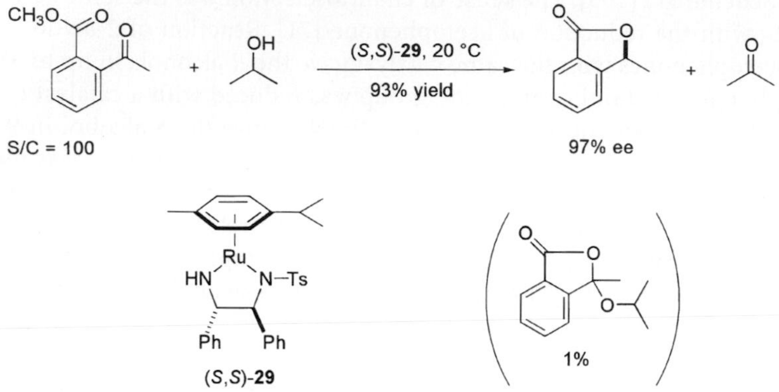

Scheme 33

2.5
Keto Esters

Ethyl benzoylacetate was hydrogenated with [RuCl$_2$(*p*-cymene)]$_2$, (*S,R*)-ephe-drine, and (CH$_3$)$_2$CHOK in 2-propanol to give the *S* hydroxy ester quantitative-ly in 94% ee (Scheme 32) [106]. Reduction of methyl 2-acetylbenzoate was per-formed with the 16-electron Ru catalyst, (*S,S*)-**29** [72], in 2-propanol to afford (*S*)-3-methylphthalide in 97% ee and in 93% yield contaminated with 1% of 3-(2-isopropoxy)-3-methylphthalide (Scheme 33) [107].

2.6
Formation of Chiral Diols and Hydroxyketones

The highly enantioselective reduction of benzils was achieved by the use of the chiral Ru complex (*S,S*)-**28** with an S/C of 1,000 in a formic acid–triethylamine mixture to give the *R,R* diol in >99% ee (Scheme 34) [108]. The sense of enan-tioselection was the same as that of the reduction of simple aromatic ketones, suggesting that the adjacent oxygen atom does not participate in the stereoreg-ulation. Introduction of electron-accepting functions at the 4 and 4′ positions increased the reaction rate, while the enantioselectivity was not affected by the electronic properties of the substituents. Use of 2-propanol as a hydride source caused both the rate and enantioselectivity to decrease. An unsymmetrical 1,2-

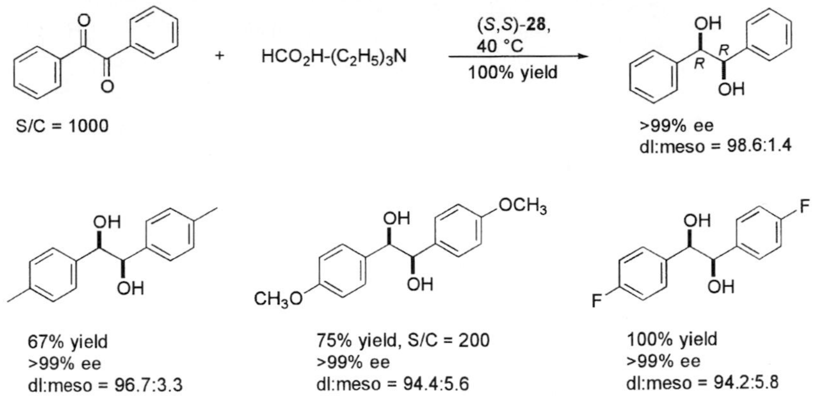
(S,S)-28,
40 °C
100% yield
HCO2H-(C2H5)3N
S/C = 1000
>99% ee
dl:meso = 98.6:1.4
67% yield
>99% ee
dl:meso = 96.7:3.3
75% yield, S/C = 200
>99% ee
dl:meso = 94.4:5.6
100% yield
>99% ee
dl:meso = 94.2:5.8

Scheme 34

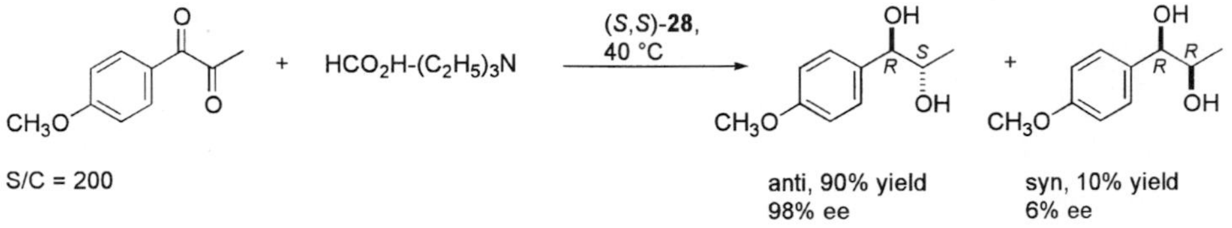
(S,S)-28,
40 °C
HCO2H-(C2H5)3N
S/C = 200
anti, 90% yield
98% ee
syn, 10% yield
6% ee

Scheme 35

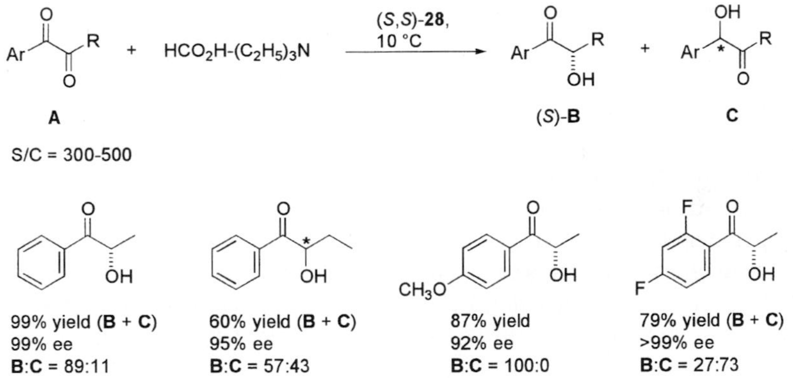
(S,S)-28,
10 °C
HCO2H-(C2H5)3N
A
(S)-B
C
S/C = 300-500
99% yield (B + C)
99% ee
B:C = 89:11
60% yield (B + C)
95% ee
B:C = 57:43
87% yield
92% ee
B:C = 100:0
79% yield (B + C)
>99% ee
B:C = 27:73

Scheme 36

diketone was reduced with the same catalyst to afford the 1*R*,2*S* (*anti*) alcohol in 98% ee in 90% yield accompanied by 10% of the 1*R*,2*R* (*syn*) alcohol in 6% ee (Scheme 35) [109].

When the unsymmetrical 1,2-diketone **A** was reduced with (*S*,*S*)-**28** and a formic acid–triethylamine mixture at 10 °C, α-hydroxy ketones, (*S*)-**B** in up to >99% ee and **C**, were obtained selectively (Scheme 36) [109]. The selectivity between **B** and **C** was highly dependent on the character of the aromatic ring of

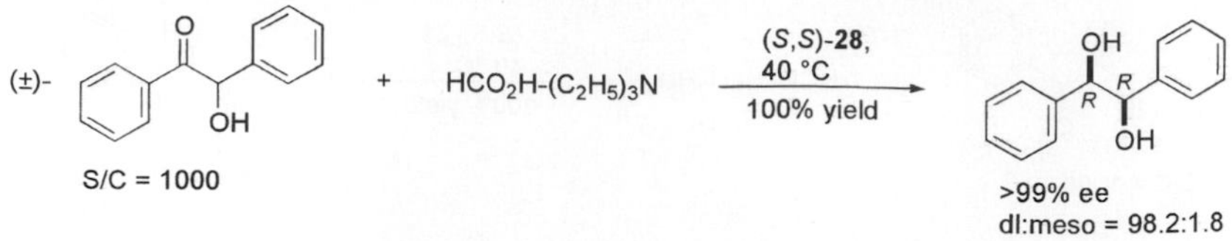

Scheme 37

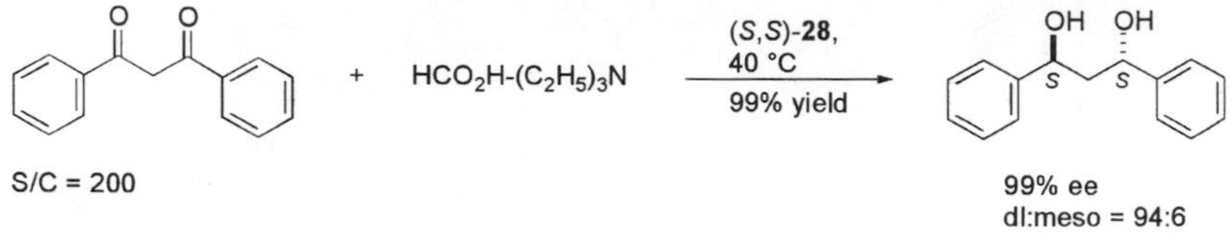

Scheme 38

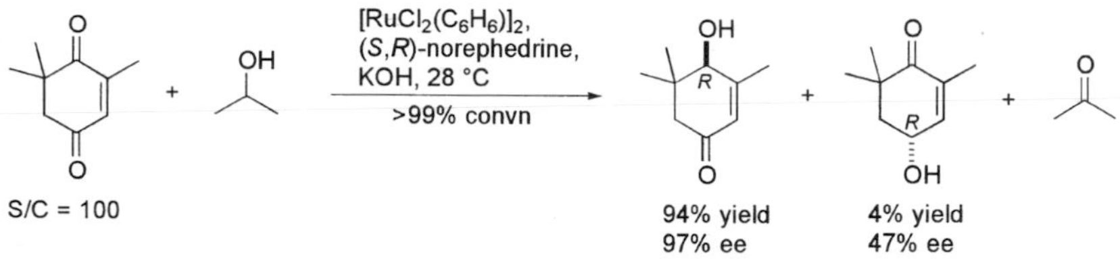

Scheme 39

substrate **A**: a 4'-methoxy substrate predominantly produced (S)-**B** in 92% ee, while a 2',4'-difluoro compound gave (S)-**B** in >99% ee and (R)-**C** in 47% ee with a 27:73 ratio.

Racemic benzoin was reduced with (S,S)-**28** in a formic acid–triethylamine mixture to give the R,R diol (dl:meso=98.2:1.8) quantitatively in >99% ee via dynamic resolution, revealing that racemization at the benzylic carbon atom occurs rapidly under transfer hydrogenation conditions (Scheme 37) [108]. The reduction rate of (R)-benzoin was calculated to be 55 times faster than the S isomer.

Reduction of 1,3-diphenyl-1,3-propanedione catalyzed by (S,S)-**28** in a formic acid–triethylamine mixture gave the S,S 1,3-diol (dl:meso=94:6) in 99% ee and 99% yield (Scheme 38) [103, 110].

Ketoisophorone was reduced selectively at the sterically hindered carbonyl function in a 1,2-fashion using a chiral catalyst consisting of [RuCl$_2$(C$_6$H$_6$)]$_2$ and (S,R)-norephedrine in a KOH containing 2-propanol to give (R)-4-hydroxyisophorone in 97% ee and 94% yield accompanied by (R)-phorenol in 47% ee and 4% yield (Scheme 39) [111]. The (R)-enol obtained is a useful intermediate for the synthesis of pharmaceuticals and natural pigments.

The combination of Ru complex-catalyzed stereomutation of secondary alcohols with enzyme-catalyzed enantioselective acylation is an efficient procedure to obtain chiral acyloxy compounds with excellent optical purity from a variety of racemic secondary alcohols via dynamic kinetic resolution [112].

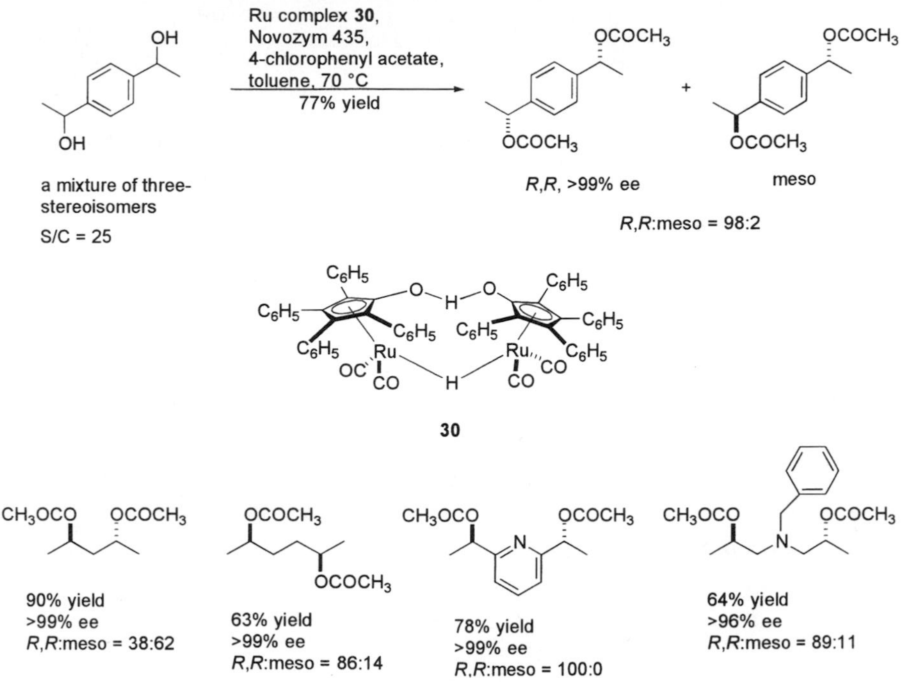

Scheme 40

This reaction was applied to a mixture of diastereoisomeric diols [113]. For instance, α,α′-dimethyl-1,4-benzenedimethanol was treated with Ru complex **30** (S/C=25), Novozyme 435, and 4-chlorophenyl acetate (3 equiv) in toluene at 70 °C to give the *R,R* diacetate in >99% ee and the *meso* isomer (*R,R:meso*=98:2) in 77% yield (Scheme 40). Reaction of 1,3-pentanediol and 1,4-hexanediol gave the corresponding *R,R* diacetate in >99% ee, while the *R,R* to *meso* ratio was lower than that of the aromatic diacetate. Nitrogen-containing substrates also gave the desired products in optically pure form.

2.7
Aldehydes

Asymmetric transfer hydrogenation of benzaldehyde-1-*d* with (*R,R*)-**28** and (CH$_3$)$_3$COK in 2-propanol gave (*R*)-benzyl-1-*d* alcohol quantitatively in 98% ee (Scheme 41) [114]. Introduction of electron-donating and electron-accepting groups at the 4′ position had little effect on the enantioselectivity. Catalytic deuteration of benzaldehydes was achieved by use of the same complex (*R,R*)-**28** and a 1:1 mixture of formic acid-2-*d* and triethylamine to give the *S* deuterio alcohols in up to 99% ee (Scheme 42) [114]. The d_1 content in the product alcohol was >99%. Only a stoichiometric amount of deuterium source is required to complete the reaction.

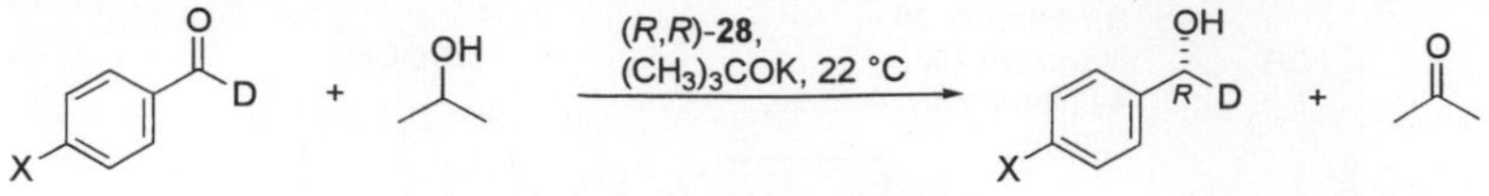

S/C = 200

X	convn [%]	ee [%]
H	100	98
CH$_3$	99	98
OCH$_3$	87	99
Br	97	96
CF$_3$	98	96

Scheme 41

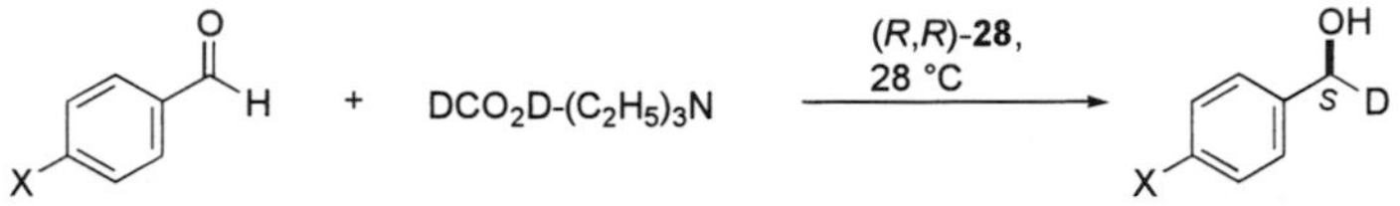

aldehyde:Ru:DCO$_2$D:(C$_2$H$_5$)$_3$N = 200:1:200:200

X	convn [%]	ee [%]
H	93	98
CH$_3$	92	98
OCH$_3$	97	99
Br	99	99
CF$_3$	95	97

Scheme 42

References

1. Noyori R (2002) Angew Chem Int Ed 41:2008
2. Noyori R, Ohkuma T (2001) Angew Chem Int Ed 40:40
3. Ohkuma T, Noyori R (1999) Hydrogenation of carbonyl groups. In: Jacobsen EN, Pfaltz A, Yamamoto H (eds) Comprehensive asymmetric catalysis, vol 1. Springer, Berlin Heidelberg New York, chap 6.1
4. Ohkuma T, Kitamura M, Noyori R (2000) Asymmetric hydrogenation. In: Ojima I (ed) Catalytic asymmetric synthesis, 2nd edn. Wiley VCH, New York, chap 1
5. Doucet H, Ohkuma T, Murata K, Yokozawa T, Kozawa M, Katayama E, England AF, Ikariya T, Noyori R (1998) Angew Chem Int Ed 37:1703
6. Ohkuma T, Koizumi M, Doucet H, Pham T, Kozawa M, Murata K, Katayama E, Yokozawa T, Ikariya T, Noyori R (1998) J Am Chem Soc 120:13529
7. Terashima K, Ohkuma T, Noyori R (2000) Jpn Patent Kokai Tokkyo Koho P2000–26344A
8. Abdur-Rashid K, Lough AJ, Morris RH (2001) Organometallics 20:1047
9. Burk MJ, Hems W, Herzberg D, Malan C, Zanotti-Gerosa A (2000) Org Lett 2:4173
10. Akotsi OM, Metera K, Reid RD, McDonald R, Bergens SH (2000) Chirality 12:514
11. Ito M, Hirakawa M, Murata K, Ikariya T (2001) Organometallics 20:379
12. Jiang Q, Jiang Y, Xiao D, Cao P, Zhang X (1998) Angew Chem Int Ed 37:1100

13. Li RX, Cheng PM, Li DW, Chen H, Li XJ, Wessman C, Wong NB, Tin KC (2000) J Mol Catal A: Chem 159:179
14. Noyori R (1994) Asymmetric catalysis in organic synthesis. Wiley, New York, chap 2
15. Kitamura M, Ohkuma T, Inoue S, Sayo N, Kumobayashi H, Akutagawa S, Ohta T, Takaya H (1988) J Am Chem Soc 110:629
16. Ohkuma T, Koizumi M, Muñiz K, Hilt G, Kabuto C, Noyori R (2002) J Am Chem Soc 124: 6508
17. Noyori R, Tokunaga M, Kitamura M (1995) Bull Chem Soc Jpn 68:36
18. Ohkuma T, Ooka H, Yamakawa M, Ikariya T, Noyori R (1996) J Org Chem 61:4872
19. Bayston DJ, Fraser JL, Ashton MR, Baxter AD, Polywka MEC, Moses E (1998) J Org Chem 63:3137
20. Ohkuma T, Takeno H, Honda Y, Noyori R (2001) Adv Synth Catal 343:369
21. ter Halle R, Schulz E, Spagnol M, Lemaire M (2000) Synlett 680
22. Yu HB, Hu QS, Pu L (2000) Tetrahedron Lett 41:1681
23. Yu HB, Hu QS, Pu L (2000) J Am Chem Soc 122:6500
24. Ohkuma T, Koizumi M, Ikehira H, Yokozawa T, Noyori R (2000) Org Lett 2:659
25. Ohkuma T, Koizumi M, Yoshida M, Noyori R (2000) Org Lett 2:1749
26. Cao P, Zhang X (1999) J Org Chem 64:2127
27. Kuroki Y, Sakamaki Y, Iseki K (2001) Org Lett 3:457
28. Nagel U, Roller C (1998) Z Naturforsch B 53:267
29. Ohkuma T, Ooka H, Ikariya T, Noyori R (1995) J Am Chem Soc 117:10417
30. Ohkuma T, Ikehira H, Ikariya T, Noyori R (1997) Synlett 467
31. Chan KK, Cohen N, De Noble JP, Specian AC Jr, Saucy G (1976) J Org Chem 41:3497
32. Takeda H, Tachinami T, Aburatani M, Takahashi H, Morimoto T, Achiwa K (1989) Tetrahedron Lett 30:363
33. Ohkuma T, Ishii D, Takeno H, Noyori R (2000) J Am Chem Soc 122:6510
34. Pasquir C, Naili S, Pelinski L, Brocard J, Mortreux A, Agbossou F (1998) Tetrahedron: Asymmetry 9:193
35. Devocelle M, Mortreux A, Agbossou F, Dormoy JR (1999) Tetrahedron Lett 40:4551
36. Saito T, Yokozawa T, Ishizaki T, Moroi T, Sayo N, Miura T, Kumobayashi H (2001) Adv Synth Catal 343:264
37. Matsumoto T, Murayama T, Mitsuhashi S, Miura T (1999) Tetrahedron Lett 40:5043
38. Ohkuma T, Ooka H, Hashiguchi S, Ikariya T, Noyori R (1995) J Am Chem Soc 117:2675
39. Ohkuma T, Doucet H, Pham T, Mikami K, Korenaga T, Terada M, Noyori R (1998) J Am Chem Soc 120:1086
40. Mikami K, Korenaga T, Terada M, Ohkuma T, Pham T, Noyori R (1999) Angew Chem Int Ed 38:495
41. Mikami K, Terada M, Korenaga T, Matsumoto Y, Ueki M, Angelaud R (2000) Angew Chem Int Ed 39:3533
42. Mikami K, Korenaga T, Ohkuma T, Noyori R (2000) Angew Chem Int Ed 39:3707
43. Abdur-Rashid K, Lough AJ, Morris RH (2000) Organometallics 19:2655
44. Abdur-Rashid K, Faatz M, Lough AJ, Morris RH (2001) J Am Chem Soc 123:7473
45. Hartmann R, Chen P (2001) Angew Chem Int Ed 40:3581
46. Ohkuma T, Noyori R (1998) Carbonyl hydrogenations. In: Beller M, Bolm C (eds) Transition metals for organic synthesis, vol 2. Wiley VCH, Weinheim, p 25
47. Pasquier C, Naili S, Mortreux A, Agbossou F, Pélinski L, Brocard J, Eilers J, Reiners I, Peper V, Martens J (2000) Organometallics 19:5723
48. Guerreiro P, Cano de Andrade MC, Henry JC, Tranchier JP, Phansavath P, Ratovelomanana-Vidal V, Genêt JP, Homri T, Youati AR, Hassine BB (1999) Compt Rendus Acad Sci Paris Ser 2 175
49. Madec J, Pfister X, Phansavath P, Ratovelomanana-Vidal V, Genêt JP (2001) Tetrahedron 57:2563
50. Zhang Z, Qian H, Longmire J, Zhang X (2000) J Org Chem 65:6223
51. Gelpke AES, Kooijman H, Spek AL, Hiemstra H (1999) Chem Eur J 5:2572

52. Pai CC, Li YM, Zhou ZY, Chan ASC (2002) Tetrahedron Lett 43:2789
53. Pai CC, Lin CW, Lin CC, Chen CC, Chan ASC (2000) J Am Chem Soc 122:11513
54. Ireland T, Tappe K, Grossheimann G, Knochel P (2002) Chem Eur J 8:843
55. ter Halle R, Colasson B, Schulz E, Spagnol M, Lemaire M (2000) Tetrahedron Lett 41:643
56. Guerreiro P, Ratovelomanana-Vidal V, Genêt JP, Dellis P (2001) Tetrahedron Lett 42:3423
57. Lamouille T, Saluzzo C, ter Halle R, Le Guyader F, Lemaire M (2001) Tetrahedron Lett 42: 663
58. Vankelecom I, Wolfson A, Geresh S, Landau M, Gottlieb M, Hershkovitz M (1999) Chem Commun 2407
59. BINAP/Ru: Schulz S (1999) Chem Commun 1239; Ganci W, Kuruüzüm A, Calis I, Rüedi P (2000) Chirality 12:139; Guerreiro P, Ratovelomanana-Vidal V, Genêt JP (2000) Chirality 12:408; Bianchini C, Barbaro P, Scapacci G, Zanobini F (2000) Organometallics 19: 2450; Kiegiel J, Józwik J, Wozniak K, Jurczak J (2000) Tetrahedron Lett 41:4959; Poupardin O, Greck C, Genêt JP (2000) Tetrahedron Lett 41:8795; Rüeger H, Stutz S, Göschke R, Spindler F, Maibaum J (2000) Tetrahedron Lett 41:10085; Makino K, Okamoto N, Hara O, Hamada Y (2001) Tetrahedron: Asymmetry 12:1757; Upadhya TT, Nikalje MD, Sudalai A (2001) Tetrahedron Lett 42:4891; Fürstner A, Dierkes T, Thiel OR, Blanda G (2001) Chem Eur J 7:5286; MeO-BIPHEP/Ru: Phansavath P, Duprat de Paule S, Ratovelomanana-Vidal V, Genêt JP (2000) Eur J Org Chem 3903; Lavergne D, Mordant C, Ratovelomanana-Vidal V, Genêt JP (2001) Org Lett 3:1909
60. Wu J, Chen H, Zhou ZY, Yeung CH, Chan ASC (2001) Synlett 1050
61. Huang HL, Liu LT, Chen SF, Ku H (1998) Tetrahedron: Asymmetry 9:1637
62. Blanc D, Ratovelomanana-Vidal V, Gillet JP, Genêt JP (2000) J Organomet Chem 603:128
63. Kuroki Y, Asada D, Iseki K (2000) Tetrahedron Lett 41:9853
64. Kitamura M, Yoshimura M, Kanda N, Noyori R (1999) Tetrahedron 55:8769
65. Bertus P, Phansavath P, Ratovelomanana-Vidal V, Genêt JP, Touati AR, Homri T, Hassine BB (1999) Tetrahedron: Asymmetry 10:1369
66. Duprat De Paule S, Piombo L, Ratovelomanana-Vidal V, Greck C, Genêt JP (2000) Eur J Org Chem 1535.
67. Noyori R, Hashiguchi S (1997) Acc Chem Res 30:97
68. Hashiguchi S, Fujii A, Takehara J, Ikaroya T, Noyori R (1995) J Am Chem Soc 117:7562
69. Takehara J, Hashiguchi S, Fujii A, Inoue S, Ikariya T, Noyori R (1996) Chem Commun 233
70. Matsumura K, Hashiguchi S, Ikariya T, Noyori R (1997) J Am Chem Soc 119:8738
71. Fujii A, Hashiguchi S, Uematsu N, Ikariya T, Noyori R (1996) J Am Chem Soc 118:2521
72. Haack KJ, Hashiguchi S, Fujii A, Ikariya T, Noyori R (1997) Angew Chem Int Ed Engl 36: 285
73. Yamakawa M, Ito H, Noyori R (2000) J Am Chem Soc 122:1466
74. Alonso DA, Brandt P, Nordin SJM, Andersson PG (1999) J Am Chem Soc 121:9580
75. Yamakawa M, Yamada I, Noyori R (2991) Angew Chem Int Ed 40:2818
76. Noyori R, Yamakawa M, Hashiguchi S (2001) J Org Chem 66:7931
77. Kenny JA, Versluis K, Heck AJR, Walsgrove T, Wills M (2000) Chem Commun 99; Petra DGI, Reek JNH, Handgraaf JW, Meijer EJ, Dierkes P, Kamer PCJ, Brussee J, Schoemaker HE, van Leeuwen PWNM (2000) Chem Eur J 6:2818; Pàmies O, Bäckvall JE (2001) Chem Eur J 7:5052; Casey CP, Singer SW, Powell DR, Hayashi RK, Kavana M (2001) J Am Chem Soc 123:1090
78. Tanaka K, Katsurada M, Ohno F, Shiga Y, Oda M (2000) J Org Chem 65:432; Rossi R, Bellina F, Biagetti M, Mannina L (1999) Tetrahedron: Asymmetry 10:1163; Suginome M, Iwanami T, Ito Y (1999) Chem Commun 2537; Yamano Y, Watanabe Y, Watanabe N, Ito M (2000) Chem Pharm Bull 48:2017
79. Gladiali S, Mestroni G (1998) Transfer hydrogenation. In: Beller M, Bolm C (eds) Transition metals for organic synthesis, vol 2. Wiley VCH, Weinheim, p 97
80. Palmer MJ, Wills M (1999) Tetrahedron: Asymmetry 10:2045

81. Aitali M, Allaoud S, Karim A, Meliet C, Mortreux A (2000) Tetrahedron: Asymmetry 11: 1367
82. Rhyoo HY, Yoon YA, Park HJ, Chung YK (2001) Tetrahedron Lett 42:5045
83. Frost CG, Mendonca P (2000) Tetrahedron: Asymmetry 11:1845
84. Petra DGI, Kamer PCJ, van Leeuwen PWNM, Goubitz K, Van Loon AM, de Vries JG, Schoemaker HE (1999) Eur J Org Chem 2335
85. Alonso DA, Guijarro D, Pinho P, Temme O, Andersson PG (1998) J Org Chem 63:2749
86. Nordin SJM, Roth P, Tarnai T, Alonso DA, Brandt P, Andersson PG (2001) Chem Eur J 7: 1431
87. Brunner H, Henning F, Weber M (2002) Tetrahedron: Asymmetry 13:37
88. Jiang Y, Jiang Q, Zhang X (1998) J Am Chem Soc 120:3817; Zhang X (1999) Enantiomer 4: 541
89. Sammakia T, Stangeland EL (1997) J Org Chem 62:6104
90. Nishibayashi Y, Takei I, Uemura S, Hidai M (1999) Organometallics 18:2291
91. Hashiguchi S, Fujii A, Haack KJ, Matsumura K, Ikariya T, Noyori R (1997) Angew Chem Int Ed Engl 36:288
92. Mashima K, Abe T, Tani K (1998) Chem Lett 1199; Mashima K, Abe T, Tani K (1998) Chem Lett 1201
93. Murata K, Ikariya T, Noyori R (1999) J Org Chem 64:2186
94. Petra DGI, Kamer PCJ, Spek AL, Schoemaker HE, van Leeuwen PWNM (2000) J Org Chem 65:3010
95. Faller JW, Lavoie AR (2001) Org Lett 3:3703
96. Palmer M, Walsgrove T, Wills M (1997) J Org Chem 62:5226
97. Rhyoo HY, Park HJ, Suh WH, Chung YK (2002) Tetrahedron Lett 43:269
98. Thorpe T, Blacker J, Brown SM, Bubert C, Crosby J, Fitzjohn S, Muxworthy JP, Williams MJ (2001) Tetrahedron Lett 42:4041; Bubert C, Blacker J, Brown SM, Crosby J, Fitzjohn S, Muxworthy JP, Thorpe T, Williams MJ (2001) Tetrahedron Lett 42:4037
99. Chen YC, Wu TF, Deng JG, Liu H, Jiang YZ, Choi MCK, Chan ASC (2001) Chem Commun 1488
100. Polborn K, Severin K (2000) Eur J Org Chem 1687; Laue S, Greiner L, Wöltinger J, Liese A (2001) Adv Synth Catal 343:711; Gao JX, Yi XD, Tang CL, Xu PP, Wan HL (2001) Polym Adv Technol 12:716; Sandee AJ, Petra DGI, Reek JNH, Kamer PCJ, van Leeuwen PWNM (2001) Chem Eur J 7:1202
101. Everaere K, Carpentier JF, Mortreux A, Bulliard M (1999) Tetrahedron: Asymmetry 10: 4083
102. Okano K, Murata K, Ikariya T (2000) Tetrahedron Lett 41:9277
103. Watanabe M, Murata K, Ikariya T (2002) J Org Chem 67:1712
104. Kawamoto A, Wills M (2000) Tetrahedron: Asymmetry 11:3257
105. Cross DJ, Kenny JA, Houson I, Campbell L, Walsgrove T, Wills M (2001) Tetrahedron: Asymmetry 12:1801
106. Everaere K, Carpentier JF, Mortreux A, Bulliard M (1999) Tetrahedron: Asymmetry 10: 4663
107. Everaere K, Scheffler JL, Mortreux A, Carpentier JF (2001) Tetrahedron Lett 42:1899
108. Murata K, Okano K, Miyagi M, Iwane H, Noyori R, Ikariya T (1999) Org Lett 1:1119
109. Koike T, Murata K, Ikariya T (2000) Org Lett 2:3833
110. Cossy J, Eustache F, Dalko PI (2001) Tetrahedron Lett 42:5005
111. Hennig M, Püntener K, Scalone M (2000) Tetrahedron: Asymmetry 11:1849
112. Persson BA, Larsson ALE, Ray ML, Bäckvall JE (1999) J Am Chem Soc 121:1645, and references therein
113. Persson BA, Huerta FF, Bäckvall JE (1999) J Org Chem 64:5237
114. Yamada I, Noyori R (2000) Org Lett 2:3425

Supplement to Chapter 6.2
Hydrogenation of Imino Groups

Takeshi Ohkuma, Ryoji Noyori

Department of Chemistry and Research Center for Materials Science, Nagoya University,
Chikusa, Nagoya 464–8602, Japan
e-mail: noyori@chem3.chem.nagoya-u.ac.jp

Keywords: Asymmetric hydrogenation, Asymmetric transfer hydrogenation, Ketimines, Oxime, Nitrones

1
Hydrogenation

1.1
Acyclic Imines

1.1.1
N-Aryl Imines

Highly enantioselective hydrogenation of acyclic ketimines is difficult to achieve [1–4], primarily due to the ease of interconversion between the E and Z isomers in solution [5]. Chiral catalysts are required to reduce one stereoisomer preferentially, or to reduce both isomers with the same sense of enantioselection.

The *N*-phenylimine of acetophenone was hydrogenated using an Ir cationic complex with a phosphinodihydrooxazole (S)-**1a** in CH_2Cl_2 under 100 atm of H_2 to give the R amine in 87% ee (Table of Scheme 1) [6]. The reaction was completed with a substrate to catalyst molar ratio (S/C) of 667. The reactivity was further increased when the reaction was conducted in supercritical CO_2 in-

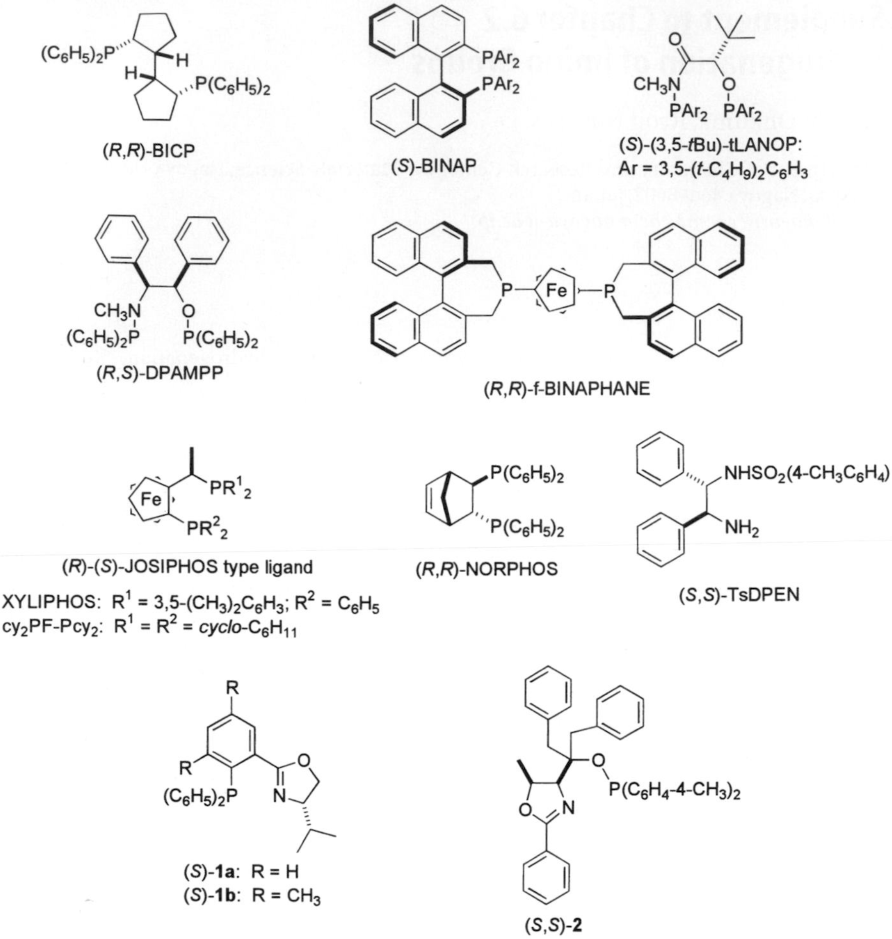

(R,R)-BICP

(S)-BINAP

(S)-(3,5-*t*Bu)-tLANOP:
Ar = 3,5-(*t*-C$_4$H$_9$)$_2$C$_6$H$_3$

(R,S)-DPAMPP

(R,R)-f-BINAPHANE

(R)-(S)-JOSIPHOS type ligand
XYLIPHOS: R^1 = 3,5-(CH$_3$)$_2$C$_6$H$_3$; R^2 = C$_6$H$_5$
cy$_2$PF-Pcy$_2$: R^1 = R^2 = *cyclo*-C$_6$H$_{11}$

(R,R)-NORPHOS

(S,S)-TsDPEN

(S)-1a: R = H
(S)-1b: R = CH$_3$

(S,S)-2

Fig. 1. Chiral ligands (in alphabetical order for named compounds

stead of in regular organic solvents, achieving a turnover number (TON, moles of product per mole of catalyst) as high as 6,830 and a turnover frequency (TOF, TON h^{-1}) of 2,820 with a slightly lower optical yield of 74% [6]. The choice of tetrakis-3,5-bis(trifluoromethyl)phenylborate (BARF) as an anion was crucial to achieve high catalytic activity. Use of **1b** as a ligand led to higher enantioselectivity [7]. A phosphiniteoxazoline ligand **2** also showed a high enantioselective ability [8]. Highly enantioselective hydrogenation of several N-aryl imines was realized by the use of a neutral f-BINAPHANE/Ir complex [9]. Ketimines with a 2,6-disubstituted N-phenyl group were hydrogenated with an S/C of 25–100 under 68 atm of H$_2$ to give the chiral amines in up to >99% ee (Table of Scheme 1). Hydrogenation of substrates with a less hindered N-phenyl or 4-methoxy-sub-

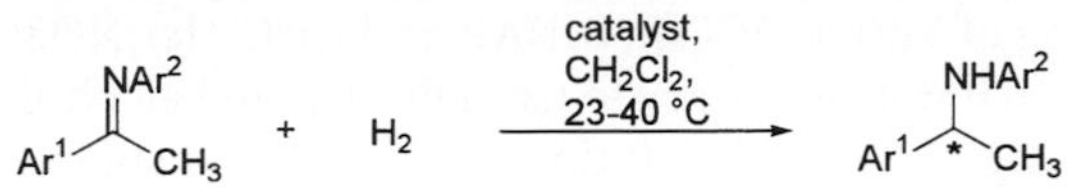

$$\underset{Ar^1}{\overset{NAr^2}{\diagdown}}\!\!\!\!\!\!\!\underset{}{\overset{}{\diagup}}CH_3 \quad + \quad H_2 \quad \xrightarrow[\text{23-40 °C}]{\substack{\text{catalyst,} \\ CH_2Cl_2,}} \quad \underset{Ar^1}{\overset{NHAr^2}{\diagdown}}\!\!\!\!\!\!\!\underset{*}{\overset{}{\diagup}}CH_3$$

Ar^1	Ar^2	Catalyst (S/C[a])	Pressure (atm)	Convn (%)	ee (%)	Con-fign
C_6H_5	C_6H_5	[Ir{(S)-1a}(COD)]BARF (667)	100	100	87	R
C_6H_5	C_6H_5	[Ir{(S)-1a}(COD)]BARF (7,143)	30[b]	96	74	R
C_6H_5	C_6H_5	[Ir{(S)-1b}(COD)]PF$_6$ (1,000)	100[c]	99	89	R
C_6H_5	C_6H_5	[Ir{(S,S)-2}(COD)]BARF (100)	50	100	80	R
C_6H_5	C_6H_5	[IrCl(COD)]$_2$-(R,R)-f-BINAPHANE+I$_2$ (25)	68[d]	100	94	–
C_6H_5	2,6-(CH$_3$)$_2$, C$_6$H$_3$	[IrCl(COD)]$_2$-(R,R)-f-BINAPHANE (25)	68	77	>99	–
C_6H_5	4-CH$_3$OC$_6$H$_4$	[IrCl(COD)]$_2$-(R,R)-f-BINAPHANE+I$_2$ (25)	68[d]	100	95	–
C_6H_5	2-CH$_3$-6-CH$_3$OC$_6$H$_3$	[IrCl(COD)]$_2$-(R,R)-f-BINAPHANE (100)	68	72	98	–
4-CH$_3$OC$_6$H$_4$	2,6-(CH$_3$)$_2$, C$_6$H$_3$	[IrCl(COD)]$_2$-(R,R)-f-BINAPHANE (25)	68	77	98	–
4-CF$_3$C$_6$H$_4$	2,6-(CH$_3$)$_2$, C$_6$H$_3$	[IrCl(COD)]$_2$-(R,R)-f-BINAPHANE (25)	68	80	99	–
1-Naphthyl	2-CH$_3$-6-CH$_3$OC$_6$H$_3$	[IrCl(COD)]$_2$-(R,R)-f-BINAPHANE (100)	68	75	96	–

[a] Substrate/catalyst molar ratio. [b] Reaction in super critical CO$_2$. [c] At 5 °C. [d] At –5 °C.

Scheme 1

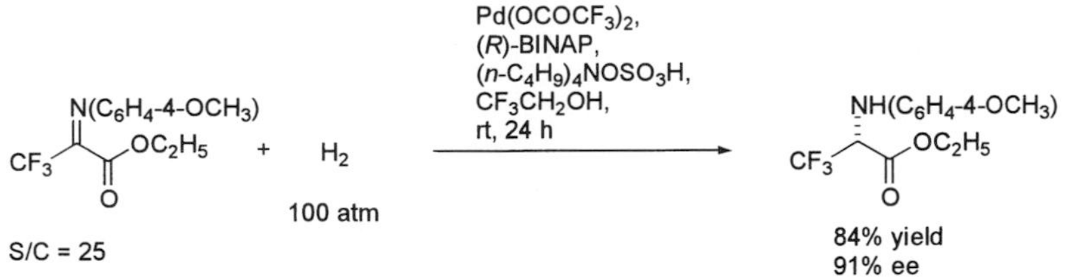

$$\underset{CF_3}{\overset{N(C_6H_4\text{-}4\text{-}OCH_3)}{\diagdown}}\!\!\!\!\!\!\!\underset{O}{\overset{}{\diagup}}OC_2H_5 \quad + \quad H_2 \quad \xrightarrow[\text{rt, 24 h}]{\substack{Pd(OCOCF_3)_2, \\ (R)\text{-}BINAP, \\ (n\text{-}C_4H_9)_4NOSO_3H, \\ CF_3CH_2OH,}} \quad \underset{CF_3}{\overset{NH(C_6H_4\text{-}4\text{-}OCH_3)}{\diagdown}}\!\!\!\!\!\!\!\underset{O}{\overset{}{\diagup}}OC_2H_5$$

S/C = 25   100 atm   84% yield   91% ee

Scheme 2

stituted N-phenyl group using the same complex in the presence of I$_2$ gave optical yields of 94% and 95%, respectively.

As shown in Scheme 2, a 3,3,3-trifluoroalanine derivative was obtained in 91% ee by asymmetric hydrogenation of the corresponding imino ester with a catalyst consisting of $Pd(OCOCF_3)_2$, BINAP, and $(n\text{-}C_4H_9)_4NOSO_3H$ in 2,2,2-trifluoroethanol [10]. It was necessary to use a fluorinated alcohol solvent in order to obtain both a high yield and enantioselectivity. For instance, reaction in ethanol gave the product in 30% ee and 29% yield. Hydrogenation of the 3,3-difluoroimino substrate gave the corresponding amine in 30% ee, even in 2,2,2-trifluoroethanol.

1.1.2
N-Diphenylphosphinyl Imines

N-Diphenylphosphinylacetophenone imines, which exclusively exist as the E isomer, were hydrogenated with a cationic $(R)\text{-}(S)\text{-cy}_2PF\text{-}Pcy_2$/Rh complex under 100 atm of H_2 to give the chiral amines in up to 99% ee (Scheme 3) [11]. Substitution of a Cl or CH_3O group at the 4' position of the substrate decreased the optical yield. The N-diphenylphosphinyl group can be removed under acidic conditions.

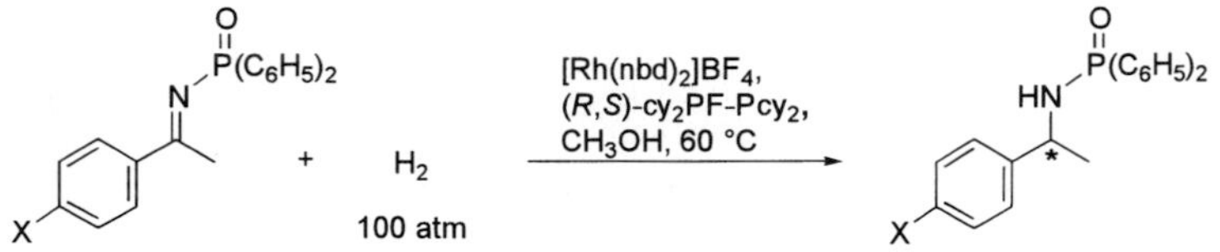

X	S/C[a]	Time (h)	Convn (%)	ee (%)	Confign
H	500	1	100	99	R
CH$_3$	100	21	100	97	–
Cl	100	17[b]	85	28	–
CF$_3$	100	18	98	93	–
OCH$_3$	100	19	100	62	–

[a] Substrate/catalyst molar ratio. [b] At 80 °C.

Scheme 3

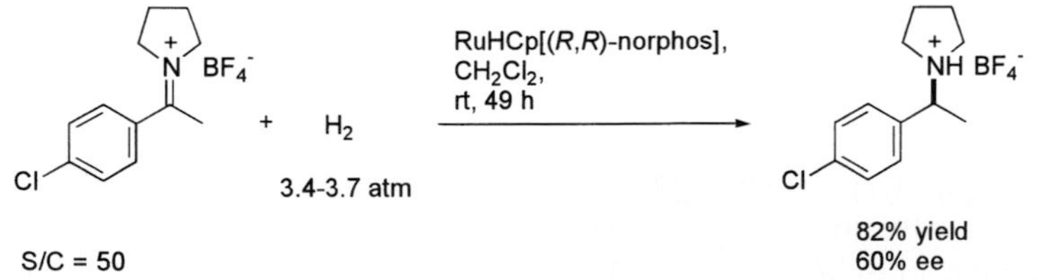

Scheme 4

1.1.3
Pyrrolidinium Salts

Hydrogenation of 4-chlorophenyl methylpyrrolidinium salt with RuHCp[(*R,R*)-NORPHOS] gave the *S* product in 60% ee (Scheme 4) [12]. The turnover-limiting step of this reaction was shown to be the hydride transfer from the catalyst to the C=N$^+$ group.

1.1.4
Oxime and Nitrones

Ethyl 4-phenyl-2-oxobutanoate oxime, which is easily prepared as a single *E* isomer, was hydrogenated with a neutral (*R,S*)-DPAMPP/Ir complex and (*n*-C$_4$H$_9$)$_4$NI under 48 atm of H$_2$ to afford the *S* amino ester in 93% ee, whereas the conversion was low (Scheme 5) [13].

Nitrones of several aryl methyl ketones were hydrogenated by a catalyst prepared in situ from [IrCl(COD)]$_2$, (*S*)-BINAP, and (*n*-C$_4$H$_9$)$_4$NBH$_4$ in THF under 80 atm of H$_2$ and at 0 °C to give the corresponding *N*-hydroxylamines in up to 86% ee (Scheme 6) [14]. Substitution of halides at the 3' or 4' position of the ar-

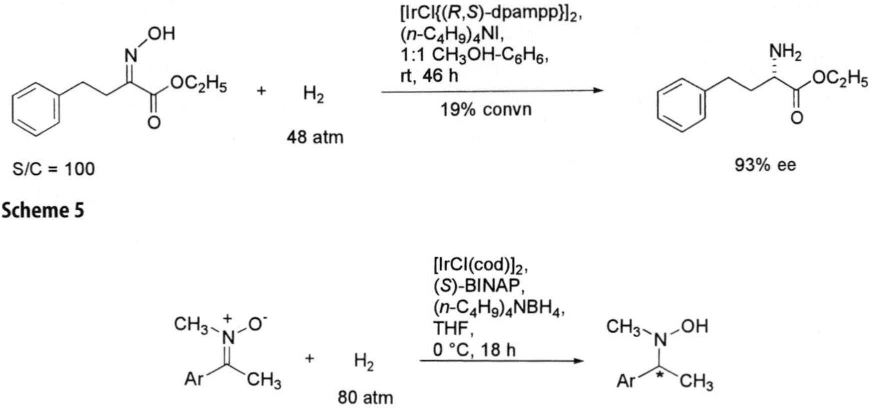

S/C = 100   93% ee

Scheme 5

S/C = 50

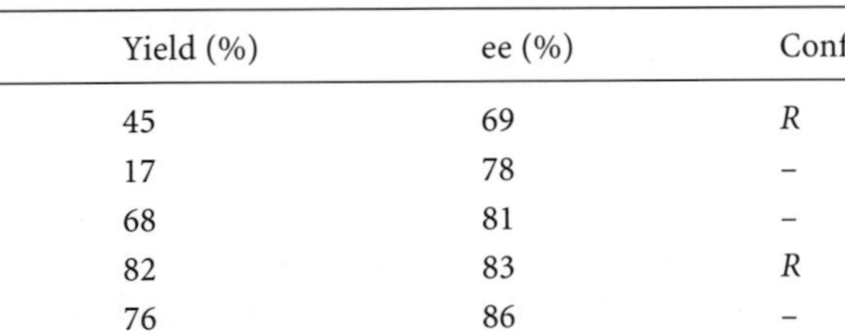

Ar	Yield (%)	ee (%)	Config'n
C$_6$H$_5$	45	69	*R*
2-ClC$_6$H$_4$	17	78	–
3-ClC$_6$H$_4$	68	81	–
4-ClC$_6$H$_4$	82	83	*R*
4-BrC$_6$H$_4$	76	86	–
2-Naphthyl	64	80	–

Scheme 6

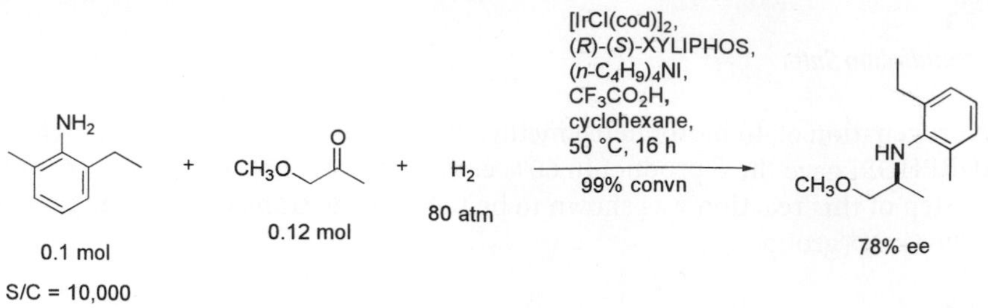

Scheme 7

omatic ring increased the optical yield (Table of Scheme 6). The 2′-chloro-substituted substrate resulted in a low yield. The N-hydroxylamines obtained can be converted to the chiral amines under reducing conditions.

1.1.5
Reductive Alkylation of Amines

Asymmetric reductive alkylation of amines with ketones without isolation of imines is a direct means of obtaining optically active amines. As shown in Scheme 7, the reaction of 2-ethyl-6-methylaniline and methoxyacetone under 80 atm of H_2 with a catalyst consisting of $[IrCl(COD)]_2$, (R)-(S)-XYLIPHOS, and $(n\text{-}C_4H_9)_4NI$ in the presence of CF_3CO_2H or CH_3SO_3H gave the desired S amine in 78% ee [15]. The high catalytic activity enabled the reaction with an S/C of 10,000 to be completed within 16 h. The chiral amine obtained is the key compound for the industrial synthesis of metolachlor, a grass herbicide. Even the productivity of this reaction could not compete with that observed in the hydrogenation of the corresponding imine, achieving a TON of >1,000,000 [16].

1.2
Cyclic Imines

Highly enantioselective hydrogenation of geometry-fixed cyclic imines has been achieved by the use of certain chiral Ti and Ir catalysts [14, 17]. In particular, a chiral titanocene catalyst developed by Buchwald possess excellent enantiodifferentiating ability for a variety of cyclic substrates [18].

A titanocene catalyst prepared in situ from (R)-3 and $n\text{-}C_4H_9Li$ effected the hydrogenation of 2-phenyl-1-pyrroline, a five-membered imine, to give (S)-2-phenylpyrrolidine in 98% ee (Scheme 8) [19]. The reaction proceeded smoothly with an S/C of 1,000.

2,3,3-Trimethylindolenine was hydrogenated with a catalyst consisting of $[IrCl(COD)]_2$, (R,R)-BICP, and phthalimide in CH_2Cl_2 under 68 atm of H_2 to afford the chiral cyclic amine in 95.1% ee (Scheme 9) [20]. The addition of phthalimide remarkably increased the optical yield [21]. Hydrogenation of a dihydroisoquinoline compound with a neutral (R)-BINAP/Ir complex in

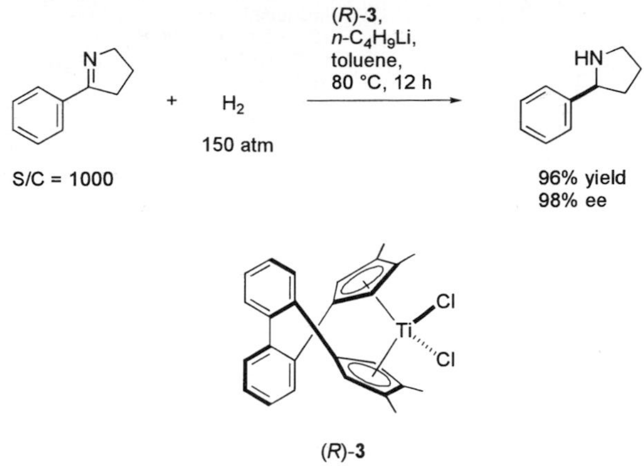

Scheme 8

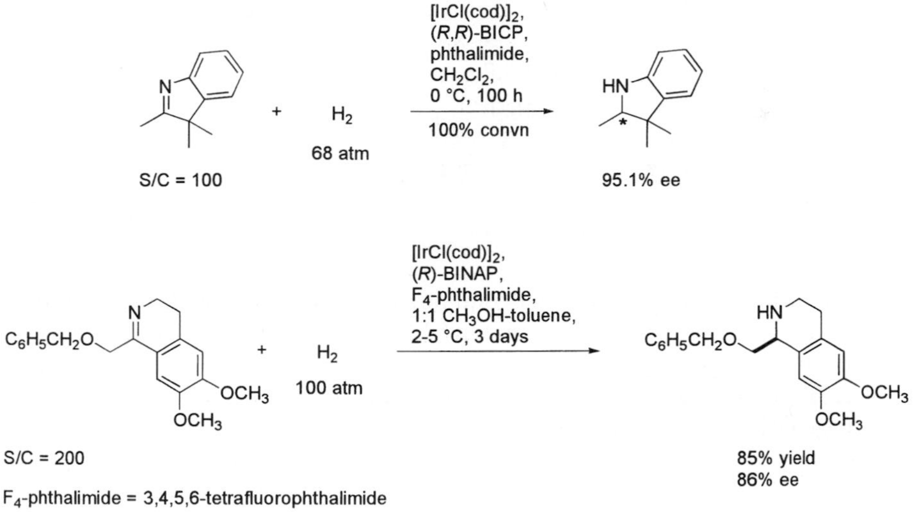

F_4-phthalimide = 3,4,5,6-tetrafluorophthalimide

Scheme 9

the presence of 3,4,5,6-tetrafluorophthalimide gave the S cyclic amine in 86% ee (Scheme 9) [22]. The reaction with the addition of phthalimide instead of tetrafluorophthalimide resulted in 75% optical yield. The removal of the benzyl group yielded (S)-calycotomine, a naturally occurring tetrahydroisoquinoline alkaloid.

A cyclic iminium salt was hydrogenated using an Ir catalyst with (S)-(3,5-tBu)-tLANOP, an amidophosphinephosphinite ligand, in the presence of (n-C_4H_9)$_4$NI under 100 atm of H_2 to give the S ammonium salt in 86% ee and in 46% yield, accompanied by 54% of the decahydroisoquinoline derivative (Scheme 10) [23].

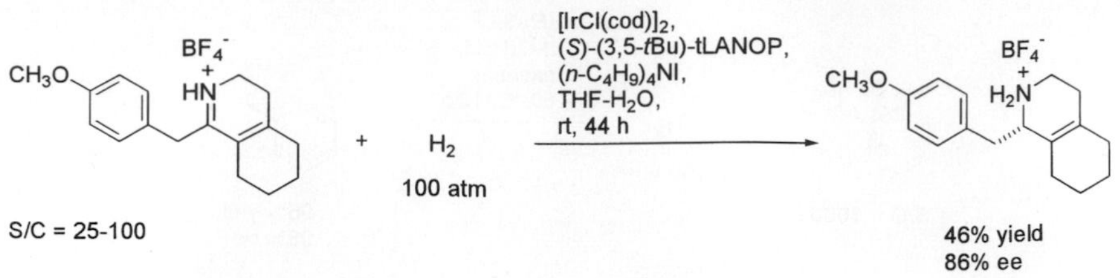

Scheme 10

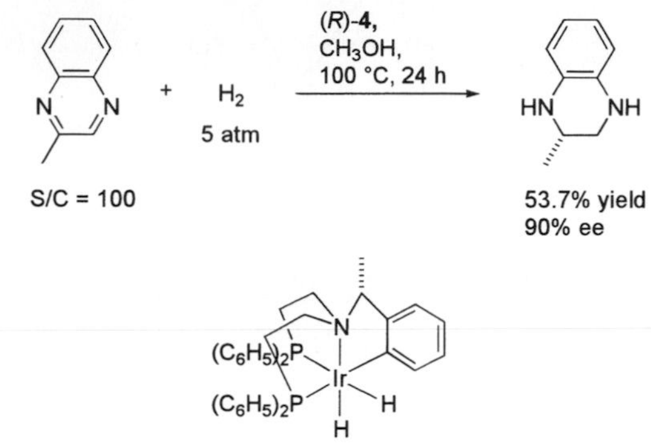

Scheme 11

Double hydrogenation of 2-methylquinoxaline using (R)-**4** with an S/C of 100 under 5 atm of H_2 afforded (S)-2-methyl-1,2,3,4-tetrahydroquinoxaline in 90% ee and in a 53.7% yield (Scheme 11) [24]. When the reaction proceeded under lower H_2 pressure, a higher optical yield was obtained. The two C=N bonds were hydrogenated at comparable rates.

2
Transfer Hydrogenation

Asymmetric transfer hydrogenation of a series of cyclic imines with high optical yield has been achieved by the use of the Ru complexes of type **5** with a formic acid–triethylamine mixture, thus providing a new and generally applicable means of obtaining natural as well as unnatural isoquinoline alkaloids (Scheme 12) [25]. This reaction has been applied to the asymmetric syntheses of morphine [26], a curare-like agent [27], an anti-influenza A virus agent [28], an asymmetric protonation agent [29], and a chiral auxiliary [30]. Examples of chiral amines obtained by reduction with (S,S)-**5** are shown in Fig. 2. The reduction of the o-bromophenyl imine [29] and the ethoxycarbonylmethyl imine [28] gave the corresponding chiral amines in 98.7% (S) and 97% (R), respectively. A cyclic sulfonimide was converted to the S sultam in 93% ee [30].

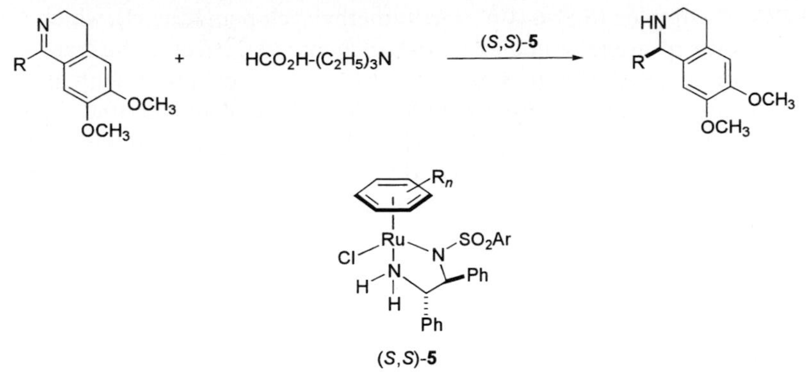

Scheme 12

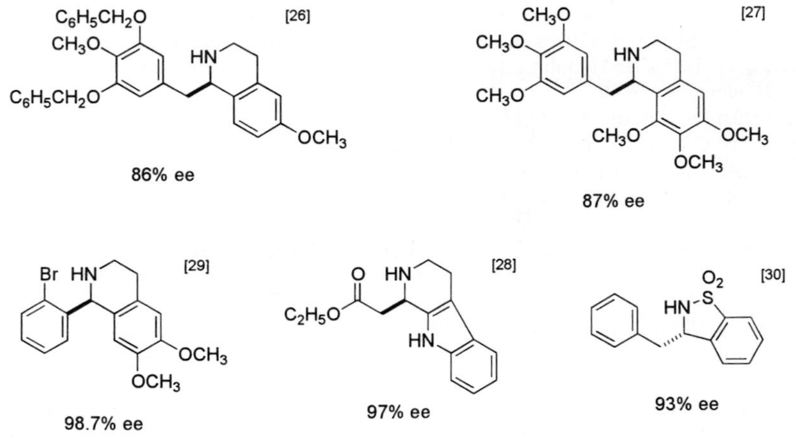

86% ee

[26]

[27]

87% ee

[29]

98.7% ee

[28]

97% ee

[30]

93% ee

Fig. 2

S/C = 200

(S,S)-**6**,
CH₂Cl₂,
20 °C, 10 min

96% yield
99% ee

95% yield
90% ee

96% yield (30 min)
81% ee

(S,S)-**6**

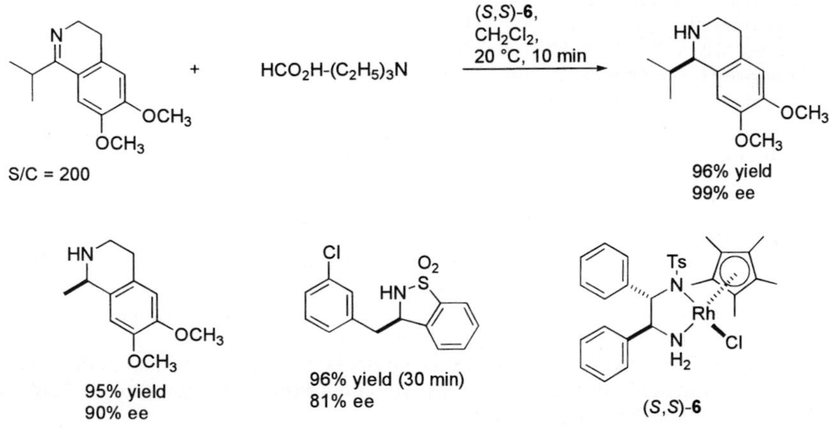

Scheme 13

A RhCp* complex (S,S)-**6** (Cp*=pentamethylcyclopentadienyl), which is isolobal with Ru(η^6-arene) complex (S,S)-**5** (Scheme 13), effected the transfer hydrogenation of a cyclic imine substituted by an isopropyl group with an S/C of 200 in the presence of a 5:2 mixture of formic acid and triethylamine to give the R amine in 99% ee (Scheme 13) [31]. When the reaction was performed with an S/C of 1,000, the optical yield decreased to 93%. The methyl imine was reduced with a 91% optical yield. Reduction of a cyclic sulfonimide resulted in the R sultam in 81% ee.

References

1. Blaser HU, Spindler F (1999) Hydrogenation of imino groups. In: Jacobsen EN, Pfaltz A, Yamamoto H (eds) Comprehensive asymmetric catalysis, vol 1. Springer, Berlin Heidelberg New York, chap 6.2
2. Ohkuma T, Kitamura M, Noyori R (2000) Asymmetric hydrogenation. In: Ojima I (ed) Catalytic asymmetric synthesis, 2nd edn. Wiley-VCH, New York, chap 1
3. Spindler F, Blaser HU (1998) Enantioselective reduction of C=N bonds and enamines with hydrogen. In: Beller M, Bolm C (eds) Transition metals for organic synthesis, vol 2. Wiley-VCH, Weinheim, p 69
4. Kobayashi S, Ishitani H (1999) Chem Rev 99:1069
5. McCarty CG (1970) *syn–anti* Isomerizations and rearrangements. In: Patai S (ed) The chemistry of the carbon–nitrogen double bonds. Wiley, London, chap 9; Bjørgo J, Boyd DR, Watson CG, Jennings WB (1974) J Chem Soc Perkin Trans 2 757; Johnson GP, Marples BA (1984) Tetrahedron Lett 25:3359
6. Kainz S, Brinkmann A, Leitner W, Pfaltz A (1999) J Am Chem Soc 121:6421
7. Schnider P, Koch G, Prétôt R, Wang G, Bohnen FM, Krüger C, Pfaltz A (1997) Chem Eur J 3:887
8. Menges F, Pfaltz A (2002) Adv Synth Catal 344:40
9. Dengming X, Zhang X (2001) Angew Chem Int Ed 40:3425
10. Abe H, Amii H, Uneyama K (2001) Org Lett 3:313
11. Spindler F, Blaser HU (2001) Adv Synth Catal 343:68
12. Magee MP, Norton JR (2001) J Am Chem Soc 123:1778
13. Xie Y, Mi A, Jiang Y, Liu H (2001) Synth Commun 31:2767
14. Murahashi S, Tsuji T, Ito S (2000) Chem Commun 409
15. Blaser HU, Buser HP. Jalett HP, Pugin B, Spindler F (1999) Synlett 867
16. Blaser HU (2002) Adv Synth Catal 344:17, and references therein
17. Yurovskaya MA, Karchava AV (1998) Tetrahedron: Asymmetry 9:3331
18. Willoughby CA, Buchwald SL (1994) J Am Chem Soc 116:8952
19. Ringwald M, Stürmer R, Brintzinger HH (1999) J Am Chem Soc 121:1524
20. Zhu G, Zhang X (1998) Tetrahedron: Asymmetry 9:2415
21. Morimoto T, Achiwa K (1995) Tetrahedron: Asymmetry 6:2661
22. Moromoto T, Suzuki N, Achiwa K (1998) Tetrahedron: Asymmetry 9:183
23. Broger EA, Burkart W, Hennig M, Scalone M, Schmid R (1998) Tetrahedron: Asymmetry 9: 4043
24. Bianchini C, Barbaro P, Scapacci G, Farnetti E, Graziani M (1998) Organometallics 17: 3308
25. Noyori R, Hashiguchi S (1997) Acc Chem Res 30: 97; Uematsu N, Fujii A, Hashiguchi S, Ikariya T, Noyori R (1996) J Am Chem Soc 118: 4916
26. Meuzelaar GJ, van Vliet MCA, Maat L, Sheldon RA (1999) Eur J Org Chem 2315
27. Kaldor I, Feldman PL, Mook, Jr. RA, Ray JA, Samano V, Sefler AM, Thompson JB, Travis BR, Boros EE (2001) J Org Chem 66: 3495
28. Tietze LF, Zhou Y, Töpken E (2000) Eur J Org Chem 2247

29. Vedejs E, Trapencieris P, Suna E (1999) J Org Chem 64:6724
30. Ahn KH, Ham C, Kim SK, Cho CW (1997) J Org Chem 62:7047
31. Mao J, Baker DC (1999) Org Lett 1:841

...

Supplement to Chapter 6.3
Hydrosilylation of Carbonyl and Imino Groups

Takeshi Ohkuma, Ryoji Noyori

Department of Chemistry and Research Center for Materials Science, Nagoya University, Chikusa, Nagoya 464–8602, Japan
e-mail: noyori@chem3.chem.nagoya-u.ac.jp

Keywords: Asymmetric hydrosilylation, optically active alcohols, amines, Chiral Titanocene Catalysts, Acyclic Imines, Cyclic Imines, Chiral Rhodium Catalysts, aromatic ketones

1
Hydrosilylation of C=O with Chiral Rhodium Catalysts

1.1
Alkyl Aryl Ketones

Asymmetric hydrosilylation of ketones catalyzed by chiral Rh complexes has been extensively studied (Scheme 1) [1–3]. The pioneering works done by Brunner [4] and Nishiyama and Itoh [5] using chiral nitrogen-based ligands are noteworthy examples. Recently, a new chiral 2-(2-pyridyl)oxazoline ligand **1** (Rh:ligand=1:4.9) was applied to the reduction of acetophenone with 1.01 equivalents of diphenylsilane followed by hydrolysis, resulting in 1-phenylethanol with 79.6% ee (Scheme 1) [6]. Use of well-designed chiral iminophosphine ligands led to high enantioselectivities [7]. A complex prepared in situ from [RhCl(COD)]₂ and (*R*,*S*)-**2** (Rh:ligand=1:2) catalyzed hydrosilylation

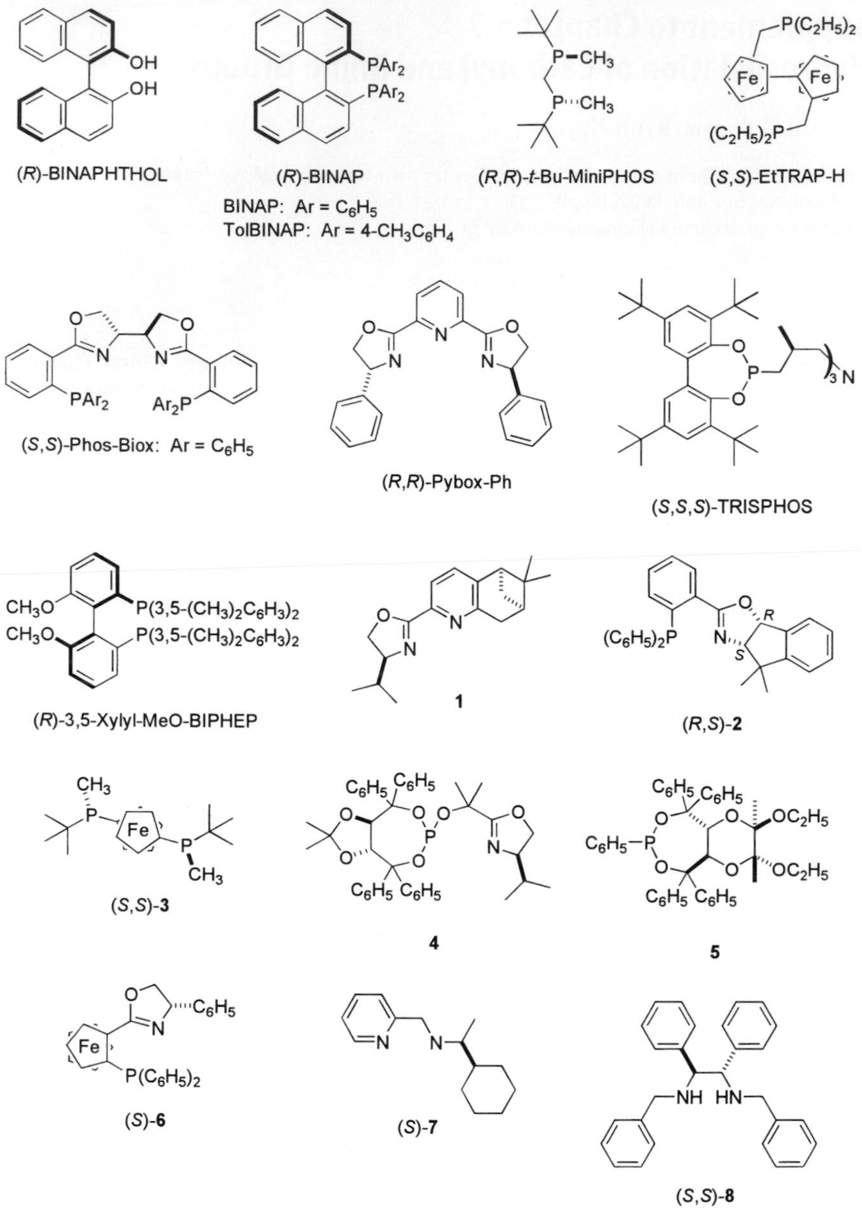
(R)-BINAPHTHOL
(R)-BINAP
BINAP: Ar = C6H5
TolBINAP: Ar = 4-CH3C6H4
(R,R)-t-Bu-MiniPHOS
(S,S)-EtTRAP-H
(S,S)-Phos-Biox: Ar = C6H5
(R,R)-Pybox-Ph
(S,S,S)-TRISPHOS
(R)-3,5-Xylyl-MeO-BIPHEP
1
(R,S)-2
(S,S)-3
4
5
(S)-6
(S)-7
(S,S)-8

Fig. 1. Chiral ligands (in alphabetical order for named compounds)

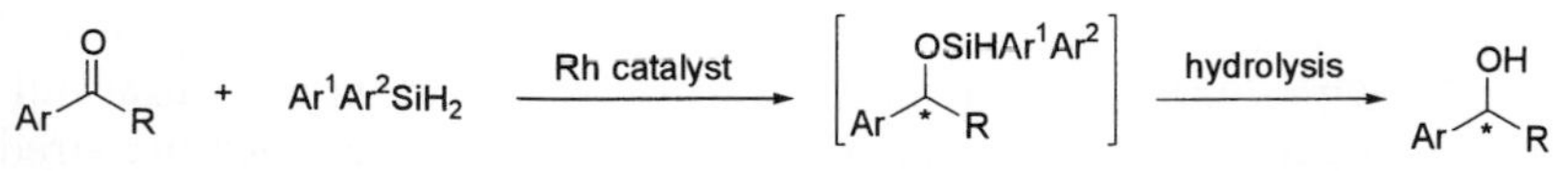
Rh catalyst
hydrolysis

Scheme 1

▶

R	Ar	Ar^{1}Ar^{2}SiH$_2$	Catalyst	S:Si:Rh:ligand[a]	Solvent	Temp. (°C)	Yield (%)	ee (%)	Con-fign
CH$_3$	C$_6$H$_5$	(C$_6$H$_5$)$_2$SiH$_2$	[RhCl(COD)]$_2$–**1**	210:212:1:4.9	CCl$_4$	0–rt	90	79.6	S
CH$_3$	C$_6$H$_5$	(C$_6$H$_5$)$_2$SiH$_2$	[RhCl(COD)]$_2$–(R,S)–**2**	50:100:1:2	Toluene	rt	84	94	R
CH$_3$	C$_6$H$_5$	(C$_6$H$_5$)$_2$SiH$_2$	[RhCl(COD)]$_2$–(S,S)-Phos-Biox	400:640:1:2	THF	0	98	97	R
CH$_3$	C$_6$H$_5$	(C$_6$H$_5$)$_2$SiH$_2$	[Rh(COD)$_2$]BF$_4$–(S,S)-EtTRAP-H	100:150:1:1.1	THF	–40	89	94	S
CH$_3$	C$_6$H$_5$	1-Naphthyl-phenylsilane	[Rh{(R,R)-t-Bu-MiniPHOS}]BF$_4$	100:150:1:1	THF	–40	86	91	R
CH$_3$	C$_6$H$_5$	1-Naphthyl-phenylsilane	[Rh{(S,S)–**3**}]BF$_4$	100:150:1:1	THF	–20	96	92	S
CH$_3$	C$_6$H$_5$	(C$_6$H$_5$)$_2$SiH$_2$	[RhCl(COD)]$_2$–**4**	100:110:1:1.2	Toluene	–20	91	88	S
CH$_3$	C$_6$H$_5$	(C$_6$H$_5$)$_2$SiH$_2$	[Rh(nbd)$_2$]ClO$_4$–(S,S,S)-TRISPHOS	50:50:1:1	Benzene	rt	65	81	R
CH$_3$	2-CH$_3$C$_6$H$_4$	1-Naphthyl-phenylsilane	[Rh{(R,R)-t-Bu-MiniPHOS}]BF$_4$	100:150:1:1	THF	–40	96	95	R
CH$_3$	4-CH$_3$C$_6$H$_4$	(C$_6$H$_5$)$_2$SiH$_2$	[RhCl(COD)]$_2$–(S,S)-Phos-Biox	400:640:1:2	THF	0	98	93	R
CH$_3$	4-CH$_3$C$_6$H$_4$	1-Naphthyl-phenylsilane	[Rh{(S,S)–**3**}]BF$_4$	100:150:1:1	THF	–20	98	84	S
CH$_3$	4-CH$_3$OC$_6$H$_4$	(C$_6$H$_5$)$_2$SiH$_2$	[RhCl(COD)]$_2$–(S,S)-Phos-Biox	400:640:1:2	THF	0	99	21	R
CH$_3$	4-CH$_3$OC$_6$H$_4$	1-Naphthyl-phenylsilane	[Rh{(S,S)–**3**}]BF$_4$	100:150:1:1	THF	–20	61	≈0	–
CH$_3$	1-Naphthyl	(C$_6$H$_5$)$_2$SiH$_2$	[RhCl(COD)]$_2$–(R,S)–**2**	50:100:1:2	Toluene	rt	90	92	R
CH$_3$	1-Naphthyl	1-Naphthyl-phenylsilane	[Rh{(R,R)-t-Bu-MiniPHOS}]BF$_4$	100:150:1:1	THF	–40	90	97	R
CH$_3$	1-Naphthyl	(C$_6$H$_5$)$_2$SiH$_2$	[RhCl(COD)]$_2$–**4**	100:110:1:1.2	Toluene	0	81	89	S
C$_2$H$_5$	C$_6$H$_5$	(C$_6$H$_5$)$_2$SiH$_2$	[RhCl(COD)]$_2$–(R,S)–**2**	50:100:1:2	Toluene	rt	91	91	R
n-C$_3$H$_7$	C$_6$H$_5$	(3-FC$_6$H$_4$)$_2$SiH$_2$	[Rh(COD)$_2$]BF$_4$–(S,S)-EtTRAP-H	100:150:1:1.1	THF	–40	93	89	S
ClCH$_2$	C$_6$H$_5$	(C$_6$H$_5$)$_2$SiH$_2$	[RhCl(COD)]$_2$–**4**	100:110:1:1.2	Toluene	0	89	83	R
Cl(CH$_2$)$_3$	C$_6$H$_5$	(3-FC$_6$H$_4$)$_2$SiH$_2$	[Rh(COD)$_2$]BF$_4$–(S,S)-EtTRAP-H	100:150:1:1.1	THF	–40	99	88	–
CH$_2$=CH–(CH$_2$)$_2$	C$_6$H$_5$	(3-FC$_6$H$_4$)$_2$SiH$_2$	[Rh(COD)$_2$]BF$_4$–(S,S)-EtTRAP-H	100:150:1:1.1	THF	–40	90	89	–
(CH$_3$)$_3$C	C$_6$H$_5$	(C$_6$H$_5$)$_2$SiH$_2$	RhCl(**5**)(NBD)	100:125:1:1	Toluene	–5	86	86	R

[a] Substrate:silane:Rh:chiral ligand molar ratio.

of several aromatic ketones with diphenylsilane in up to 94% optical yield [8]. A neutral Rh catalyst with a C_2-symmetric bisimidophosphine Phos-Biox ligand achieved an ee of 97% in the reduction of acetophenone with a substrate to Rh molar ratio of 400 [9]. 4'-Methylacetophenone was reduced with 93% optical yield, while the selectivity was decreased to 21% in the reaction of the corresponding 4'-methoxy ketone.

Highly enantioselective hydrosilylation of ketones with chiral diphosphine/metal complexes has proven difficult to achieve [10], in contrast to the fruitful results for the corresponding asymmetric hydrogenation [11]. A cationic Rh complex with a planar chiral diphosphine EtTRAP-H (Rh:ligand=1:1.1) catalyzed the hydrosilylation of acetophenone with a 1.5 equivalents of diphenylsilane at -40 °C to give the desired product in 94% ee (Scheme 1) [12]. When di(3-fluorophenyl)silane was used for the reduction in place of diphenylsilane, several alkyl phenyl ketones were reduced with 88–89% optical yield. The Cl and olefinic groups on the side chain were left intact. Reduction of aryl methyl ketones with a 1.5 equivalents of 1-naphthylphenylsilane in the presence of [Rh{(R,R)-t-Bu-MiniPHOS}]BF$_4$ with a substrate to Rh molar ratio of 100 at -40 °C followed by hydrolysis afforded the R alcohols in up to 97% ee [13]. Propiophenone was reduced with 83% optical yield. A cationic Rh complex with a P-chiral diphosphine (S,S)-3 catalyzed the hydrosilylation of acetophenone with 1-naphthylphenylsilane at -20 °C to give the S product in 92% ee [14]. 4'-Methylacetophenone was reduced with 84% optical yield, while the hydrosilylation of the 4'-methoxy analogue resulted in the racemic product. The tendency was similar to that observed in the reduction with the Rh/Phos-Biox complex [9] (vide supra).

The cyclic phosphite and phosphonite ligands derived from $\alpha,\alpha,\alpha',\alpha'$-tetraaryl-1,3-dioxolane-4,5-dimethanol (TADDOL) [15] have been reported as effective chiral monodentate ligands for the asymmetric hydrosilylation of aromatic ketones by Seebach [16]. The chiral bidentate-iminophosphite ligand 4 is effective for the Rh complex-catalyzed asymmetric hydrosilylation of a series of aromatic ketones with diphenylsilane [17]. Alkyl aryl ketones were reduced with up to 89% optical yield (Scheme 1). Phenacyl chloride was converted to (R)-2-chloro-1-phenylethanol in 83% optical yield. A Rh/TRISPHOS complex prepared in situ (Rh:ligand=1:1) catalyzed the hydrosilylation of acetophenone with 81% optical yield [18]. For the reduction of pivalophenone with diphenylsilane, use of a chiral cyclic phosphite 5 prepared from a TADDOL analogue having a 1,4-dioxane backbone gave the R product in 86% ee [19].

A cyclic ketone, 1-tetralone, was reduced with diphenylsilane in the presence of the Rh/(S,R)-2 complex (substrate:Rh:ligand=50:1:2) in toluene at room temperature followed by hydrolysis to give the S alcohol in 92% optical yield (Scheme 2) [8]. Use of the Rh/4 complex (substrate:Rh:ligand=100:1:1.2) at 0 °C yielded the S product in 80% optical yield [17]. The reduction of 4-chromanone with the same complex gave an optical yield of 87%.

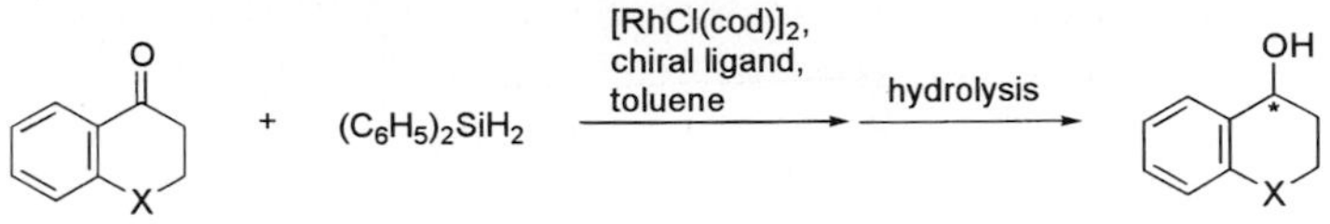

X	Ligand	S:Si:Rh:ligand[a]	Temp. (°C)	Yield (%)	ee (%)	Config[n]
CH_2	(S,R)–2	50:100:1:2	rt	97	92	S
CH_2	4	100:110:1:1.2	0	70	80	S
O	4	100:110:1:1.2	0	86	87	S

[a] Substrate:silane:Rh:chiral ligand molar ratio.

Scheme 2

1.2
Alkyl Methyl Ketones

Enantioselective reduction of simple aliphatic ketones is one of the most challenging of the currently unresolved problems in this field. The Rh/4 complex catalyzed the hydrosilylation of 2-butanone with diphenylsilane at 0 °C, which after hydrolysis gave (S)-2-butanol in 56% ee (Scheme 3) [17]. 2-Octanone and 4-phenyl-2-butanone were reduced with diphenylsilane in the presence of the cationic Rh/EtTRAP-H at -50 °C and gave optical yields of 77% and 81%, respectively [12]. The cationic Rh/(R,R)-t-Bu-MiniPHOS was also effective for the reduction of 4-phenyl-2-butanone with 1-naphthylphenylsilane at -20 °C, affording the R product in 80% ee [13]. 3-Methyl-2-butanone was reduced using the Rh/4 complex with 76% optical yield [17]. Hydrosilylation of cyclohexyl methyl ketone with the Rh/(R,S)-2 complex followed by hydrolysis afforded the R alcohol in 87% ee [8]. Highly enantioselective hydrosilylation of pinacolone with diphenylsilane at -20 °C was achieved by means of the Rh/4 complex and yielded the desired R product in 95% ee [17].

1.3
Keto Esters

Ketopantolactone was reduced with diphenylsilane in the presence of the Rh/4 complex in THF at 0 °C followed by desilylation with an acidic methanol affording (S)-pantoyl lactone in 84% optical yield (Scheme 4) [17]. (S)-Ethyl 4-hydroxypentanoate was obtained in 80% ee by the reduction of ethyl 4-oxopentanoate with the Rh/4 complex in toluene followed by cleavage of the silyl ether with a neutral methanol [17].

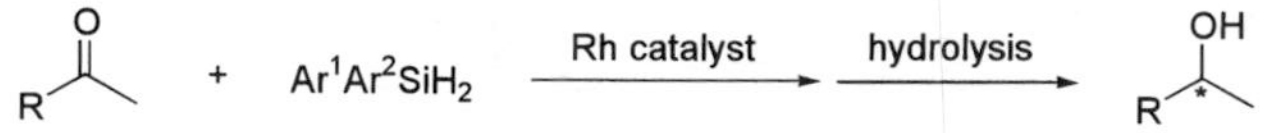

$$\underset{R}{\overset{O}{\|}}\!\!\diagup + \; Ar^1Ar^2SiH_2 \xrightarrow{\text{Rh catalyst}} \xrightarrow{\text{hydrolysis}} \underset{R}{\overset{OH}{\diagup}}\!\!\overset{*}{\diagup}$$

R	$Ar^1Ar^2SiH_2$	Catalyst	S:Si:Rh:ligand[a]	Solvent	Temp. (°C)	Yield (%)	ee (%)	Config^n
C_2H_5	$(C_6H_5)_2SiH_2$	[RhCl(COD)]$_2$–4	100:110:1:1.2	Toluene	0	–[b]	56	S
n-C_6H_{13}	$(C_6H_5)_2SiH_2$	[Rh(COD)$_2$]BF$_4$–(S,S)-EtTRAP-H	100:150:1:1.1	THF	–50	82	77	S
$C_6H_5(CH_2)_2$	$(C_6H_5)_2SiH_2$	[Rh(COD)$_2$]BF$_4$–(S,S)-EtTRAP-H	100:150:1:1.1	THF	–50	94	81	S
$C_6H_5(CH_2)_2$	1-Naphthylphenylsilane	Rh{(R,R)-t-Bu-MiniPHOS}]BF$_4$	100:150:1:1	THF	–20	93	80	R
$(CH_3)_2CH$	$(C_6H_5)_2SiH_2$	[RhCl(COD)]$_2$–4	100:110:1:1.2	Toluene	0	–[b]	76	S
$cyclo$-C_6H_{12}	$(C_6H_5)_2SiH_2$	[RhCl(COD)]$_2$–(R,S)-2	50:100:1:2	Toluene	rt	85	87	R
$(CH_3)_3C$	$(C_6H_5)_2SiH_2$	[RhCl(COD)]$_2$–4	100:110:1:1.2	Toluene	–20	–[b]	95	R

[a] Substrate:silane:Rh:chiral ligand molar ratio. [b] >99% Conversion with no detectable by-product.

Scheme 3

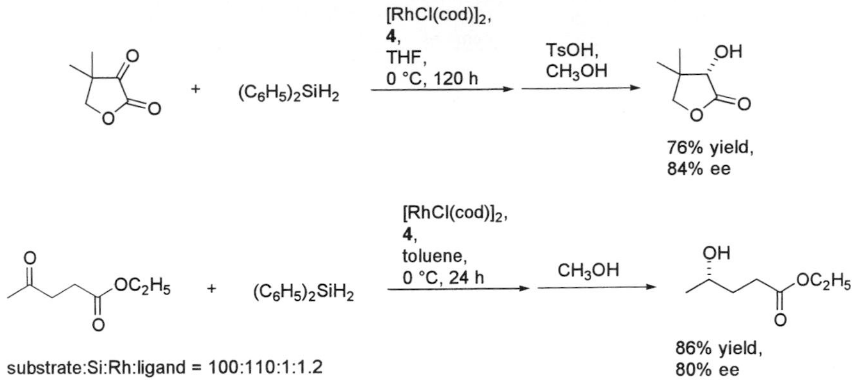

Scheme 4

2
Hydrosilylation of C=O with Chiral Titanocene Catalysts

2.1
Ketones

A chiral titanocene catalyst (S)-**10** prepared from the corresponding complex with 1,1'-binaphth-2,2'-diolate (S,S,S)-**9**, n-C$_4$H$_9$Li, and polymethylhydrosiloxane (PMHS) promoted asymmetric hydrosilylation of a variety of simple ketones with very high enantioselectivity, although relatively high catalyst loading (substrate:Ti=10–22:1) was required to achieve a reasonable reaction rate (Scheme 5) [20–22]. Recently, a titanocene complex (R,R)-(EBTHI)–TiF$_2$ with five equivalents of phenylsilane in the presence of pyrrolidine and methanol in THF at 60 °C afforded the active catalyst (R)-**11** for the hydrosilylation with PMHS as a reducing agent (Scheme 6) [23]. Slow addition of methanol (3–7 equivalents) to the reaction system significantly increased the reaction rate. Propiophenone was reduced with a substrate to Ti molar ratio of 200:1 to give the S product in 98% ee (Scheme 6). More sterically hindered isobutyrophenone was reduced with 99% optical yield. An unconjugated trisubstituted olefin was not affected under the reaction conditions. Acetyl cycloalkenes were reduced preferentially at the carbonyl function and after base treatment gave the allylic alcohols in up to 97% ee. Hydrosilylation of simple aliphatic ketones gave moderate enantioselectivity. 2-n-Pentylcyclopentenone, a cyclic α,β-unsaturated ketone, was also converted to the allylic alcohol in 84% yield (Scheme 7) [23].

2.2
Acyclic Imines

Asymmetric hydrosilylation of several N-alkyl ketimines with PMHS was effectively promoted by the chiral titanocene catalyst (S)-**11** (Scheme 8) [24, 25]. The

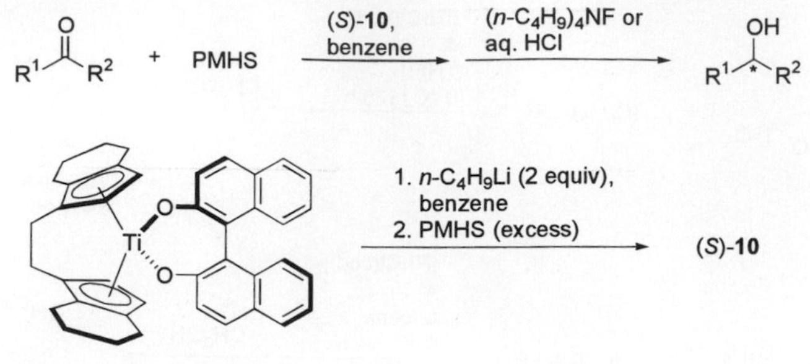

$$R^1 \overset{O}{\underset{}{\bigg\Vert}} R^2 \; + \; PMHS \xrightarrow[\text{benzene}]{(S)\text{-}10,} \xrightarrow{(n\text{-}C_4H_9)_4NF \text{ or} \atop \text{aq. HCl}} R^1 \overset{OH}{\underset{*}{\bigg|}} R^2$$

$$(S,S,S)\text{-}9 \xrightarrow[\text{2. PMHS (excess)}]{1.\ n\text{-}C_4H_9Li\ (2\ equiv),\ \text{benzene}} (S)\text{-}10$$

PMHS = polymethylhydrosiloxane

Scheme 5

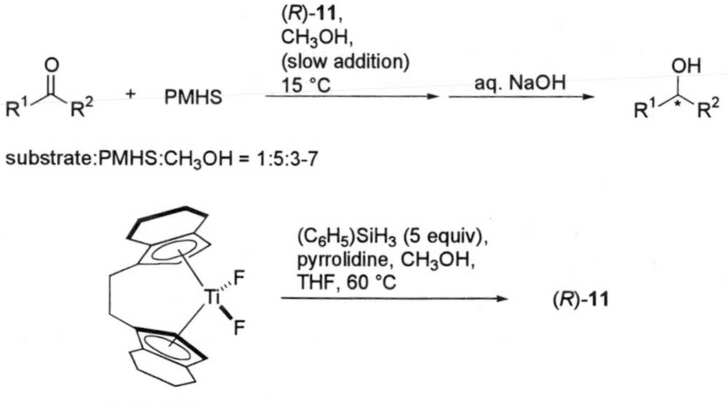

$$R^1 \overset{O}{\underset{}{\bigg\Vert}} R^2 \; + \; PMHS \xrightarrow[\substack{\text{(slow addition)}\\ 15\ ^\circ\text{C}}]{\substack{(R)\text{-}11,\\ CH_3OH}} \xrightarrow{\text{aq. NaOH}} R^1 \overset{OH}{\underset{*}{\bigg|}} R^2$$

substrate:PMHS:CH$_3$OH = 1:5:3-7

$$(R,R)\text{-(EBTHI)-TiF}_2 \xrightarrow[\substack{\text{pyrrolidine, CH}_3\text{OH},\\ \text{THF, 60 }^\circ\text{C}}]{(C_6H_5)SiH_3\ (5\ equiv),} (R)\text{-}11$$

R^1	R^2	S:Ti[a]	Time (h)	Yield (%)	ee (%)	Config^n
C$_6$H$_5$	C$_2$H$_5$	200:1	13	86	98	S
C$_6$H$_5$	CH(CH$_3$)$_2$	100:1	8	86	99	S
C$_6$H$_5$	(CH$_3$)$_2$C=CH(CH$_2$)$_2$	50:1	6	84	97	S
4-CH$_3$C$_6$H$_4$	CH$_3$	100:1	5	87	98	S
1-Cyclohexenyl	CH$_3$	50:1	5	87	96	S
2-Methyl-1-Cyclopentenyl	CH$_3$	50:1	4.5	90	97	–
Cyclohexyl	CH$_3$	100:1	6	>98	23	R
(CH$_3$)$_3$C	CH$_3$	100:1	17	30	53	–

[a] Substrate:Ti molar ratio.

Scheme 6

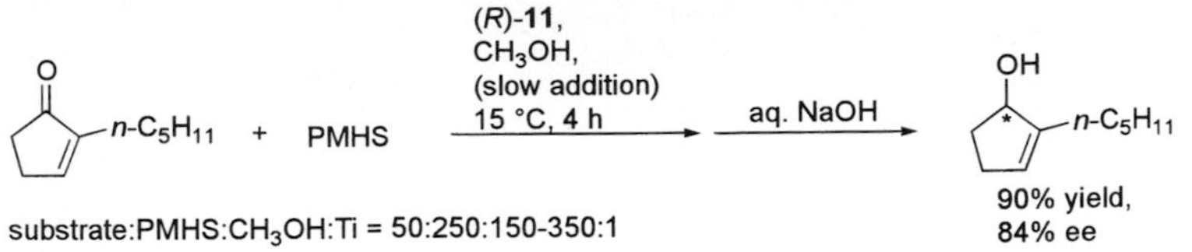

substrate:PMHS:CH$_3$OH:Ti = 50:250:150-350:1

Scheme 7

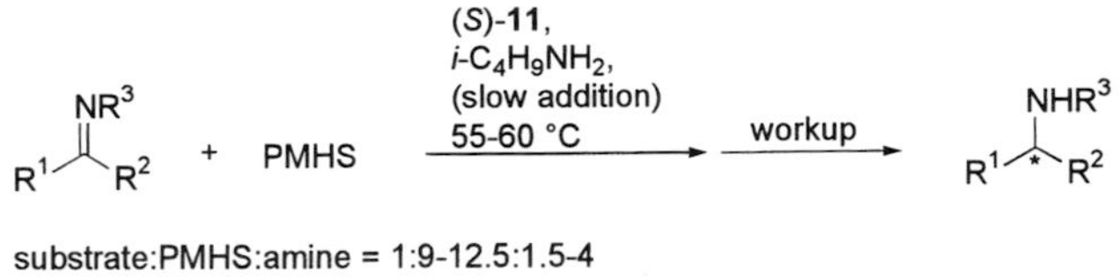

substrate:PMHS:amine = 1:9-12.5:1.5-4

R^1	R^2	R^3	E:Z Ratio	S:Ti^a	Yield (%)	ee (%)	Con-fign
C$_6$H$_5$	CH$_3$	C$_6$H$_5$CH$_2$	15:1	200:1	95	98	S
C$_6$H$_5$	n-C$_4$H$_9$	4-CH$_3$OC$_6$H$_4$CH$_2$	2.5:1	20:1	86	93	S
4-ClC$_6$H$_4$	CH$_3$	4-CH$_3$OC$_6$H$_4$CH$_2$	15:1	200:1	92	99	–
4-ClC$_6$H$_4$	CH$_3$	n-C$_3$H$_7$	18:1	100:1	97	98	–
3-CH$_3$OC$_6$H$_4$	CH$_3$	2-ClC$_6$H$_4$(CH$_2$)$_3$	–	96:1	83	97	R^b
4-CH$_3$OC$_6$H$_4$	(CH$_3$)$_2$CHCH$_2$	n-C$_3$H$_7$	1.8:1	100:1^c	90	97	–
Cyclohexyl	CH$_3$	n-C$_3$H$_7$	20:1	2,000:1	95	98	–
n-Hexyl	CH$_3$	C$_6$H$_5$CH$_2$	3.5:1	100:1	96	69	–

^{a} Substrate:Ti molar ratio. ^{b} (R)-11 was used. ^{c} Phenylsilane was used instead of PMHS.

Scheme 8

reactivity and enantioselectivity were greatly enhanced when i-C$_4$H$_9$NH$_2$ (1.5–4 equivalents) was slowly added to the reaction mixture at about 60 °C, while less hindered N-methylimines were smoothly reduced without any additives [26]. Alkyl aryl ketimines were reduced with up to 99% ee (Scheme 8) [24]. The high enantioselectivity was not affected by the E:Z ratio of the imines. For example, a 1.8:1 E:Z mixture of the N-propylimine of 4'-methoxy-3-methylbutyrophenone was converted to the desired product in 97% optical yield. Hydrosilylation of the N-propylimine of cyclohexyl methyl ketone with a substrate to Ti molar ratio of 2,000:1 was completed to give the product in 98% ee [24]. N-Benzylimine of 2-octanone, a simple aliphatic ketimine, was reduced with 69% optical yield. The reduction of N-benzyl-1-indanimine gave the corresponding amine in 92% ee (Scheme 9) [24].

Highly enantioselective hydrosilylation of N-aryl imines derived from aliphatic ketones was achieved by the use of (S)-11 as a chiral catalyst (Scheme 10) [27]. N-Phenylimine of cyclohexyl methyl ketone was hydrosilylated using PMHS (12 equivalents) as a reducing agent with a slow addition of i-C$_4$H$_9$NH$_2$

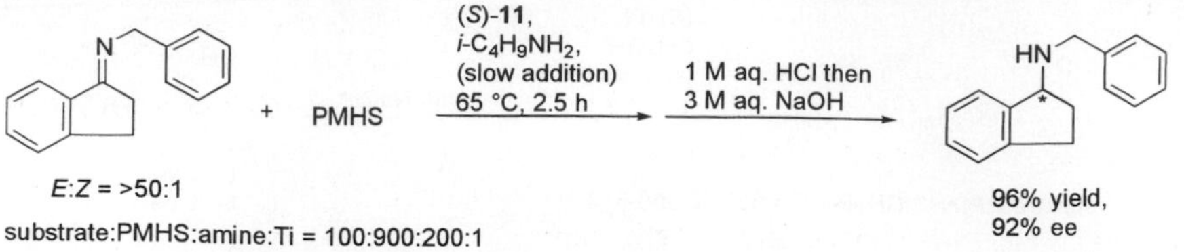

Scheme 9

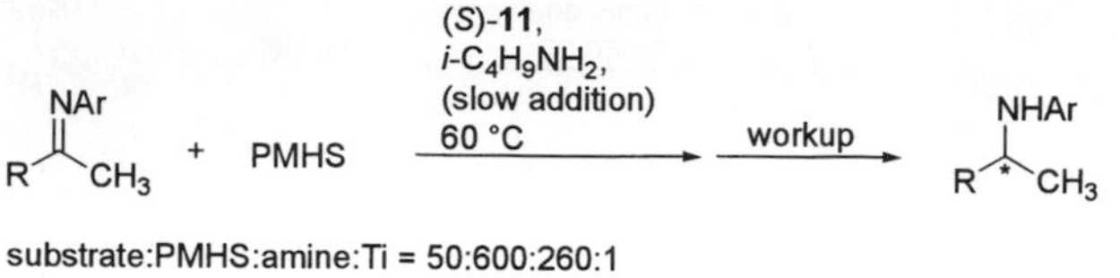

R	Ar	Yield (%)	ee (%)	Confign
C_6H_5	C_6H_5	100[a]	13	
Cyclohexyl	C_6H_5	63	99	S
Cyclohexyl	$4-CH_3OC_6H_4$	79	99	–
n-Hexyl	$4-CH_3OC_6H_4$	70	88	–
$(CH_3)_2C=CH(CH_2)_2$	$4-CH_3OC_6H_4$	68	90	–

[a] % Conversion.

Scheme 10

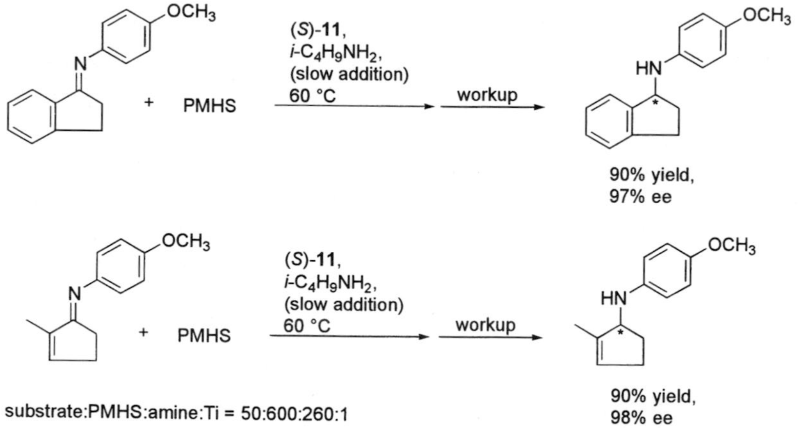

Scheme 11

(5.2 equivalents) at 60 °C followed by a workup process to give the *S* amine in 99% ee (Scheme 10). Interestingly, the enantioselectivity was dramatically decreased to 13% in the reduction of the corresponding imine of acetophenone, an acyclic aromatic ketone. The *N*-(4-methoxy)phenylimine of cyclohexyl methyl ketone was also reduced with a high enantioselectivity. The *N*-(2-methyl)phenyl analogue did not convert at all. *N*-(4-Methoxy)phenylimines derived from 2-octanone and 6-methyl-5-hepten-2-one were reduced with optical yields of 88% and 90%, respectively. Reduction of *N*-(4-methoxy)phenylimines of 1-indanone and 2-methylcyclopentanone under the same conditions led to the chiral amines in 97% and 98% ee, respectively (Scheme 11) [27].

Racemic *N*-methylimines derived from 4-substituted 1-tetralones were kinetically resolved by asymmetric hydrosilylation with phenylsilane (1 equivalent) as a reducing agent using the titanocene catalyst (*R*)-**11** (substrate:Ti=100:1) at 13 °C, followed by a workup procedure to afford the corresponding chiral ketones and chiral *cis* amines with very high enantio- and diastereoselectivity (Scheme 12) [28]. The extent of the enantiomeric differentiation, k_{fast}/k_{slow}, was calculated to be up to 114. The *cis*-selectivity of this reaction was

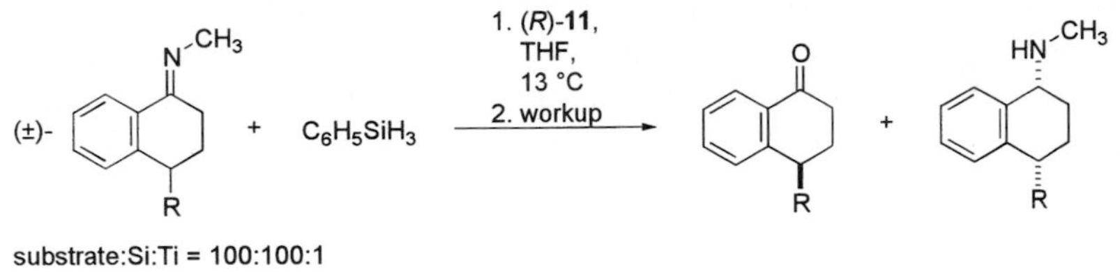

substrate:Si:Ti = 100:100:1

R	Time (h)	Convn (%)	Ketone ee (%)	k_{fast}/k_{slow}	Amine cis:trans	ee of cis (%)	ee of trans (%)
CH₃	24	54	99	60.8	96:4	93	99
Cyclopentyl	48	44	75.5	114	98.5:1.5	98	99
C₆H₅	15	53	99	80.1	95.5:4.5	96	–
3,4-Cl₂C₆H₃	24ᵃ	44	71	–	96.5:3.5	97ᵇ	–

[a] (*S*)-**11** was used at rt (substrate:Ti = 40:1). [b] The 1*S*,4*S* product was sertraline.

Scheme 12

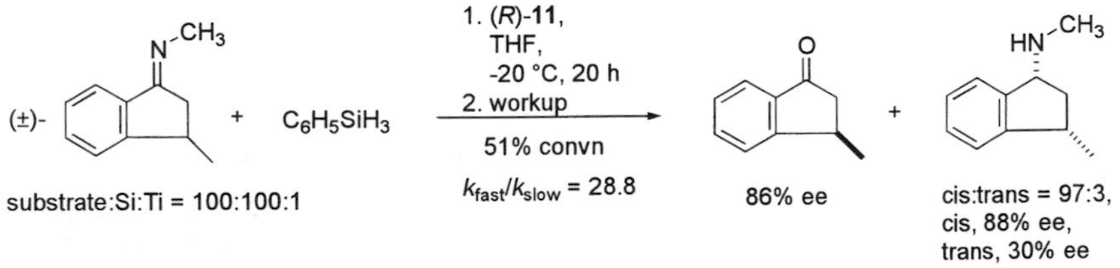

substrate:Si:Ti = 100:100:1

Scheme 13

much higher than that observed in the reduction with NaBH$_4$. This reaction was applied to the synthesis of an antidepressant sertraline [28]. The racemic imine derived from 3-methylindanone was also efficiently resolved with the same catalyst (Scheme 13) [28].

2.3
Cyclic Imines

Several cyclic imines were reduced with phenylsilane as a reducing agent in the presence of the chiral titanocene catalyst 11 followed by a workup process to give the corresponding cyclic amines in excellent ee [26]. The hydrosilylation of 2-propyl-3,4,5,6-tetrahydropyridine with (R)-11 (substrate:Ti=100:1) in THF at room temperature was completed in about 6 h (Scheme 14) [29]. The reaction mixture was treated with an acid and then with an aqueous base to afford (S)-coniine, the poisonous hemlock alkaloid, in 99% ee.

Racemic 2,5-disubstituted 1-pyrrolines were successfully resolved by hydrosilylation with PMHS in the presence of 11. For instance, reduction of 5-methyl-2-phenyl-1-pyrroline with a five equivalents of PMHS using (R)-11 (substrate:Ti=40:1) in the presence of a three equivalents of i-C$_4$H$_9$NH$_2$ in THF at 70 °C followed by purification by chromatography resulted in the S unreacted imine in 98.7% ee (42% yield) and the 2R,5R amine in 98.5% ee (43% yield) accompanied by a small amount of (R)-2-methyl-5-phenyl-1-pyrroline in 98% ee (Scheme 15)

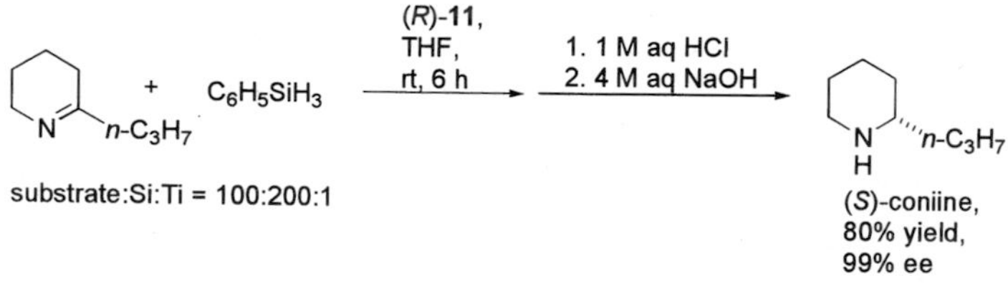

Scheme 14

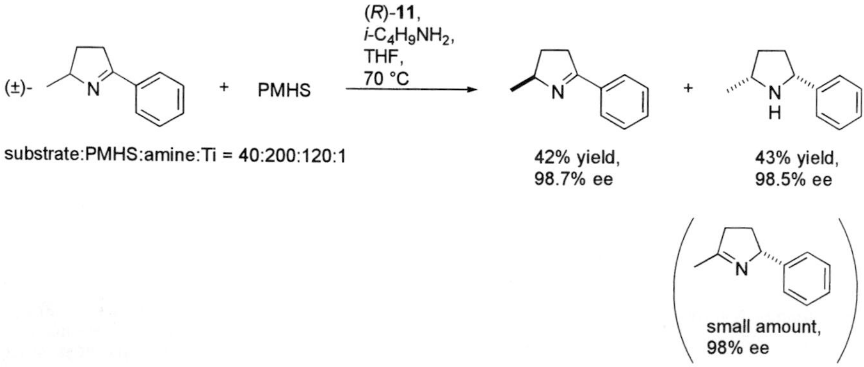

Scheme 15

[30]. The enantiomeric purity of the byproduct indicated that the chiral catalyst also promoted isomerization of the C=N bond, although the isomerization rate was much slower than that of hydrosilylation.

3
Hydrosilylation of C=O and C=N with Miscellaneous Metal Catalysts

A Ru complex with a chiral oxazolinylferrocenylphosphine (S)-6 (substrate: Ru=100:1) with Cu(OSO$_2$CF$_3$)$_2$ promoted hydrosilylation of acetophenone using diphenylsilane (2 equivalents) in ether at 0 °C to give (R)-1-phenylethanol in 95% ee and in 59% yield after hydrolysis (Scheme 16) [31]. Propiophenone was reduced with 97% optical yield. This complex was also effective for the hydrosilylation of 2-phenyl-1-pyrroline in toluene at 0 °C to afford the S chiral amine in 88% ee [31]. RuCl$_2$[(R)-TolBINAP][(S)-7] and AgOSO$_2$CF$_3$ catalyzed the hydrosilylation of acetophenone (substrate:Ru:Ag=100:1:4) with diphenylsilane to give the R product in 82% ee [32, 33].

A neutral Ir complex consisting of [IrCl(COD)]$_2$ and (S)-6 catalyzed the hydrosilylation of the N-methylimine of acetophenone (substrate:Ir=100:1) with two equivalents of diphenylsilane in ether at 0 °C to give the S product in 89% optical yield (Scheme 17) [34]. The corresponding N-phenylimine was reduced with 23% optical yield. 2-Phenyl-1-pyrroline was reduced with the same catalyst to afford the S cyclic amine in 88% ee. The enantiomeric excess was decreased to 7% in the hydrosilylation of the corresponding 6-membered cyclic imine.

A complex prepared in situ from CuCl, (R)-3,5-Xyl-MeO-BIPHEP, and t-C$_4$H$_9$ONa promoted the hydrosilylation of several alkyl aryl ketones (substrate: Cu:ligand:base=33:1:1:1) with PMHS in toluene at -50 or -78 °C to afford the corresponding R products with high optical purity (Scheme 18) [35]. The reduction of propiophenone gave 97% ee. 4'-Trifluoromethylacetophenone and 2'-acetonaphthone were converted to the corresponding R alcohols in 95% ee. 1-Te-

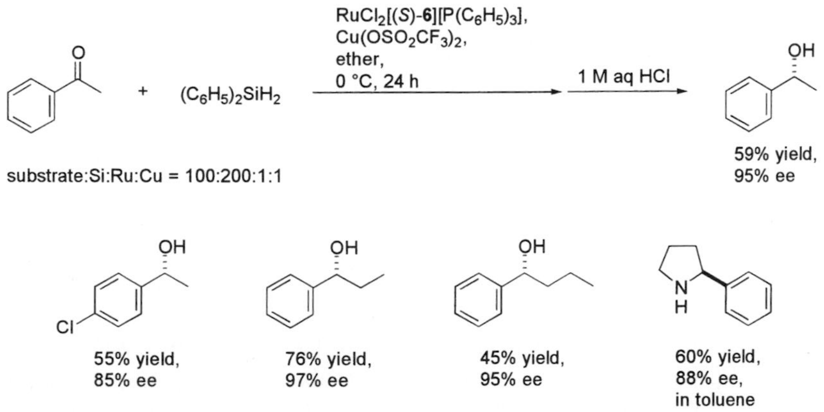

substrate:Si:Ru:Cu = 100:200:1:1

RuCl$_2$[(S)-6][P(C$_6$H$_5$)$_3$],
Cu(OSO$_2$CF$_3$)$_2$,
ether,
0 °C, 24 h

1 M aq HCl

59% yield,
95% ee

55% yield,
85% ee

76% yield,
97% ee

45% yield,
95% ee

60% yield,
88% ee,
in toluene

Scheme 16

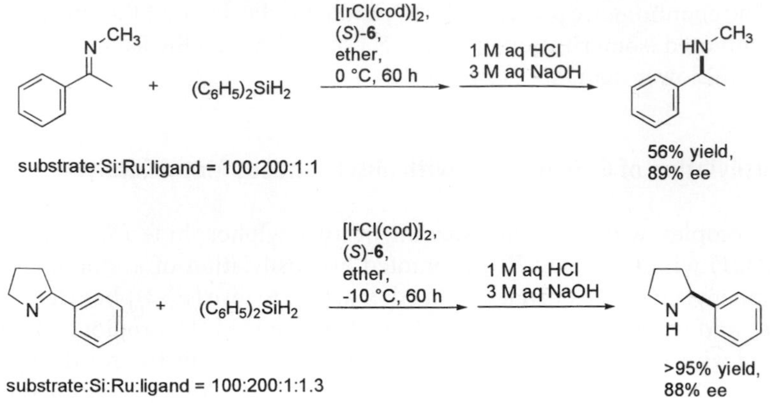

substrate:Si:Ru:ligand = 100:200:1:1

56% yield,
89% ee

substrate:Si:Ru:ligand = 100:200:1:1.3

>95% yield,
88% ee

Scheme 17

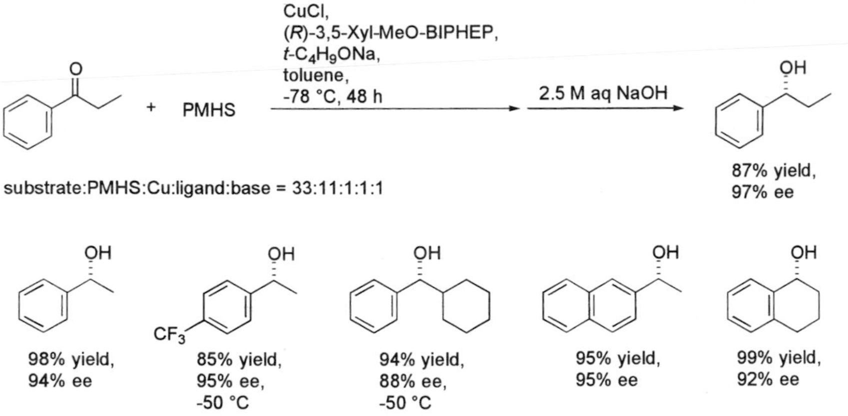

substrate:PMHS:Cu:ligand:base = 33:11:1:1:1

87% yield,
97% ee

98% yield,
94% ee

85% yield,
95% ee,
-50 °C

94% yield,
88% ee,
-50 °C

95% yield,
95% ee

99% yield,
92% ee

Scheme 18

tralone, a cyclic ketone, was also reduced with high enantioselectivity. The reduction of acetophenone with a substrate to Cu to ligand molar ratio of 20,000:600:1 proceeded without loss of enantioselectivity. Hydrosilylation of butyrophenone with phenylsilane catalyzed by a neutral Cu complex consisting of CuF_2 and (S)-BINAP (substrate:Cu:ligand=100:1:1) in toluene at room temperature followed by hydrolysis resulted in the S alcohol in 92% ee (Scheme 19) [36]. Interestingly, the reaction under air proceeded much faster than in argon atmosphere. Use of other copper halides did not work at all. Several 4'-substituted acetophenones were reduced with high enantioselectivity.

A Zn catalyst prepared from diethylzinc and a chiral diamine (S,S)-8 catalyzed the hydrosilylation of acetophenone (substrate:Zn:ligand=50:1:1) with PMHS in toluene at room temperature to afford the R product in 88% ee (Scheme 20) [37]. A catalyst system consisting of zinc diethylacetate, (S,S)-8,

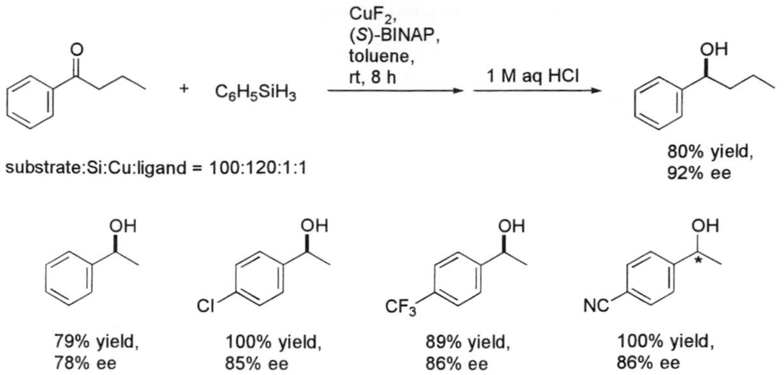

Scheme 19

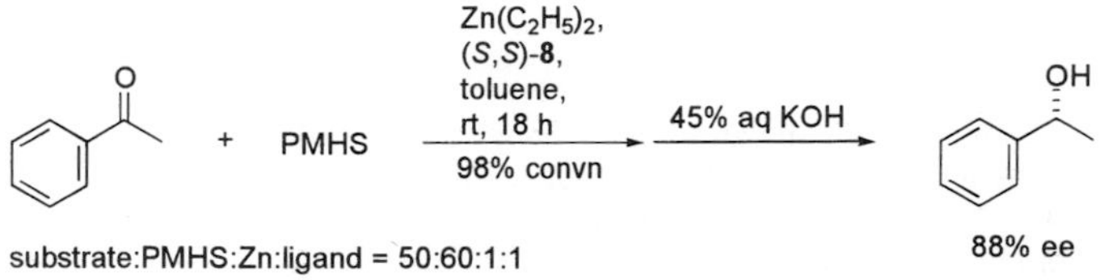

Scheme 20

and $NaAlH_2[O(CH_2)_2OCH_3]_2$ gave the same result. This reaction could be performed at the 1-kg scale without scale-up problems.

Several alkyl aryl ketones were reduced with PMHS in the presence of a complex prepared in situ consisting of $Sn(OSO_2CF_3)_2$ and a nitrogen-based ligand Pybox-Ph [5] (substrate:Sn:ligand=10:1:1) in methanol at room temperature to give the chiral products in up to 58% ee [38].

4
Hydrosilylation of C=O with a Hypervalent Alkoxysilane

Asymmetric reduction of alkyl aryl ketones with trialkoxysilanes is promoted by a catalytic amount of chiral nucleophiles [39]. The reactive species is a transiently prepared hypervalent silicon hydride. 2′,4′,6′-Trimethylacetophenone was reduced with equimolecular amounts of trimethoxysilane in the presence of the monolithio salt of (R)-BINAPHTHOL (substrate:Li=20:1) in a 30:1 ether–TMEDA mixed solvent at 0 °C to afford the R product in 90% ee (Scheme 21) [40]. The presence of TMEDA was crucial to achieve high yield and enantioselectivity. Reduction of less hindered ketonic substrates preferentially gave the S configured products. 1-Tetralone was converted to the R alcohol in 93% optical yield.

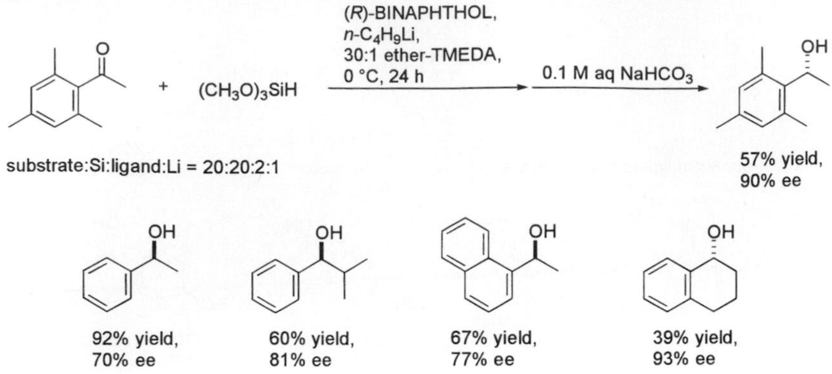
(R)-BINAPHTHOL,
n-C4H9Li,
30:1 ether-TMEDA,
0 °C, 24 h
(CH3O)3SiH
0.1 M aq NaHCO3
substrate:Si:ligand:Li = 20:20:2:1
57% yield, 90% ee
92% yield, 70% ee
60% yield, 81% ee
67% yield, 77% ee
39% yield, 93% ee

Scheme 21

References

1. Brunner H (1998) Hydrosilylation of carbonyl compounds. In: Beller M, Bolm C (eds) Transition metals for organic synthesis, vol 2. Wiley-VCH, Weinheim, p 131
2. Nishiyama H (1999) Hydrosilylation of carbonyl and imino groups. In: Jacobsen EN, Pfaltz A, Yamamoto H (eds) Comprehensive asymmetric catalysis, vol 1, chap 6.3. Springer, Berlin Heidelberg New York
3. Nishiyama H, Itoh K (2000) Asymmetric hydrosilylation and related reactions. In: Ojima I (ed) Catalytic asymmetric synthesis, chap 2, 2nd edn. Wiley-VCH, New York
4. Brunner H, Becker R, Ripel G (1984) Organometallics 3:1354
5. Nishiyama H, Kondo M, Nakamura T, Itoh K (1991) Organometallics 10:500
6. Brunner H, Störiko R, Nuber B (1998) Tetrahedron: Asymmetry 9:407
7. Nishibayashi Y, Segawa K, Ohe K, Uemura S (1995) Organometallics 14:5486; Nishibayashi Y, Segawa K, Takada H, Ohe K, Uemura S (1996) Chem Commun 847
8. Sudo A, Yoshida H, Saigo K (1997) Tetrahedron: Asymmetry 8:3205
9. Lee S, Lim CW, Song CE, Kim IO (1997) Tetrahedron: Asymmetry 8:4027
10. Sawamura M, Kuwano R, Ito Y (1994) Angew Chem Int Ed Engl 33:111
11. Ohkuma T, Noyori R (1999) Hydrogenation of carbonyl groups. In: Jacobsen EN, Pfaltz A, Yamamoto H (eds) Comprehensive asymmetric catalysis, vol 1, chap 6.1. Springer, Berlin Heidelberg New York; Ohkuma T, Kitamura M, Noyori R (2000) Asymmetric hydrogenation. In: Ojima I (ed) Catalytic asymmetric synthesis, chap 1, 2nd edn. Wiley-VCH, New York
12. Kuwano R, Uemura T, Saitoh M, Ito Y (1999) Tetrahedron Lett 40:1327
13. Yamanoi Y, Imamoto T (1999) J Org Chem 64:2988
14. Tsuruta H, Imamoto T (1999) Tetrahedron: Asymmetry 10:877
15. Seebach D, Beck AK, Heckel A (2001) Angew Chem Int Ed 40:92
16. Sakaki J, Schweizer WB, Seebach D (1993) Helv Chim Acta 76:2654
17. Heldmann DK, Seebach D (1999) Helv Chim Acta 82:1096
18. Pastor SD, Shum SP (1998) Tetrahedron: Asymmetry 9:543
19. Haag D, Runsink J, Scharf HD (1998) Organometallics 17:398
20. Carter MB, Schiøtt B, Gutiérrez A, Buchwald SL (1994) J Am Chem Soc 116:11667
21. Rahimian K, Harrod JF (1998) Inorg Chim Acta 270:330
22. Halterman RL, Ramsey TM, Chen Z (1994) J Org Chem 59:2642
23. Yun J, Buchwald SL (1999) J Am Chem Soc 121:5640
24. Verdaguer X, Lange UEW, Buchwald SL (1998) Angew Chem Int Ed 37:1103
25. Hansen MC, Buchwald SL (1999) Tetrahedron Lett 40:2033

26. Verdaguer X, Lange UEW, Reding MT, Buchwald SL (1996) J Am Chem Soc 118:6784
27. Hansen MC, Buchwald SL (2000) Org Lett 2:713
28. Yun J, Buchwald SL (2000) J Org Chem 65:767
29. Reding MT, Buchwald SL (1998) J Org Chem 63:6344
30. Yun J, Buchwald SL (2000) Chirality 12:476
31. Nishibayashi Y, Takei I, Uemura S, Hidai M (1998) Organometallics 17:3420
32. Moreau C, Frost CG, Murrer B (1999) Tetrahedron Lett 40:5617
33. Noyori R, Ohkuma T (2001) Angew Chem Int Ed 40:40
34. Takei I, Nishibayashi Y, Akiyama Y, Uemura S, Hidai M (1999) Organometallics 18:2271
35. Lipshutz BH, Noson K, Chrisman W (2001) J Am Chem Soc 123:12917
36. Sirol S, Courmarcel J, Mostefai N, Riant O (2001) Org Lett 3:4111
37. Minoun H, de Saint Laumer JY, Giannini L, Scopelliti R, Floriani C (1999) J Am Chem Soc 121:6158
38. Lawrence NJ, Bushell SM (2000) Tetrahedron Lett 41:4507
39. Kohra S, Hayashida H, Tominaga Y, Hosomi A (1988) Tetrahedron Lett 29:89
40. Schiffers R, Kagan HB (1997) Synlett 1175

Supplement to Chapter 14
Heck Reaction

Masakatsu Shibasaki, Erasmus M. Vogl, Takashi Ohshima

Graduate School of Pharmaceutical Sciences, The University of Tokyo, 7–3–1 Hongo, Bunkyo-ku, Tokyo 113–0031, Japan
e-mail: mshibasa@mol.f.u.tokyo.ac.jp

Keywords: Carbopalladation, Asymmetric Heck reaction, Vinylation, Arylation

1
Supplements

Among palladium-catalyzed carbon–carbon bond-forming reactions, the Heck reaction enjoys considerable current popularity because of its versatility and tolerance of functionality [102]. Moreover, the intramolecular variant has been regarded as one of the most effective methods for the asymmetric construction of molecules with quaternary carbon stereocenters [103]. Thus, the Heck reaction continues to be utilized in the key steps of many total syntheses of complex natural products, for example, as illustrated by Link and Overman [102]. Notable progress in the asymmetric Heck reaction (AHR) has been made towards the development of new chiral ligands in order to overcome unsolved problems, such as the regioisomer problem, low reaction rates, high catalyst loading, and difficulty in recovery of the catalyst.

1.1
Ligands

1.1.1
P,N-Ligands

Given the success and popularity of phosphinooxazoline ligands **102**, many types of P,N-ligands, which contain an oxazole unit as a nitrogen ligand in most cases, were synthesized and applied to the AHR. New chiral phosphinooxazoline ligands (*S*,*S*p)- and (*S*,*R*p)-2-[4-(isopropyl)oxazole-2-yl]-2′-diphenylphosphino-1,1′-binaphthyls (**114a** and **114b**) were synthesized by Ikeda et al. [104] and Hayashi et al. [105] independently. One of their structural characteristics is that **114a** and **114b** have two independent chiral elements, the binaphthyl axial chirality and the carbon central chirality on the oxazoline ring. The latter authors demonstrated the AHR of dihydrofuran **48** with phenyl triflate using **114a** or **114b** as a chiral ligand to afford **55** in a highly enantio- and regioselective manner (Scheme 1). The regioisomer **54**, which was a major product in the AHR catalyzed by Pd-BINAP, was not detected at all. This regioselectivity is similar to that in Pfaltz's report. As shown, ligands **114a** and **114b** induced opposite configurations in the product **55**. This observation indicates that the axial chirality plays more important role in the enantiocontrol than the carbon central chirality on the oxazolines. The X-ray crystal structure studies supported this argument [105]. The same tendency was also observed in the palladium-catalyzed asymmetric allylic alkylation [104, 105].

Two further phosphinooxazoline ligands, **115** and **116**, which were synthesized by Kündig et al. [106] and Hashimoto et al. [107] respectively, were utilized for the AHR of dihydrofuran **48** with phenyl triflate (Fig. 1). In both cases, the absolute configuration of the product is controlled by the absolute configuration of the 4-position on the oxazolines, meaning (4*S*)-ligands gave (*R*)-products. These results are consistent with those using **102**.

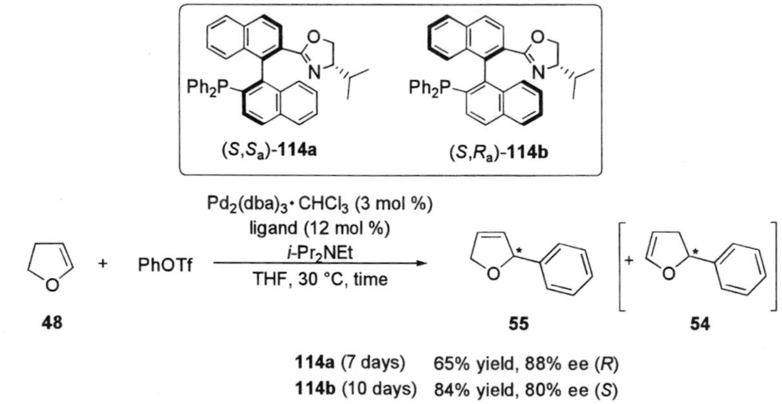

(*S*,*S*a)-**114a**   (*S*,*R*a)-**114b**

$Pd_2(dba)_3 \cdot CHCl_3$ (3 mol %)
ligand (12 mol %)
i-Pr$_2$NEt
THF, 30 °C, time

48   +   PhOTf   →   **55**   +   [**54**   ]

114a (7 days)	65% yield, 88% ee (*R*)
114b (10 days)	84% yield, 80% ee (*S*)

Scheme 1

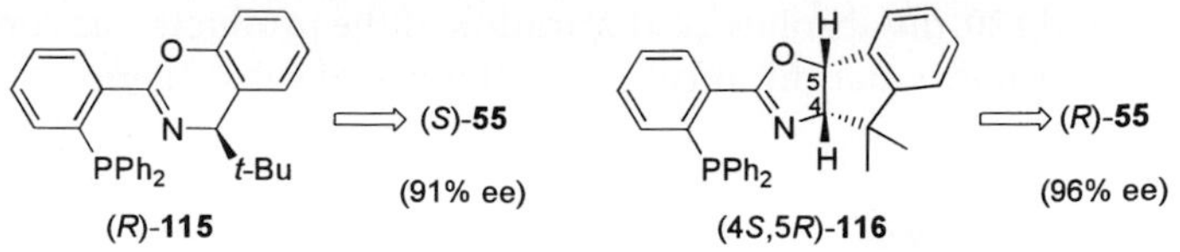

Fig. 1

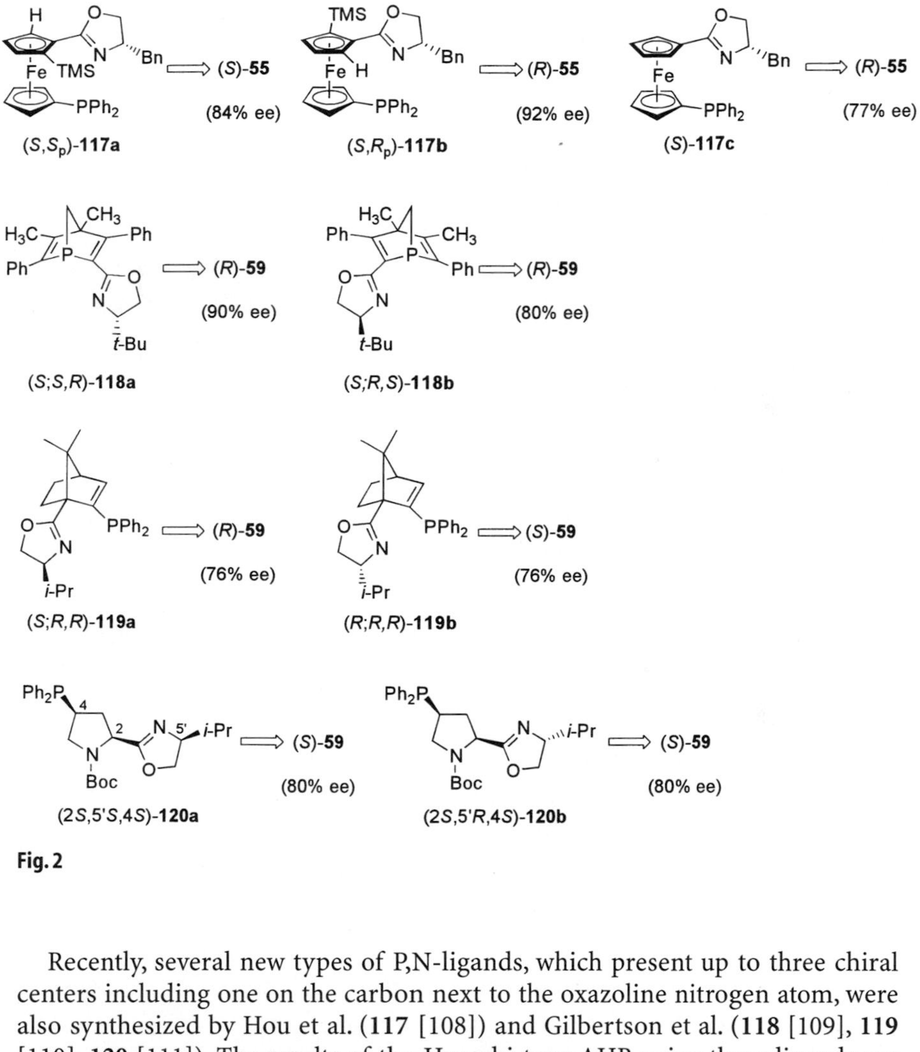

Fig. 2

Recently, several new types of P,N-ligands, which present up to three chiral centers including one on the carbon next to the oxazoline nitrogen atom, were also synthesized by Hou et al. (**117** [108]) and Gilbertson et al. (**118** [109], **119** [110], **120** [111]). The results of the Hayashi-type AHR using these ligands are summarized in Fig. 2. In the cases of **118** and **119**, the absolute configuration of the products was controlled by the chirality of the 4-position on the oxazolines as described above, although **118a** and **118b** gave opposite enantioselectivity in the palladium-catalyzed asymmetric allylic alkylation. On the other hand, in the cases of **117** and **120**, the absolute configuration of the products was controllable by changing the additional chirality and/or the size of substituents.

1.1.2
P,P-Ligands and Derivatives

Many BINAP-type bidentate phosphine ligands have been applied to a variety transition metal-catalyzed asymmetric transformations with a remarkable degree of success. Although the great majority of them reported to date are electron-rich phosphine ligands, less electron-rich ligands, such as tri-2-furylphosphine (TFP) and triphenylarsine, are advantageous for some transition metal-mediated organic reactions. As mentioned in above, the new diarsine ligand (BINAs [26]) is a very effective ligand for the AHR. In this direction several new biaryl bidentate ligands were developed (Fig. 3). 2-Diphenylarsino-2′-diphenyl-

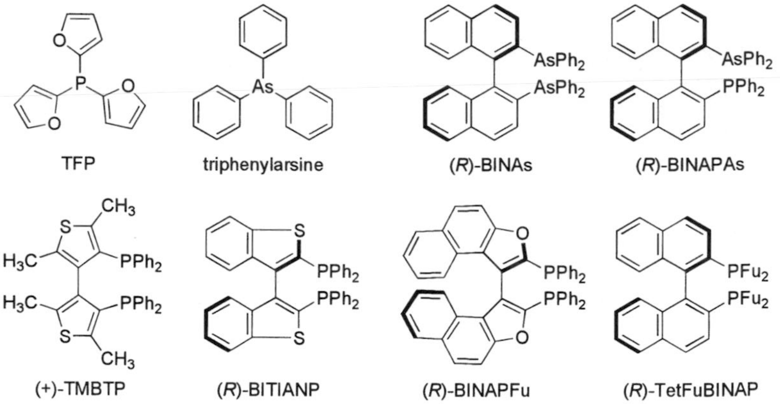
TFP
triphenylarsine
(R)-BINAs
(R)-BINAPAs
(+)-TMBTP
(R)-BITIANP
(R)-BINAPFu
(R)-TetFuBINAP

Fig. 3

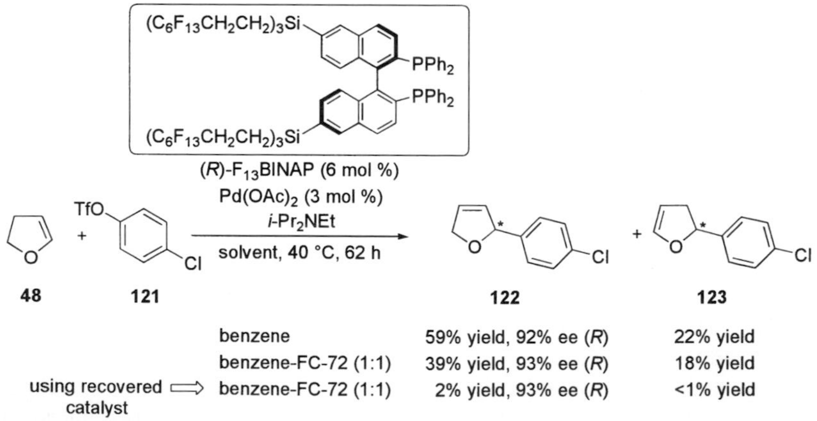
(R)-F13BINAP (6 mol %)
Pd(OAc)2 (3 mol %)
i-Pr2NEt
solvent, 40 °C, 62 h
48
121
122
123
benzene 59% yield, 92% ee (R) 22% yield
benzene-FC-72 (1:1) 39% yield, 93% ee (R) 18% yield
using recovered catalyst benzene-FC-72 (1:1) 2% yield, 93% ee (R) <1% yield

Scheme 2

phosphino-1,1'-binaphthyl (BINAPAs) was synthesized by Shibasaki et al. and successfully applied in AHRs of a system similar to **91** with superior reactivity compared to BINAP [112]. Thienyl-type ligands (TMBTP and BITIANP by Tietze et al. [113]) and furyl-type ligands (BINAPFu and TetFuBINAP by Keay et al. [114]) also show several advantages over BINAP in some AHRs.

As mentioned above, recovery of the catalyst, even that of the chiral ligand, is usually not practical in the AHR process. Recently, fluorous chiral BINAP (F_{13}BINAP), which can be recycled in an organic–fluorous biphasic system, was synthesized by Nakamura et al. and applied to the AHR (Scheme 2) [115]. F_{13}BINAP had good solubility in fluorinated solvents. However, F_{13}BINAP was easily oxidized by a trace amount of oxygen in the fluorous phase during the reaction; as a result, recycling of the catalyst has not succeeded so far.

1.1.3
Other Types of Ligands

Recently, the efficient intramolecular AHR of cyclohexadienones **126** using *monodentate* phosphoramidite ligand **124** was developed by Feringa et al. [116]. Very interestingly, in this reaction system monodentate ligand **124** gave higher enantiomeric excess than bidentate ligand **125** (3). For comparison, BINAP was also examined in this system and **127** was obtained in 0–50% yield and 0–5% ee. Preliminary mechanistic studies of this reaction indicated a possible "neutral" pathway.

In 1995 Herrmann discovered highly efficient palladacycle catalysts in Heck and related reactions of aryl halides with catalyst turnover numbers (TONs) up to 500,000 [117]. Later, TONs of the intermolecular Heck reaction reached up to 8,900,000 [118]. On the other hand, few syntheses of chiral palladacycle catalysts were envisioned and most of these attempts failed. Recently, the first AHR using a chiral phosphapalladacycle catalyst was reported by Buono et al. [119]. The chiral phosphapalladacycle catalyst **129**, which was prepared from Pd(OAc)$_2$

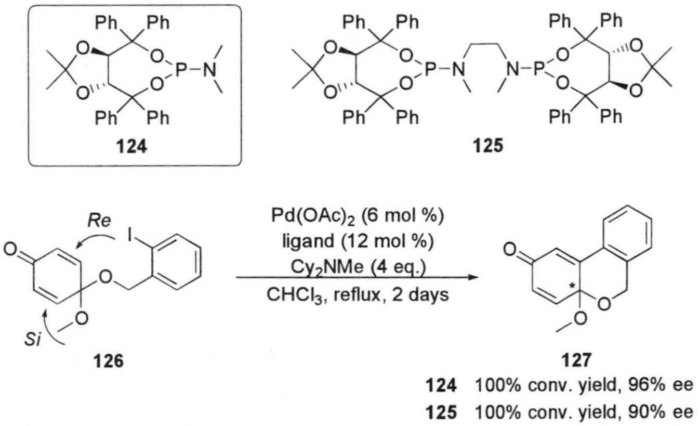

124   **125**

126   **127**

124 100% conv. yield, 96% ee
125 100% conv. yield, 90% ee

Scheme 3

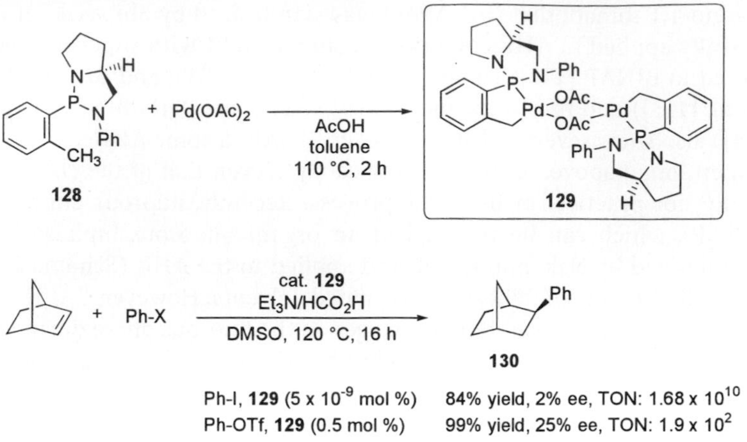

Scheme 4

and the chiral *o*-tolyldiazaphospholidine ligand **128**, promoted the AHR of nor-bornene with phenyl triflate or iodobenzene to afford *exo*-phenylnorbornane **130** (Scheme 4). Although the enantiomeric excesses are low (up to 25% ee), ex-cellent TONs (up to 16,800,000,000) were achieved. Isolation of the Pd(IV) inter-mediate suggests a mechanism for the Heck reaction involving a Pd(II)-Pd(IV) catalytic cycle.

1.2
Methodological Developments

Following a general trend in organic chemistry [99, 120], the AHR can also be integrated in tandem, cascade, or domino reaction sequences. This methodolo-gy allows the formation of complex compounds starting from simple substrate in very few steps. As described above, Keay et al. reported an elegant catalytic asymmetric total synthesis of (+)-xestoquinone, in which a highly efficient cas-cade-type AHR was developed [98]. The one-pot transformation of triflate **112** under typical cationic conditions provided the pentacyclic product **113** with a respectable 68% ee (Scheme 33). Interestingly, the bromide analogue gives little or no asymmetric induction (<15% ee), even in the presence of silver salts. As aryl halides can be prepared more readily than the corresponding triflates, the cascade AHR of the bromide analogue was further improved (in up to 63% ee) by adjustment of silver source (Ag-exchanged zeolite) and the amount of the sil-ver salt (1.0 equivalent) [121].

Bräse reported an intramolecular asymmetric Heck–intermolecular Heck cascade reaction of 1,3-bis(enolnonaflates) **131** to the highly congested bicyclic compound **132** [122] (Scheme 5). Although the level of asymmetric induction is low (up to 52% ee), this result shows that the concept of two leaving groups in the desymmetrization reaction can be applied.

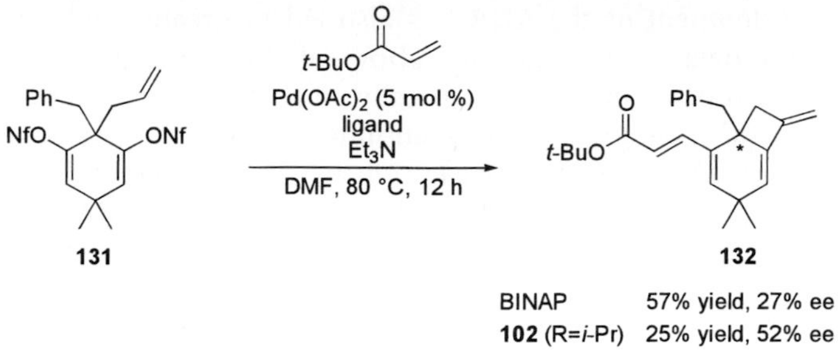

Scheme 5

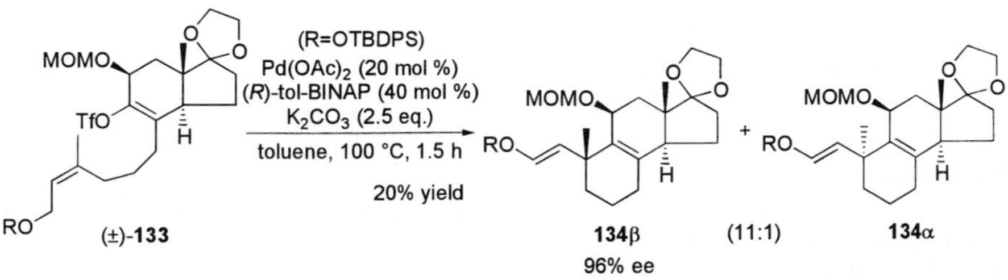

Scheme 6

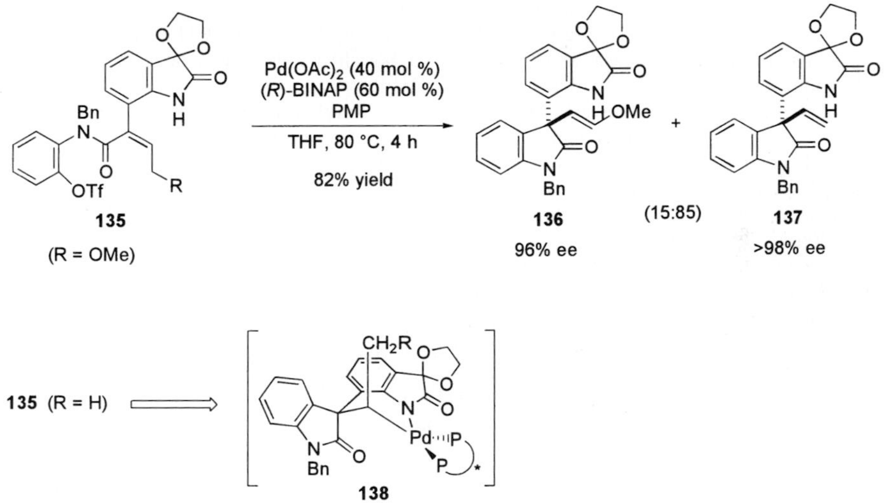

Scheme 7

Kinetic resolution using the AHR was achieved for the first time by Shibasaki et al. [123]. The AHR of racemic **133** catalyzed by the Pd-(R)-tol-BINAP complex provided the desired product **134β**, a potential synthetic intermediate for (+)-wortmannin, in 20% yield (**134β**:**134α**=11:1) and 96% ee (Scheme 6).

Recent development of the AHR has also led to greater understanding of its mechanistic details. As mentioned above, allylic ether substrates **2** (R^3 or R^4=OR) usually provide thermodynamically more stable enol ether compounds **3b** (R^3 or R^4=OR) through β'-hydride elimination (Scheme 1). Recently, Overman et al. reported the interesting exception, in which preferentially β-methoxide elimination proceeded rather than β-hydride elimination [124]. When compound **135** (R=OMe) was used for the AHR, two products were formed in an 85: 15 ratio and 82% yield (Scheme 7). The quaternary carbon centers of both oxindoles **136** and **137** were formed in high enantiomeric ratios. Surprisingly, the AHR of the simpler substrate **135** (R=H), which can be expected to provide **137** more readily, provided only traces of **137** under identical conditions together with unreacted starting material. When **135** (R=H) was treated with 100 mol % of Pd(OAc)$_2$ and 150 mol % of BINAP, a stable σ-alkylpalladium complex **138** (R=H) was isolated, suggesting that β-methoxide elimination proceeded via the palladacycle intermediate **138** (R=OMe).

References

102. For recent reviews, see Bräse S, de Meijere A (1998) Metal-catalyzed cross coupling reactions. Wiley-VCH, Weinheim; Link JT, Overman LE (1998) Metal-catalyzed cross coupling reactions. Wiley-VCH, Weinheim; Donde Y, Overman LE (2002) Catalytic asymmetric synthesis, 2nd edn. Wiley-VCH, New York
103. Corey EJ, Guzman-Perez A (1998) Angew Chem Int Ed Engl 37:388
104. Imai Y, Zhang W, Kida T, Nakatsuji Y, Ikeda I (1998) Tetrahedron Lett 39:4343
105. Ogasawara M, Yoshida K, Kamei H, Kato K, Uozumi Y, Hayashi T (1998) Tetrahedron: Asymmetry 9:1779; Ogasawara M, Yoshida K, Hayashi T (1998) Heterocycles 52:195
106. Kündig EP, Meier P (1999) Helv Chim Acta 82:1360
107. Hashimoto Y, Horie Y, Hayashi M, Saigo K (2000) Tetrahedron: Asymmetry 11:2205
108. Deng WP, Hou XL, Dai LX, Dong XW (2000) Chem Commun 1483
109. Gilbertson SR, Genov DG, Rheingold AL (2000) Org Lett 2:2885
110. Gilbertson SR, Fu Z (2001) Org Lett 3:161
111. Gilbertson SR, Fu Z, Xie D (2001) Tetrahedron Lett 42:365; Gilbertson SR, Xie D, Fu Z (2001) J Org Chem 66:7240
112. Cho SY, Shibasaki M (1998) Tetrahedron Lett 39:1773
113. Tietze LF, Thede K, Schimpf, Sannicolò (2000) Chem Commun 583
114. Andersen NG, Parvez M, Keay BA (2000) Org Lett 2:2817; Andersen NG, McDonald R, Keay BA (2001) Tetrahedron: Asymmetry 12:263
115. Nakamura Y, Takeuchi S, Zhang S, Okumura K, Ohgo Y (2002) Tetrahedron Lett 43:3053
116. Imbos R, Minnaard AJ, Feringa BL (2002) J Am Chem Soc 124:184
117. Herrmann WA, Brossmer C, Reisinger CP, Riermeir TH, Öfele K, Beller M (1997) Chem Euro J 3:1357; Beller M, Fisher H, Herrmann WA, Öfele K, Brossmer C (1995) Angew Chem Int Ed Engl 34:1848; Herrmann WA, Brossmer C, Öfele K, Reisinger CP, Priermeier T, Beller M, Fischer H (1995) Angew Chem Int Ed Engl 34:1844
118. Miyazaki F, Yamaguchi K, Shibasaki M (1999) Tetrahedron Lett 40:7379
119. Brunel JM, Hirlemann MH, Heumann A, Buono G (2000) Chem Comm 1869
120. For a recent review of domino reaction in organic synthesis, see Tietze LF, Haunert F (2000) Stimulating concepts in chemistry. Wiley-VCH, Weinheim
121. Miyazaki F, Uotsu K, Shibasaki M (1998) Tetrahedron 54:13073
122. Bräse S (1999) Synlett 10:1654
123. Honzawa S, Mizutani T, Shibasaki M (1999) Tetrahedron Lett 40:311

124. Oestreich M, Dennison PR, Kodanko JJ, Overman LE (2001) Angew Chem Int Ed Engl 40: 1439

Chapter 16.4
C–H Insertion Reactions, Cycloadditions, and Ylide Formation of Diazo Compounds

Huw M. L. Davies

Department of Chemistry, University at Buffalo, State University of New York, Buffalo, NY 14260–3000, USA
e-mail: hdavies@acsu.buffalo.edu

Keywords: Asymmetric C–H insertion, C–H activation, [3+4] cycloaddition, [3+2] cycloaddition, 1,3-dipolar cycloaddition, Rhodium-catalyzed diazo decomposition

1
Introduction

The metal-catalyzed decomposition of diazo compounds generates transient metal carbenoid intermediates, which have numerous applications in asymmetric synthesis [1]. The most broadly applied reaction is the asymmetric cyclopropanation of alkenes [2] which has been extensively covered in Vol. II, Chap. 16 of this series [3]. Another very important reaction is the metal carbenoid-induced asymmetric C–H insertion. The C–H insertion leads to the functionalization of unactivated C–H bonds and is an example of a C–H activation process. Due to the challenges associated with controlling the chemoselectivity of such a process, the early breakthroughs were in intramolecular versions of this reaction [4] and these were reviewed in Vol. II, Chap. 16 of this series [3]. Since 1998, major advances have been made in controlling the chemoselectivity of intermolecular C–H insertions and these advances will be described in this review. Recently, several examples have appeared of other catalytic enantioselective cycloadditions of diazo compounds rather than the well-established cyclopropanation and these advances will also be summarized. Another emerging methodology is the asymmetric carbenoid transformations involving ylide intermediates [5, 6], and the recent developments in this field will also be described.

2
Chiral Catalysts

A vast array of chiral catalysts have been developed for the enantioselective reactions of diazo compounds but the majority has been applied to asymmetric
cyclopropanations of alkyl diazoacetates [2]. Prominent catalysts for asymmetric intermolecular C–H insertions are the dirhodium tetraprolinate catalysts,
$Rh_2(S\text{-TBSP})_4$ (1a) and $Rh_2(S\text{-DOSP})_4$ (1b), and the bridged analogue $Rh_2(S\text{-}$
$biDOSP)_2$ (2) [7] (Fig. 1). A related prolinate catalyst is the amide 3 [8]. Another catalyst that has been occasionally used in intermolecular C–H activations is
$Rh_2(S\text{-MEPY})_4$ (4) [9]. The most notable catalysts that have been used in enantioselective ylide transformations are the valine derivative, $Rh_2(S\text{-BPTV})_4$ (5)
[10], and the binaphthylphosphate catalysts, $Rh_2(R\text{-BNP})_4$ (6a) and $Rh_2(R\text{-}$
$DDNP)_4$ (6b) [11]. All of the catalysts tend to be very active in the decomposition of diazo compounds and generally, carbenoid reactions are conducted with
1 mol % or less of catalyst loading [1–3].

3
Intermolecular C–H Insertions

Carbenoid intermediates are highly reactive and capable of functionalization of
unactivated C–H bonds [1]. In order for this reaction to be a practical transformation, however, the chemoselectivity of the reaction needed to be enhanced.
Not only would it be necessary to distinguish between different C–H bonds, but

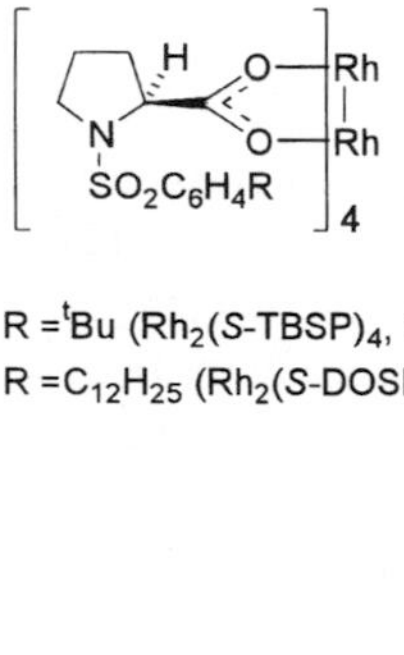

$R = ^t Bu$ ($Rh_2(S\text{-TBSP})_4$, 1a)
$R = C_{12}H_{25}$ ($Rh_2(S\text{-DOSP})_4$, 1b)

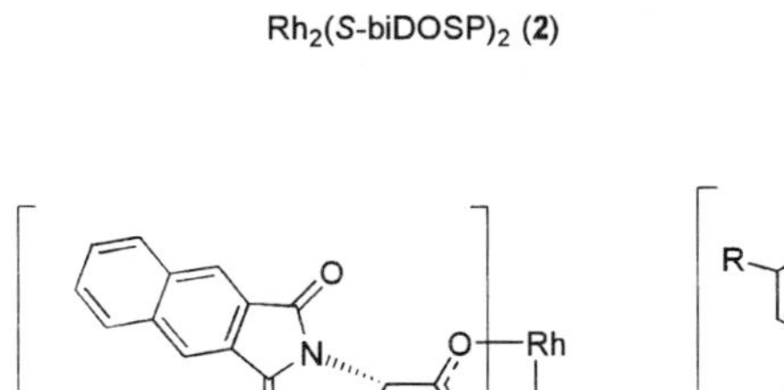

$Ar = p\text{-}C_{12}H_{25}C_6H_4$

$Rh_2(S\text{-biDOSP})_2$ (2)

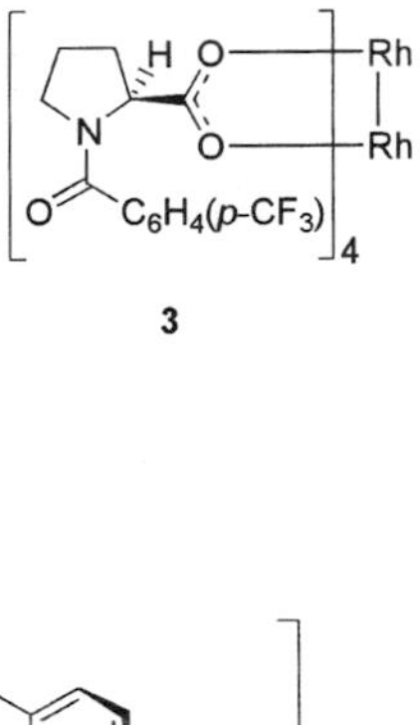

3

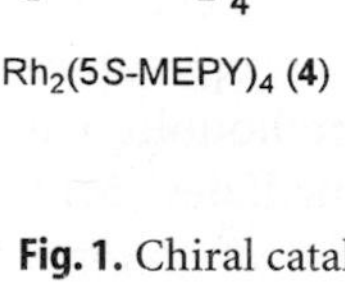

$Rh_2(5S\text{-MEPY})_4$ (4)

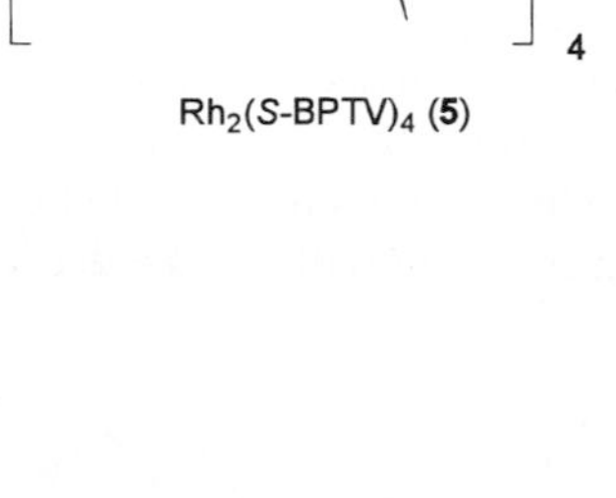

$Rh_2(S\text{-BPTV})_4$ (5)

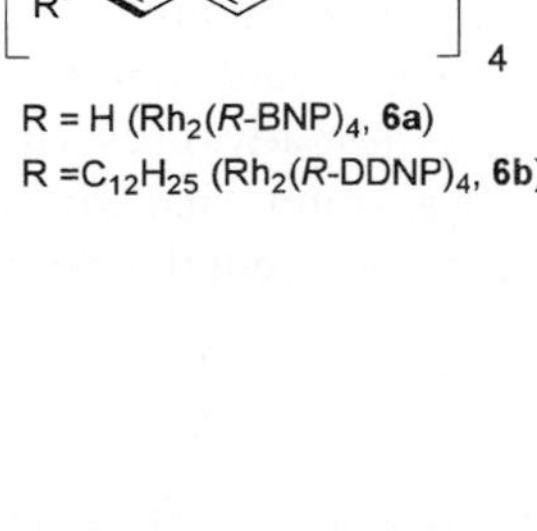

$R = H$ ($Rh_2(R\text{-BNP})_4$, 6a)
$R = C_{12}H_{25}$ ($Rh_2(R\text{-DDNP})_4$, 6b)

Fig. 1. Chiral catalysts

also, the problems of carbene dimerization would need to be overcome. The early approach to control this chemistry was to carry out the reactions intramolecularly [3, 4].

At the time of the previous review [3], only one communication had been published describing high enantioselectivity in intermolecular C–H insertions [12]. Since then, this field has experienced explosive growth. The traditional carbenoids derived from diazo compounds functionalized with either one or two electron-withdrawing groups are simply too reactive for highly chemoselective intermolecular C–H insertions [13]. Some improvement in chemoselectivity and avoidance of carbene dimerization with these traditional carbenoids can be achieved by using sterically crowded catalysts [14]. A major breakthrough in this field, however, was the discovery that carbenoids containing both electron-donating (such as aryl or vinyl) and electron-withdrawing groups are much more chemoselective and less prone to carbene dimerization than the traditional carbenes containing only electron-withdrawing group(s) [15]. The remarkable difference between the two carbenoid systems is readily seen in the dirhodium tetrapivalate-catalyzed reaction with cyclohexane. The reactions with ethyl diazoacetate gave only a 10% yield of the C–H insertion product 7, while the reaction with methyl phenyldiazoacetate under identical conditions gave a 94% yield of the C–H insertion product 8 [16].

$$\text{(1)}$$

$$\text{(2)}$$

Dirhodium tetraprolinate complexes such as $Rh_2(S\text{-DOSP})_4$ have been shown to be excellent chiral catalysts for the reactions of donor–acceptor-substituted carbenes [7]. Application of this catalyst to intermolecular C–H insertions of alkanes by carbenes derived from aryldiazoacetates results in highly chemoselective and enantioselective reactions. Highly efficient C–H functionalization of cyclohexane is possible resulting in the formation of 8 in 80% yield and 95% ee (Eq. 3) [17]. A similar reaction with 2-methylbutane generates a single C–H insertion product 9 in 68% ee (Eq. 4) [17]. The clean reaction with 2-methylbutane is notable because the reaction of ethyl diazoacetate with 2-methylbutane generates a mixture of all four C–H insertion products [13]. In general, there is a delicate balance between C–H insertions at methine or methylene sites: methine sites are favored on electronic grounds, while methylene sites are favored on steric grounds [17]. As a consequence of these steric and electronic effects, an inert solvent for this chemistry is 2,2-dimethylbutane which contains only neopentyl methylene and methyl sites.

$$(3)$$

$$(4)$$

The C–H activation of benzylic positions further underscores the interesting interplay that exists between electronic and steric influences in this chemistry. The C–H activation of 4-ethyltoluene results in clean insertions at the benzylic methylene site to form a 4:1 diastereomeric mixture of **10a** and **10b** [Eq. (5)] [18]. A Hammett study using competition reactions between various substituted ethylbenzenes revealed that the C–H activation is strongly favored in systems containing electron-donating *para* substituents (ϱ=-1.2) [18].

$$(5)$$

One of the most exciting features of these intermolecular C–H insertions is that the functionalization of unactivated C–H bonds can be efficiently achieved, leading to new strategies for the synthesis of chiral molecules. An example of this is the asymmetric synthesis of (+)-indatraline (**13**) shown in Eq. (6) [19]. $Rh_2(S\text{-}DOSP)_4$ catalyzed reaction of **11** with 1,4-cyclohexadiene generated **12** in 93% ee, which was then readily converted to (+)-indatraline (**13**).

$$(6)$$

A second example of the synthetic utility of this chemistry is the enantiose-lective synthesis of (+)-cetidil (**16**) described in Eq. (7) [20]. The key C–H inser-tion step between the thienyldiazoacetate **14** and 1,4-cyclohexadiene to form **15** demonstrates the potentially wide scope of this chemistry because it is compati-ble with an alkyl halide functionality and a nucleophilic thiophene ring.

$$(7)$$

An interesting aspect of the allylic C–H insertion is that the products are γ,δ-unsaturated esters. Traditionally, γ,δ-unsaturated esters are most common-ly prepared by a Claisen rearrangement, especially if stereocontrol is required. Diastereocontrol is also possible in the C–H insertion as long as the reaction oc-curs at a methylene site where there is good size differentiation between the two substituents [21]. An example is the reaction between **17** and the silylcyclohex-ene **18** which forms the C–H insertion product **19** in 88% de and 97% ee [21]. Other catalysts such as Rh$_2$(*R*-BNP)$_4$ and Rh$_2$(*S*-MEPY)$_4$ have been explored for allylic C–H activation of cyclohexene but none were was as effective as Rh$_2$(*S*-DOSP)$_4$ [22].

$$(8)$$

A related allylic C–H insertion that has considerable promise for strategic organic synthesis is the reaction with enol silyl ethers [23]. The resulting silyl-protected 1,5-dicarbonyls would otherwise typically be formed by means of a Michael addition. Even though with ethyl diazoacetates vinyl ethers are readily cyclopropanated [1], such reactions are generally disfavored in trisubstituted vi-nyl ethers with the sterically crowded donor/acceptor carbenoids [23]. Instead, C–H insertion predominates. Again, if sufficient size differentiation exists at the C–H activation site, highly diastereoselective and enantioselective reactions can be achieved as illustrated in the reaction of **20** with **17** to form **21** [23].

$$(9)$$

C–H insertion α to oxygen would generate β-hydroxy or β-alkoxy esters. Thus, the reaction could be considered as a surrogate of the aldol reaction. In Vol. II, Chap. 16 of this series, the preliminary studies on C–H activation of tetrahydrofuran were summarized [3]. Since then this reaction has been optimized such that the major C–H activation product **22a** can be obtained in 97% ee [17]. The optimum reaction conditions (2 equivalents of THF in hexane at -50 °C) demonstrate the regioselectivity that is possible with this chemistry because no reaction with the solvent was observed under these mild conditions.

$$\tag{10}$$

22a/22b 2.8 : 1

Since the early studies with tetrahydrofurans, new substrates have been explored that are capable of highly diastereoselective reactions. *trans*-Allyl silyl ethers are exceptional substrates as illustrated by the conversion of **23** to **24** [Eq. (11)] [24]. The reactions are highly diastereoselective (96–98% de) and in many respects outperform the standard aldol reaction. Even better substrates are tetraalkoxysilanes as shown for the reaction of tetraethoxysilane (**25**) [25]. In this case, both the enantioselectivity and the diastereoselectivity for the formation of C–H activation product **26** are >90% de and ee.

$$\tag{11}$$

80% ee, 96% de, 72% yield

$$\tag{12}$$

95% ee, >94% de, 70% yield

The subtle interplay between steric and electronic effects is revealed in competition reactions between different tetraalkoxylsilanes [24]. Only C–H insertion into tetraethoxysilane occurred in competition reactions between tetramethoxysilane (**27**) and tetraethoxysilane (**25**). A similar selectivity occurred in the reaction between tetraethoxysilane (**25**) and tetraisopropoxysilane (**28**). Thus, even though the tetraisopropoxysilane may be electronically the most favored, it is sterically too crowded. Due to the competing steric and electronic influences, tetraethoxysilane is the most reactive substrate.

$$
\begin{array}{cc}
\text{Si(OEt)}_4 & + & \text{Si(OMe)}_4 \\
\mathbf{25} & & \mathbf{27} \\
\text{(4 equiv)} & & \text{(4 equiv)}
\end{array}
\quad\xrightarrow[\text{Rh}_2\text{(R-DOSP)}_4,\ 23\ °\text{C}]{}\quad
\begin{array}{c}
\text{MeO}_2\text{C} \cdots \text{Ph} \\
\text{ÖSi(OEt)}_3 \\
\text{63\% yield} \\
\text{(exclusive C-H insertion product)}
\end{array}
\tag{13}
$$

$$
\begin{array}{cc}
\text{Si(OEt)}_4 & + & \text{Si(O}^i\text{Pr)}_4 \\
\mathbf{25} & & \mathbf{28} \\
\text{(4 equiv)} & & \text{(4 equiv)}
\end{array}
\quad\xrightarrow[\text{Rh}_2\text{(R-DOSP)}_4,\ 23\ °\text{C}]{}\quad
\begin{array}{c}
\text{MeO}_2\text{C} \cdots \text{Ph} \\
\text{ÖSi(OEt)}_3 \\
\text{60\% yield} \\
\text{(exclusive C-H insertion product)}
\end{array}
\tag{14}
$$

C–H Insertion α to nitrogen is a useful process because it represents a direct method for the asymmetric synthesis of β-amino esters. An especially attractive example of this is the reaction with *N*-Boc-protected amine **29** [26]. Even though electronically functionalization at the benzylic position would be highly favored, this position is sterically too crowded and the reaction occurs cleanly at the *N*-methyl site. The resulting β-amino ester **30**, formed in 96% ee, represents a potentially useful precursor to novel β-peptides.

$$
\begin{array}{c}
\text{Ph} \diagup \text{N} \diagup \\
\quad\quad\underset{\mathbf{29}}{|}\ \text{Boc}
\end{array}
+
\begin{array}{c}
\text{Ph} \diagdown \diagup \text{CO}_2\text{Me} \\
\quad\quad \underset{\mathbf{(2\ equiv.)}}{\|}\ \text{N}_2
\end{array}
\quad\xrightarrow[\text{2. TFA}]{\text{1. Rh}_2\text{(S-DOSP)}_4}\quad
\begin{array}{c}
\text{Ph} \diagup \text{N} \diagup \diagup \text{Ph} \\
\quad\quad \text{H} \quad \text{CO}_2\text{Me} \\
\mathbf{30} \\
\text{67\% yield, 96\% ee}
\end{array}
\tag{15}
$$

Even though it is difficult to achieve a C–H insertion at methylene sites α to an *N*-Boc group in acyclic systems, many cyclic amines are excellent substrates for the C–H insertion. The reaction with *N*-Boc-pyrrolidine (**31**) is a spectacular example [27]. The Rh$_2$(S-DOSP)$_4$-catalyzed reaction of methyl phenyldiazoacetate with **31** at -50 °C generates the C–H insertion product **32** in 94% ee and 92% de [Eq. (16)].

$$
\begin{array}{c}
\underset{\substack{|\\ \text{Boc}\\ \mathbf{31}\\ \text{(4 equiv)}}}{\overset{\displaystyle\bigcirc}{\text{N}}}
\end{array}
+
\begin{array}{c}
\text{N}_2 {=} \diagdown \overset{\text{Ph}}{\underset{\text{CO}_2\text{Me}}{}}
\end{array}
\quad\xrightarrow[\text{2. TFA}]{\text{1. Rh}_2\text{(S-DOSP)}_4\ \ -50\ °\text{C}}\quad
\begin{array}{c}
\underset{\substack{|\\ \text{H}}}{\overset{\displaystyle\bigcirc}{\text{N}}} \diagdown \overset{\text{Ph}}{\underset{\text{H}\ \ \text{CO}_2\text{Me}}{}} \\
\mathbf{32} \\
\text{94\% ee, 92\% de, 72\% yield}
\end{array}
\tag{16}
$$

One of the most impressive examples of the chemoselectivity that is possible in these C–H insertions is the reaction of (±)-silylated alcohol 33 [Eq. (17) [28]. Even though this compound contains three positions that might be expected to be susceptible to C–H insertion, only the methylene position adjacent to the N-Boc group is reactive as the other two sites are too sterically encumbered. Furthermore, this reaction is highly diastereoselective and displays a high level of kinetic resolution such that the C–H insertion product 34 is obtained essentially as a single diastereomer in 85% yield and 98% ee.

$$(17)$$

98% ee, >94% de, 85% yield

The reaction with N-Boc-pyrrolidine may be taken a step further by inducing a double C–H insertion sequence [27]. This results in the formation of the elaborate C_2-symmetric amine 35 as a single diastereomer with control of stereochemistry at four stereogenic centers. The enantiomeric purity of 35 is higher than that obtained for the single C–H insertion products, presumably because kinetic resolution is occurring in the second C–H insertion step.

$$(18)$$

97% ee, 78% yield

The reaction of methyl phenyldiazoacetate with N-Boc-piperidine (36) is a good illustration of the potential of this chemistry because it leads to the direct synthesis of *threo*-methylphenidate (37) [27]. The most efficient rhodium carboxylate catalyst for carrying out this transformation is Rh$_2$(S-biDOSP)$_2$ (2), which results in the formation of a 71:29 mixture of the readily separable *threo* and *erythro* diastereomers. The *threo* diastereomer 37 is produced in 52% isolated yield and 86% ee [Eq. (19)]. Other catalysts have also been explored for this reaction. Rh$_2$(R-DOSP)$_4$ gives only moderate stereoselectivity while Rh$_2$(R-MEPY)$_4$ gave the best diastereoselectivity in this reaction (94% de) [29].

$$(19)$$

threo-methylphenidate (37)

86% ee, 52% yield

The reaction of vinyldiazoacetates at allylic C–H sites does not result in the predominant formation of the products of a simple C–H insertion [25]. Instead, combined C–H insertion/Cope rearrangement products are formed with very high enantioselectivity as illustrated by the example shown in Eq. (20) [25]. This process has been applied to a very short asymmetric synthesis of (+)-sertraline (**38**).

$$ (20) $$

(99% ee, 60 % yield)   (+)-sertraline (**38**)

4
Cycloadditions

The [2+1] cycloaddition between metal carbenoid intermediates and alkenes is a very powerful method for the stereoselective synthesis of cyclopropanes [1–3]. Indeed, the vast majority of chiral catalysts developed for carbenoid chemistry were specifically designed for asymmetric cyclopropanation [1–3]. In recent years, however, a number of other enantioselective cycloadditions have been reported.

The reaction of electron-rich alkenes with diazo compounds containing two electron-withdrawing groups does not always result in cyclopropane formation. Instead, a [3+2] cycloaddition can occur, presumably by means of a stepwise process [1, 2]. One report has appeared which demonstrates that the *N*-amidoprolinate **3** can catalyze a [3+2] cycloaddition between diazocyclohexane-1,3-dione (**39**) and 2,3-dihydrofuran (**40**) to form the tricyclic system **41** with very high enantioselectivity [Eq. (21)] [8].

$$ (21) $$

96% ee, 66% yield

A different [3+2] cycloaddition occurs on reaction of certain vinyldiazoacetates with styryl ethers [30]. Although the mechanistic details are not known, the reaction is considered to be a concerted process and furthermore, it displays very high levels of stereocontrol. As illustrated in the Rh$_2$(*S*-DOSP)$_4$-catalyzed reaction shown in Eq. (22), the [3+2] cycloadduct **42** is formed as the all-*cis* diastereomer in 98% ee. Cuprate addition to **42** stereoselectively generates the cyclopentane **43** with five contiguous stereocenters.

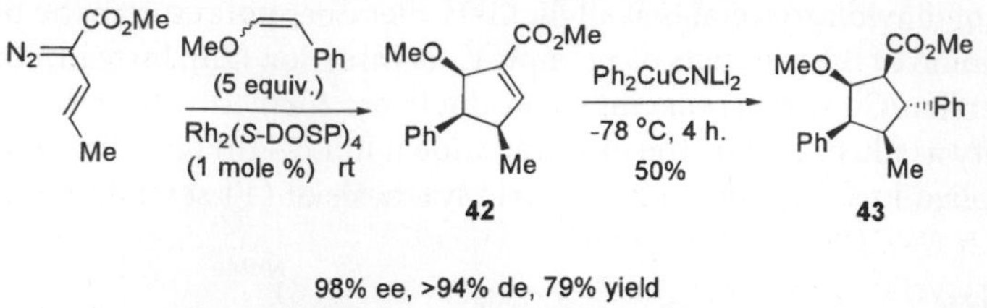

$$(22)$$

98% ee, >94% de, 79% yield

A highly stereoselective [3+4] cycloaddition occurs in the rhodium(II) car-boxylate-catalyzed reaction of vinyldiazoacetates in the presence of dienes [31]. The [3+4] cycloaddition proceeds via a divinylcyclopropane, which then under-goes a Cope rearrangement in a stereodefined manner, leading to cyclohepta-dienes with stereocontrol at up to three stereogenic centers. When these reac-tions are catalyzed by $Rh_2(S\text{-DOSP})_4$, highly enantioselective reactions are ob-tained as illustrated for (E) and (Z)-1,3-pentadiene (Eqs. 23 and 24) [32]. Enan-tioselective [3+4] cycloadditions are possible between vinylcarbenoids and a wide range of acyclic and cyclic dienes, including furans [33], N-Boc-pyrroles [34], dihydropyridines [35], and pyridones [35]. Intramolecular versions of the reactions are also known [36].

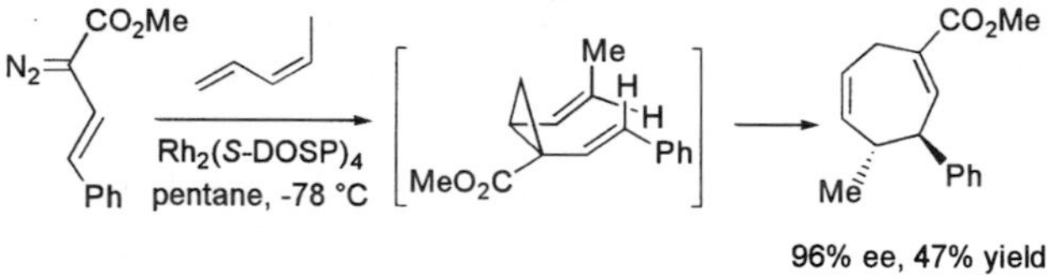

$$(23)$$

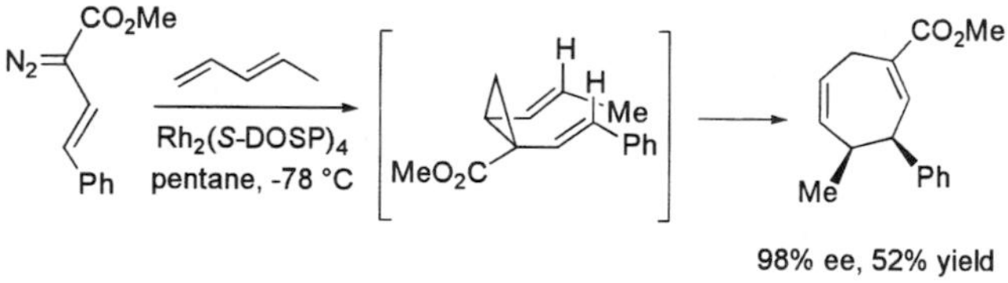

96% ee, 47% yield

$$(24)$$

98% ee, 52% yield

5
Ylide Formation

A promising synthetic transformation is the reaction of carbenoid interme-diates with heteroatoms to form ylides that are capable of undergoing further transformations [5, 6]. Enantioselective transformations in which the ylide in-termediates undergo either 1,2- or 2,3-sigmatropic rearrangement were brief-ly reviewed in the previous issue (Vol. II, pp. 531–532) and several recent exam-ples have appeared [37]. A major breakthrough has been made in the enanti-oselective transformation of carbonyl ylides derived from capture of the metal carbenoid intermediates by carbonyl groups. The carbonyl ylides have been extensively employed in various elaborate 1,3-dipolar cycloadditions [5], but until the recent landmark studies of Hodgson [38] and Hashimoto [39], highly enantioselective variants had not been developed.

The potential of the enantioselective intramolecular 1,3-dipolar cycloaddition was first described by Hodgson in the intramolecular version shown in Eq. (25). The $Rh_2(S\text{-DOSP})_4$-catalyzed reaction of the diazoacetatoacetate **44** generated the tricyclic product **45** in 53% ee [38], but in a more recent study using the binaphthylphosphate catalyst $Rh_2(R\text{-DDNP})_4$ (**6b**) the tricyclic product was formed in 90% ee [11].

$$(25)$$

$Rh_2(S\text{-DOSP})_4$,   53% ee

$Rh_2(R\text{-DDNP})_4$,   90% ee

Hashimoto has shown that the the valine-derived catalyst $Rh_2(S\text{-BPTV})_4$ (**5**) is effective in intermolecular tandem cyclization/intermolecular cycloaddition resulting in the formation of **46** in 92% ee (Eq. (26) [10]. More recent studies have broadened the range of substrates that can be used in the reaction although the enantioselectivity is variable [38, 39].

$$(26)$$

92% ee

In summary, the metal-catalyzed decomposition of diazo compounds results in a broad array of opportunities for the development of new asymmetric catalytic transformations. In the last few years considerable advances have been made in enantioselective intermolecular C–H insertion, novel cycloadditions, and tandem cyclization/cycloadditions. These new transformations offer new strategies for the rapid enantioselective construction of complex structures.

References

1. Doyle MP, McKervey MA, Ye T (1998) In: Modern catalytic methods for organic synthesis with diazo compounds: from cyclopropanes to ylides. Wiley, New York
2. Davies HML, Antoulinakis EG (2001) Org Reacts 59:1–302
3. Pfaltz A (1999) In: Jacobsen EN, Pfaltz A, Yamamoto H (eds) Comprehensive asymmetric catalysis, vol II. Springer, Berlin Heidelberg New York, pp 513–538; Lydon KM, McK-

ervey MA (1999) In: Jacobsen EN, Pfaltz A, Yamamoto H (eds) Comprehensive asymmetric catalysis, vol II. Springer, Berlin Heidelberg New York, pp 539–580; Charette AB, Lebel H (1999) In: Jacobsen EN, Pfaltz A, Yamamoto H (eds) Comprehensive asymmetric catalysis, vol II. Springer, Berlin Heidelberg New York, pp 581–603

4. Sulikowski GA, Cha KL, Sulikowski MM (1988) Tetrahedron: Asymmetry 9:3145
5. Padwa A, Hornbuckle SF (1991) Chem Rev 91:263
6. Hodgson DM, Pierard FYTM, Stupple PA (2001) Chem Soc Rev 30:50
7. Davies HML (1999) Eur J Org Chem 2459; Davies HML (1997) Aldrichim Acta 30:105
8. Ishitani H, Achiwa K (1997) Heterocycles 46:153
9. Doyle MP (1996) Aldrichim Acta 29:3
10. Kitagakii S, Yasugahira M, Anada M, Nakajima M, Hashimoto S (2000) Tetrahedron Lett 41:5931
11. Hodgson DM, Stupple PA, Johnstone C (1999) Chem Commun 2185
12. Davies HML, Hansen T (1997) J Am Chem Soc 119:9075
13. Domenceau A, Noels AF, Costa JL, Hubert A (1990) J Mol Catal 58:21; Demonceau A, Noels AF, Hubert AJ, Teyssie P (1984) Bull Soc Chim Belg 93:945
14. Diaz-Requejo MM, Belderrain TR, Nicasio MC, Trofimenko S, Perez PJ (2002) J Am Chem Soc 124:896
15. Davies HML, Antoulinakis EG (2001) J Organomet Chem 617–618:45
16. Davies HML, Hodges LM, Matasi JJ, Hansen T, Stafford DG (1998) Tetrahedron Lett 39:4417
17. Davies HML, Hansen T, Churchill MR (2000) J Am Chem Soc 122:3063
18. Davies HML, Jin Q, Ren P, Kovalevsky AY (2002) J Org Chem 67:4165
19. Davies HML, Gregg TM (2002) Tetrahedron Lett 43:4951
20. Davies HML, Walji AM, Townsend RJ (2002) Tetrahedron Lett 43:4981
21. Davies HML, Ren P, Jin Q (2001) Org Lett 3:3587
22. Muller P, Tohill S (2000) Tetrahedron 56:1725
23. Davies HML, Ren P (2001) J Am Chem Soc 123:2071
24. Davies HML, Antoulinakis EG, Hansen T (1999) Org Lett 1:383
25. Davies HML, Antoulinakis EG (2000) Org Lett 2:4153
26. Davies HML, Venkataramani C (2002) Angew Chem Int Ed Engl 41:2197
27. Davies HML, Hansen T, Hopper D, Panaro SA (1999) J Am Chem Soc 121:6509
28. Davies HML, Venkataramani C (2001) Org Lett 3:1773
29. Axten JM, Ivy R, Krim L, Winkler JD (1999) J Am Chem Soc 121:6511
30. Davies HML, Xiang B, Kong N, Stafford DG (2001) J Am Chem Soc 123:7461
31. Davies HML (1999) In: Haramata ME (ed) Advances in cycloaddition. JAI Press 5:119–164
32. Davies HML, Stafford DG, Doan BD, Houser JH (1998) J Am Chem Soc 120:3326
33. Davies HML, Ahmed G, Churchill MR (1996) J Am Chem Soc 118:10774
34. Davies HML, Matasi JJ, Hodges LM, Huby NJS, Thornley C, Kong NX Houser JH (1997) J Org Chem 62:1095
35. Davies HML, Hodges LM (2002) J Org Chem 67:5683
36. Davies HML, Doan BD (1999) J Org Chem 64:8501
37. Kitagaki S, Yanamoto Y, Tsutsui H, Anada M, Nakajima M, Hashimoto S (2001) Tetrahedron Lett 42:6361; Kitagaki S, Yanamoto Y, Okubo H, Nakajima M, Hashimoto S (2001) Heterocycles 54:623; Doyle MP, Forbes DC, Vasbinder MM, Peterson CS (1998) J Am Chem Soc 120:7653; Calter MA, Sugathapala PM (1998) Tetrahedron Lett 39:8813; Fukuda T, Irie R, Katsuki T (1999) Tetrahedron 55:649; McMillen DW, Varga N, Reed BA, King C (2000) J Org Chem 65:2532
38. Hodgson DM, Stupple PA, Johnstonne C (1997) Tetrahedron Lett 38:6471; Hodgson DM, Stupple PA, Pierard FYTM, Labande AH, Johnstone C (2001) Chem Eur J 7:4465; Hodgson DM, Glen R, Redgrave AJ (2002) Tetrahedron Lett 43:3927
39. Kitagaki S, Anada M, Kataoka O, Matsuno K, Umeda C, Watanabe N, Hashimoto S (1999) J Am Chem Soc 1999 121:1417

Supplement to Chapter 26.1
Alkylation of C=O

Kenso Soai*, Takanori Shibata

*Department of Applied Chemistry, Faculty of Science, Tokyo University of Science,
  Kagurazaka, Shinjuku-ku, 162–8601 Tokyo, Japan
e-mail: soai@rs.kagu.tus.ac.jp

Department of Chemistry, School of Science & Engineering, Waseda University, 3-4-1 Okubo,
Shinjuku-ku, Tokyo 169-8555, Japan
e-mail: tshibata@waseda.jp

Keywords: amino alcohol, dialkylzinc, alkylation, alcohol, asymmetric autocatalysis

1
Introduction

One of the most important methods for the synthesis of optically active *sec*-alcohols is the catalytic enantioselective alkylation of aldehydes by organometallic reagents. Among organometallic reagents, dialkylzinc is utilized most frequently in the catalytic enantioselective alkylation of aldehydes [1–3]. Since the publication of our review [4], a number of chiral catalysts have been developed. In this chapter, we describe selected examples of chiral catalysts for the enantioselective alkylation of aldehydes. Enantioselective alkynylation of aldehydes and arylation of ketones are also described. Asymmetric autocatalysis of nitrogen-containing heterocyclic alkanols such as 5-pyrimidyl alkanol in the enantioselective isopropylation of nitrogen-containing heterocyclic aldehydes such as pyrimidine-5-carbaldehyde has been achieved with amplification of enantiomeric excess [5]. Alkylation of aldehydes with additional functional groups and alkylation with functionalized zinc reagents are beyond the scope of this review.

2
Asymmetric Alkylation of Aldehydes Catalyzed by Chiral Lewis Bases

Recently, new types of chiral catalysts are disclosed and selected examples are listed in Scheme 1. Various chiral catalysts containing a ferrocenyl moiety are highly enantioselective catalysts [3]. Among them, ferrocene 1 and 2-amino-substituted 1-sulfinylferrocene 2 are a new family of planar chiral catalysts [6, 7]. Tetracoordinate complex 3, which is easily prepared from phenylalanine, is an efficient catalyst for the enantioselective addition of Me_2Zn and ethylation of aliphatic aldehyde [8]. Cu-bis(oxazoline) complexes have been comprehensively examined as chiral catalysts for enantioselective 1,4-additions. Bis(oxazoline) 4 is also an efficient catalyst for the enantioselective addition of Et_2Zn to aldehydes [9]. Various bicyclic tertiary amines bearing N-chiral bridgehead nitrogen atoms are readily accessible by a combination of primary amines, cyclic amino

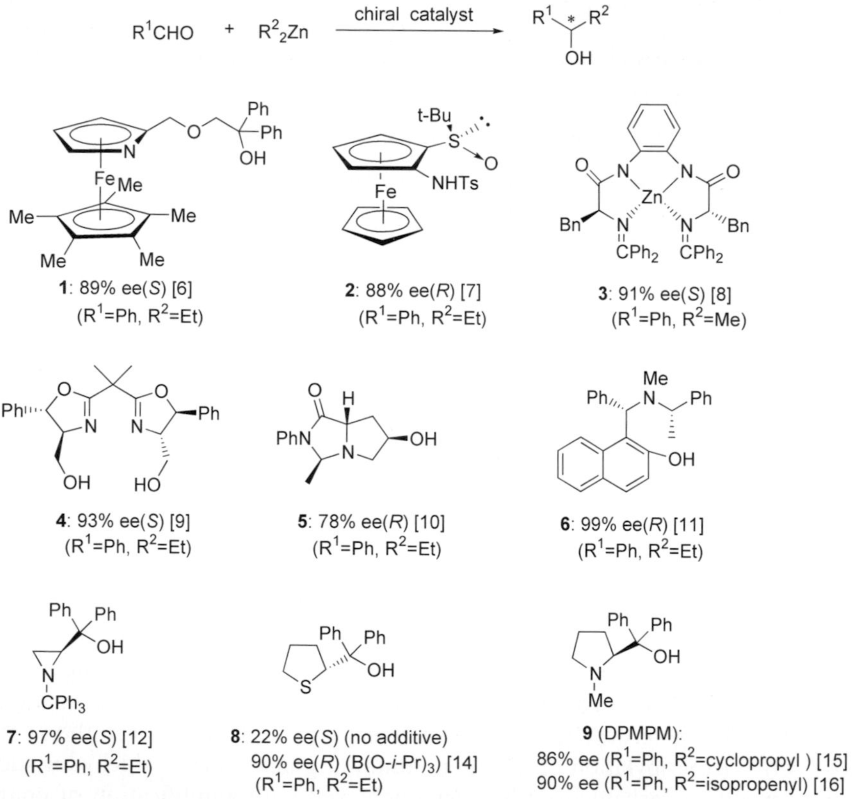

$$R^1CHO \quad + \quad R^2_2Zn \quad \xrightarrow{\text{chiral catalyst}} \quad R^1 \underset{OH}{\overset{*}{\diagdown}} R^2$$

1: 89% ee(S) [6]
(R^1=Ph, R^2=Et)

2: 88% ee(R) [7]
(R^1=Ph, R^2=Et)

3: 91% ee(S) [8]
(R^1=Ph, R^2=Me)

4: 93% ee(S) [9]
(R^1=Ph, R^2=Et)

5: 78% ee(R) [10]
(R^1=Ph, R^2=Et)

6: 99% ee(R) [11]
(R^1=Ph, R^2=Et)

7: 97% ee(S) [12]
(R^1=Ph, R^2=Et)

8: 22% ee(S) (no additive)
90% ee(R) (B(O-i-Pr)$_3$) [14]
(R^1=Ph, R^2=Et)

9 (DPMPM):
86% ee (R^1=Ph, R^2=cyclopropyl) [15]
90% ee (R^1=Ph, R^2=isopropenyl) [16]

Scheme 1

acids, and aldehydes. Enantioselectivity of amino alcohol 5 is not very high but this parallel library examination provides a new approach for the optimization of chiral catalysts [10]. Aminonaphthol 6 is also a good catalyst for the enantioselective ethylation of aryl aldehydes [11]. Aziridine-based amino alcohol 7 also induces very high enantioselectivity [12]. Chiral alcohol 8, the sulfur analogue of DPMPM [diphenyl(1-methylpyrrolidin-2-yl)methanol] 9 [1, 4, 13], by itself induces only low enantioselectivity, but addition of a catalytic amount of metal alkoxide improves the ee dramatically [14].

We also reported that DPMPM 9 can operate as an efficient catalyst in the enantioselective addition of dicyclopropylzinc and diisopropenylzinc [15, 16].

3
Asymmetric Alkylation of Aldehydes Catalyzed by Chiral Lewis Acids

Lewis acid-promoted asymmetric addition of dialkylzincs to aldehydes is also an acceptable procedure for the preparation of chiral secondary alcohol. Various chiral titanium complexes are highly enantioselective catalysts [4]. C_2-Symmetric disulfonamide, chiral diol (TADDOL) derived from tartaric acid, and chiral thiophosphoramidate are efficient chiral ligands. C_2-Symmetric chiral diol 10, readily prepared from 1-indene by Brown's asymmetric hydroboration, is also a good chiral source (Scheme 2) [17]. Even a simple α-hydroxycarboxylic acid 11 can achieve a good enantioselectivity [18].

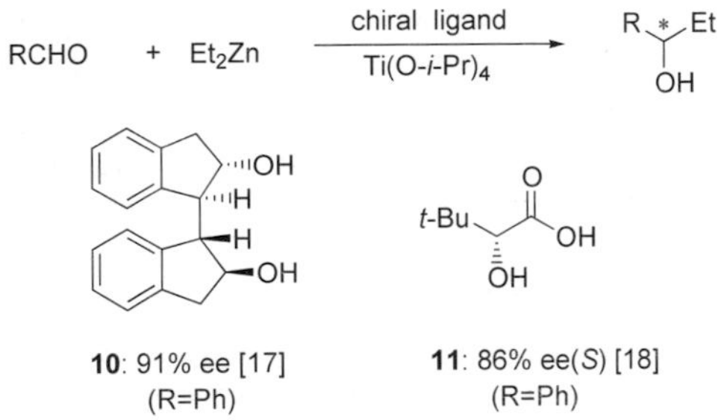

Scheme 2

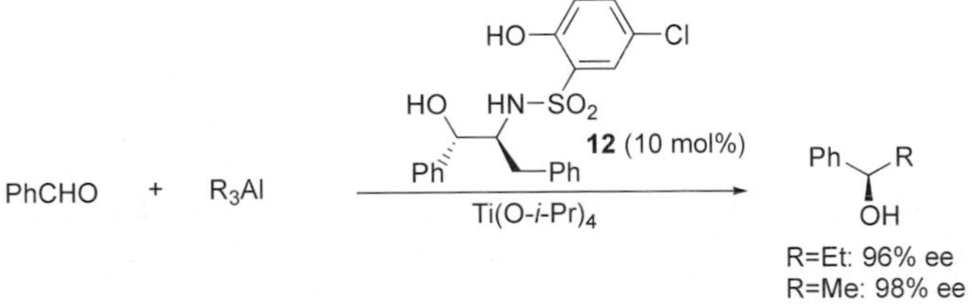

Scheme 3

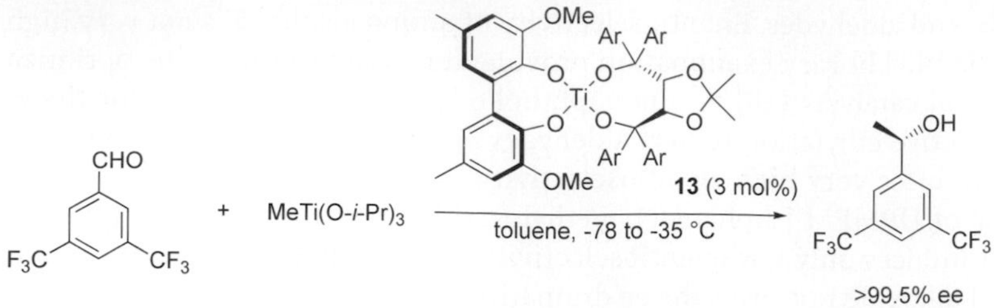

Scheme 4

Trialkylaluminiums are also suitable alkylating reagents in the presence of a Ti(IV) complex. *N*-Sulfonylated amino alcohol **12** is an especially appropriate ligand not only for asymmetric ethylation but also for methylation; this is noteworthy, since it is less easy to attain high ee in the latter process (Scheme 3) [19]. Moreover, titanium 3,3'-modified-biphenolate complex **13** which is atropisomerically controlled by TADDOL gave very high enantioselectivity when a methyltitanium complex was used as the alkylating reagent (Scheme 4) [20].

4
Asymmetric Alkylation of Aldehydes using Polymer and Dendritic Catalysts

When considering the easy recovery and reuse of chiral catalysts, or simple separation process of the product from chiral catalyst, polymer-supported catalysts are very attractive [1, 3]. For the enantioselective ethylation using dialkylzinc, Fréchet and Itsuno's group and our group developed polystyrene-supported amino alcohols [1].

Recently, dendrimers, which are hyperbranched macromolecules, were found to be an appropriate support for polymer catalysts, because chiral sites can be designed at the peripheral region of the dendrimers (Scheme 5). Seebach synthesized chiral dendrimer **14**, which has TADDOLs on its periphery and used an efficient chiral ligand in the Ti(IV)-promoted enantioselective alkylation [21]. We developed chiral hyperbranched hydrocarbon chain **15** which has six β-amino alcohols [22]. It catalyzes the enantioselective addition of diethylzinc to aldehydes. We also reported dendritic chiral catalysts with flexible carbosilane backbones [23].

In 1997 Pu reported a new type of main chain chiral polymer derived from BINOLs [24]. Polymer **16** catalyzed enantioselective ethylation using diethylzinc to give secondary alcohols in up to 94% ee. It is noteworthy that **16** is a derivative of chiral BINOL but the addition of Ti(IV) is unnecessary unlike other reported chiral monomeric diols. In 1998, Pu reported that polymer **17**, which has a phenylene spacer between two BINOL moieties, results in better ees of up to 98% [24].

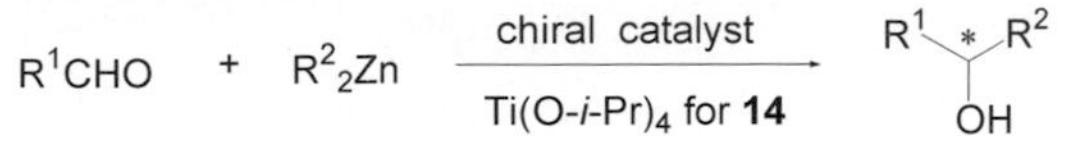

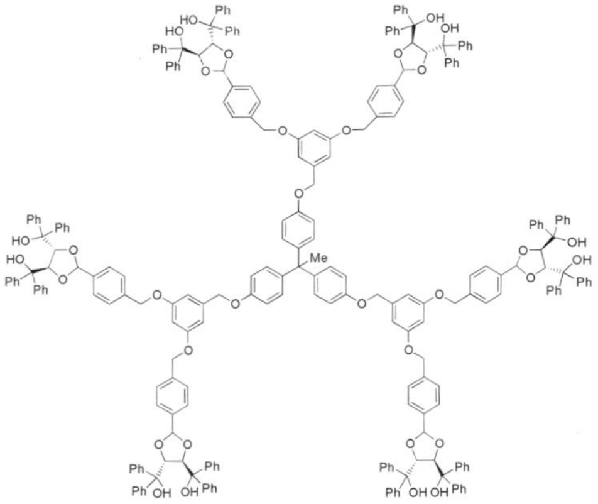

$$R^1CHO \quad + \quad R^2_2Zn \quad \xrightarrow[\text{Ti(O-}i\text{-Pr)}_4 \text{ for } \mathbf{14}]{\text{chiral catalyst}} \quad R^1 \underset{OH}{\overset{*}{\diagdown}} R^2$$

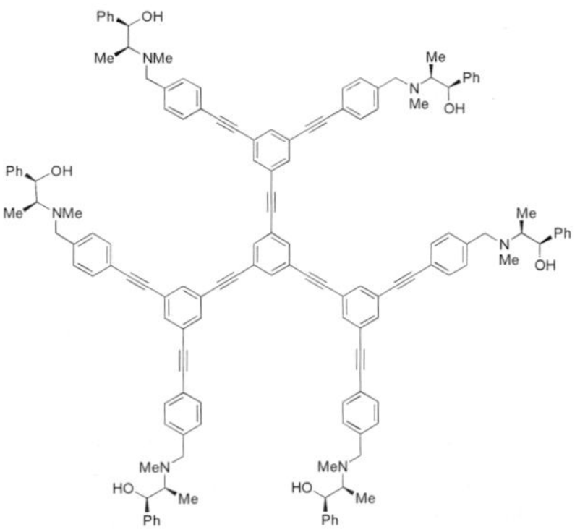

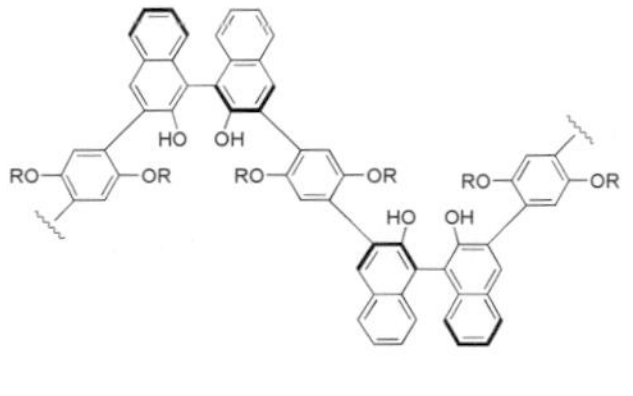

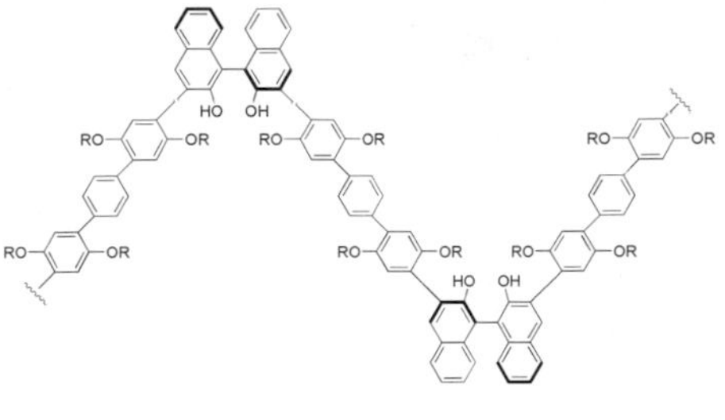

Scheme 5

5
Asymmetric Alkynylation of Aldehydes and Ketones

Asymmetric alkynylation of carbonyl compounds gives chiral propargyl alcohols. We reported catalytic and enantioselective alkynylation of aldehydes by dialkynylzincs, which are prepared in situ from the corresponding acetylenes and diethylzinc using N,N-dibutylnorephedrine (DBNE) (**18**) [1, 4, 25] as a chiral catalyst (Scheme 6) [26]. Corey reported highly enantioselective addition of alkynylboranes using substoichiometric amounts of chiral oxazaborolidines derived from proline [27]. An highly enantioselective addition of alkynyllithium to activated ketones using a stoichiometric amount of 1-phenyl-2-pyrrolidinyl-1-propanol (**19**) was reported [28]. We previously showed **19** to be an efficient chiral catalyst for catalytic and enantioselective ethylation [29]. The prop-

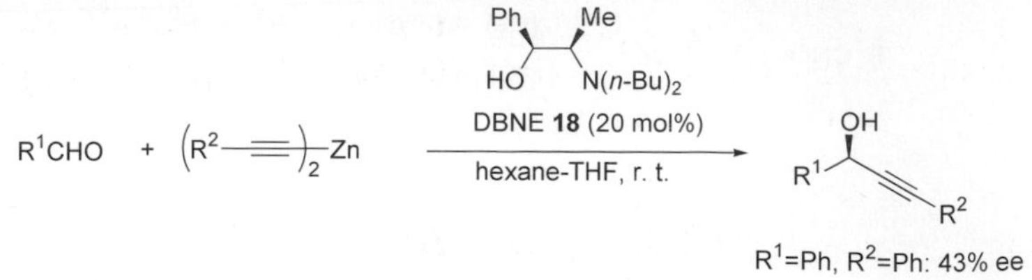

R¹CHO + $\left(R^2-\!\!\!\equiv\!\!\!-\right)_2$Zn

Ph, Me
HO N(n-Bu)₂
DBNE **18** (20 mol%)
───────────────→
hexane-THF, r. t.

$R^1\!\!\overset{\displaystyle OH}{\underset{\displaystyle \equiv}{\diagup}}\!\!R^2$

R¹=Ph, R²=Ph: 43% ee

Scheme 6

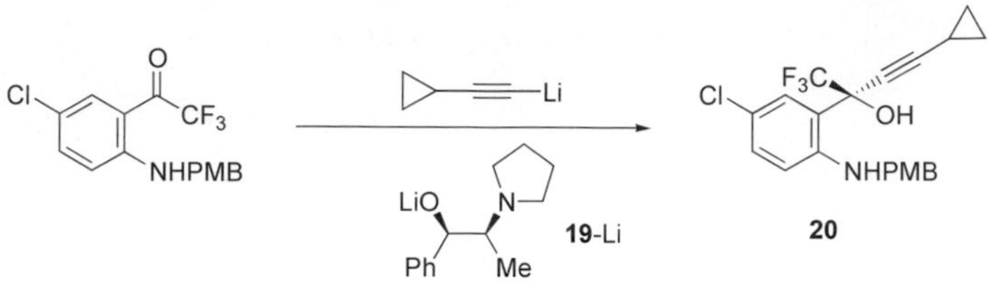

Scheme 7

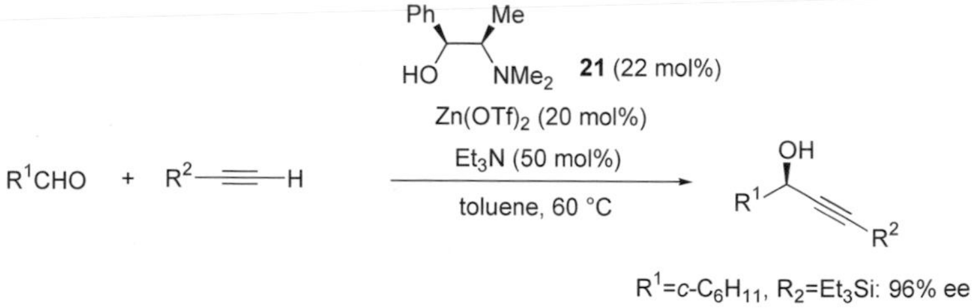

R¹CHO + R²─≡─H

Ph Me
HO NMe₂ **21** (22 mol%)
Zn(OTf)₂ (20 mol%)
Et₃N (50 mol%)
───────────────→
toluene, 60 °C

$R^1\!\!\overset{\displaystyle OH}{\underset{\displaystyle \equiv}{\diagup}}\!\!R^2$

R¹=c-C₆H₁₁, R₂=Et₃Si: 96% ee

Scheme 8

RCHO + Ph─≡─H

Me₂Zn   (R)-H₈-BINOL **22**
─────────── ───────────────→
               THF, r.t.

$R\!\!\overset{\displaystyle OH}{\underset{\displaystyle \equiv}{\diagup}}\!\!Ph$

R=Ph: 92% ee

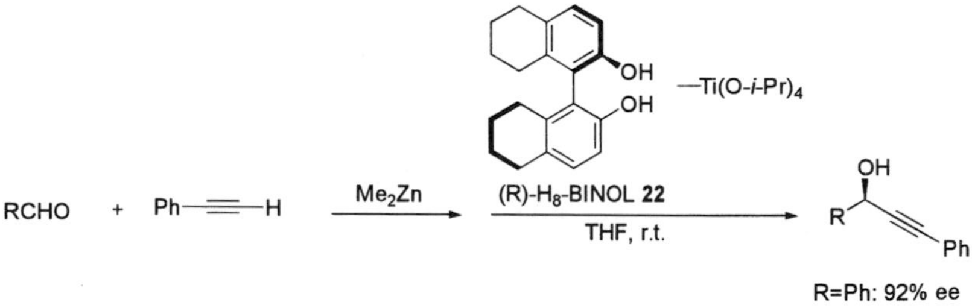

Scheme 9

argyl alcohol **20** is an important synthetic intermediate of Efavirenz, a drug for HIV approved by US FDA (Scheme 7).

Recently Carreira reported the first catalytic and highly enantioselective alkynylation. Thus, in the presence of a catalytic amount of Zn(II) salt and N-methylephedrine (**21**), the reaction of aldehydes and terminal acetylenes proceeds to give various chiral propargyl alcohols with high ee (Scheme 8) [30]. A titanium–chiral diol complex was also found to be an efficient catalyst for asymmetric alkynylation. In this reaction, H_8-BINOL **22** is a better chiral ligand than BINOL (Scheme 9) [31].

6
Asymmetric Phenylation of Aldehydes and Ketones

Compared with asymmetric ethylation, reports on asymmetric phenylation are limited. We disclosed the enantioselective phenylation using diphenylzinc prepared in situ from phenyl Grignard reagent and zinc chloride. This method needs a stoichiometric amount of chiral amino alcohol DBNE **18** but good ee of 82% was achieved [32]. A catalytic phenylation was examined using planar chiral compound **1** based on ferrocene, and chiral diaryl carbinols of moderate ee were provided from diphenylzinc and 4-chlorobenzaldehyde (Scheme 10) [33]. A catalytic and highly enantioselective phenylation was realized by binaphthyl-based chiral catalyst **23**. In this reaction, the addition of 2 equivalents of diethylzinc against catalyst increases the yield and ee [34]. Recently, chiral amino alcohol DPMPM **9** was also reported to be an efficient catalyst for asymmetric phenylation [35].

Bolm found that chiral ferrocenyl hydroxy oxazoline **24** is also a good catalyst (88% ee). When the zinc species is prepared from a 1:2 mixture of diphenylzinc and diethylzinc, enantioselectivity is dramatically improved (97% ee) (Scheme 11) [36]. Extremely high enantioselectivity was also achieved by using planar chiral η^5-cyclopentadienylrhenium tricarbonyl complex **25** [37].

Fu reported the first example of a catalytic and enantioselective phenylation of ketones. 3-*exo*-(Dimethylamino)isoborneol (DAIB) (**26**) is an appropriate catalyst but the addition of 1.5 equivalents of MeOH is vital to ensure a better yield and enantioselectivity (Scheme 12) [38].

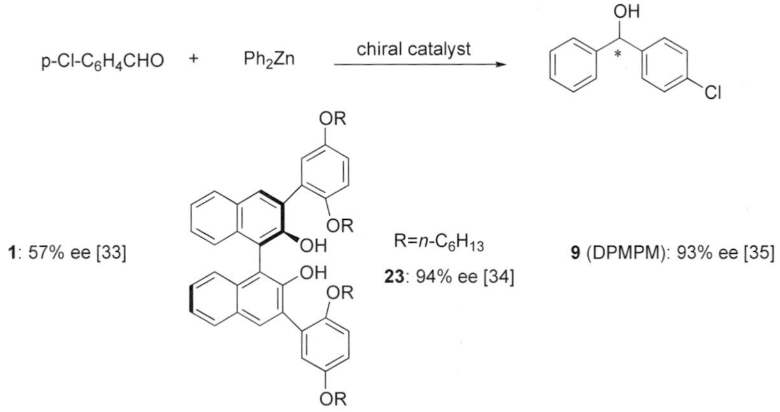

1: 57% ee [33]   R=*n*-C_6H_{13} **9** (DPMPM): 93% ee [35]
23: 94% ee [34]

Scheme 10

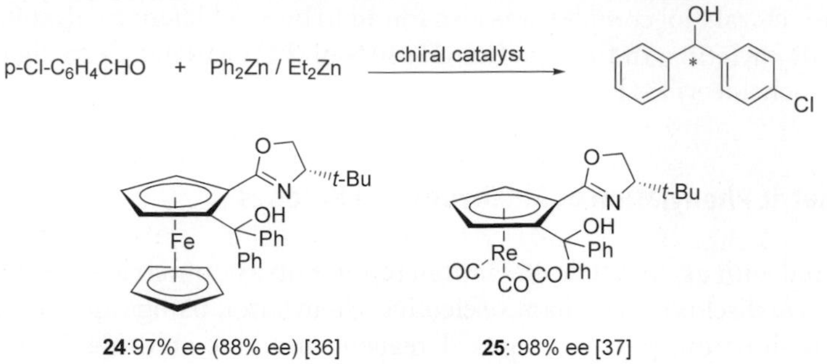

Scheme 11

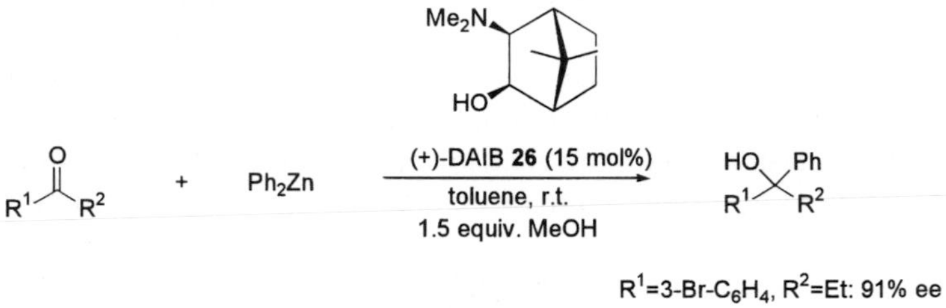

R^1=3-Br-C$_6$H$_4$, R^2=Et: 91% ee

Scheme 12

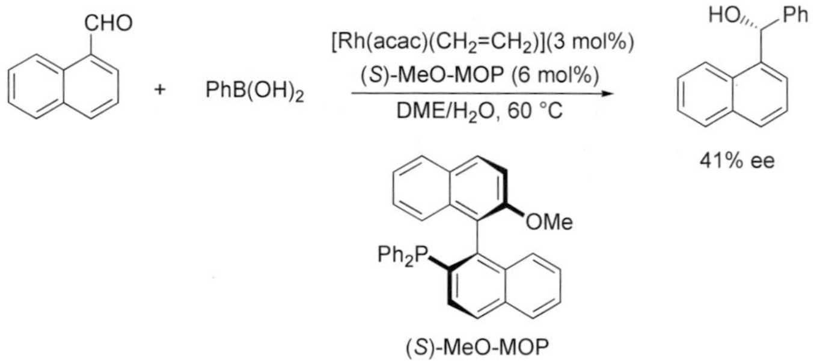

41% ee

(S)-MeO-MOP

Scheme 13

A rhodium–chiral phosphine complex catalyzes the enantioselective addition of phenylboronic acid to 1-naphthaldehyde to give a chiral diaryl carbinol but with moderate ee (Scheme 13) [39]. When considering the introduction of functionalized aryl groups, arylboronic acid is a promising alternative arylating reagent to diarylzinc compounds.

7
Asymmetric Autocatalysis in the Alkylation of Aldehydes

Asymmetric autocatalysis, in which the chiral product acts as a chiral catalyst for its own production, is a very intriguing system from both a scientific and synthetic standpoint [40].

3-Pyridyl alkanol [41], diol [42], and ferrocenyl alcohol [43] were the first asymmetric autocatalysts found by Soai and co-workers in the enantioselective alkylation of pyridine-3-carbaldehyde, dialdehyde, and ferrocenecarbaldehyde, respectively, with dialkylzincs.

Soai et al. established highly enantioselective asymmetric autocatalysis in the asymmetric isopropylation of pyrimidine-5-carbaldehyde 27 (Scheme 14) [44], quinoline-3-carbaldehyde [45], and 5-carbamoylpyridine-3-carbaldehyde [46]. Among these, 2-alkynyl-5-pyrimidyl alkanol is a practically perfect asymmetric autocatalysis [47]. When 0.2 equivalents of 2-alkynyl-5-pyrimidyl alkanol 28b with >99.5% ee was employed as an asymmetric autocatalyst in the isopropylation of 2-alkynylpyrimidine-5-carbaldehyde 27b, it automultiplies in a yield of >99% without any loss of ee (>99.5% ee). When the product was used as an asymmetric autocatalyst for the next run, pyrimidyl alkanol 28b with >99.5% ee was obtained in >99%. Even after tenth round, pyrimidyl alkanol 28b with >99.5% ee was formed in a yield of >99% [47].

Moreover, when these alkanols with low ee are utilized as asymmetric autocatalysts, 5-pyrimidyl alkanol 28 [48], 3-quinolyl alkanol [49], and 5-carbamoyl-3-pyridyl alkanols [50] with higher ees were obtained. The successive reactions were performed in order to make the best use of the autocatalysis, that is, the products of one round served as the asymmetric autocatalysts for the next. In the case of pyrimidyl alkanol, staring from (S)-alkanol 28a with only 2% ee, the ee reached almost 90% after four rounds [48] without the assistance of any other chiral auxiliary (Scheme 14). 2-Alkynyl-5-pyrimidyl alkanol 28b [5] and 2-

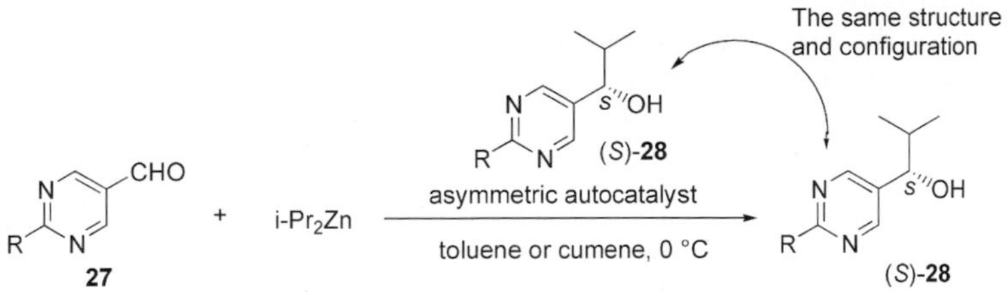

R	initial catalyst	product
H (28a)	2% ee	89% ee
t-BuC≡C (28b)	>99.5% ee	>99.5% ee
	0.6% ee	>99.5% ee
(E)-t-BuCH=CH (28c)	7% ee	99% ee

Scheme 14

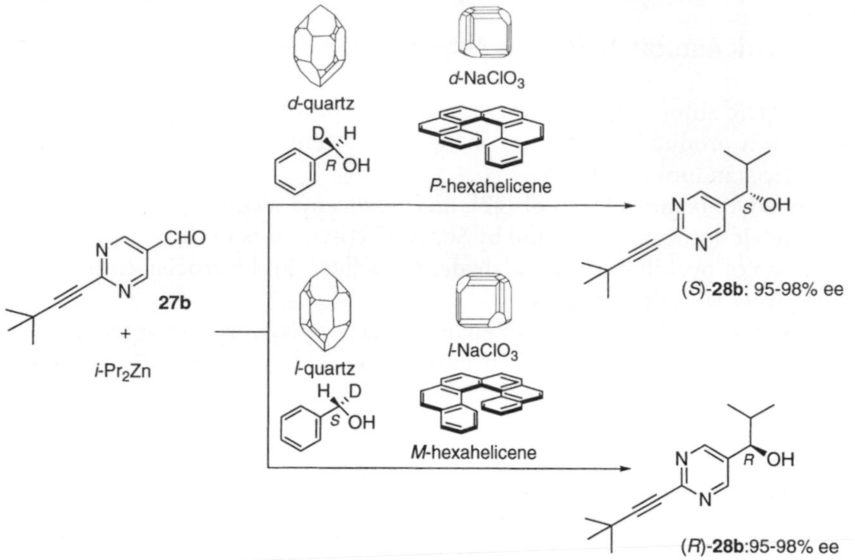

Scheme 15

alkenyl-5-pyrimidyl alkanol **28c** [51] with 0.6% ee and 7% ee, respectively, enhance their ee up to >99.5% ee and 99% ee, respectively, during asymmetric autocatalysis [5].

The origin of chirality of organic compounds has been a puzzle [52]. Soai et al. found that chiral inorganic crystals such as d- and l-quartz and sodium chlorate ($NaClO_3$) serve as chiral initiators of asymmetric autocatalysis (Scheme 15). In the presence of d-quartz or d-$NaClO_3$, reaction of 2-alkynylpyrimidine-5-carbaldehyde **27b** with diisopropylzinc afford (S)-pyrimidyl alkanol **28b** with 95–98% ee, whereas in the presence of l-quartz [53] or l-$NaClO_3$ [54], the opposite enantiomer (R)-alkanol **28b** was formed with 95–98% ee. These results correlate, for the first time, the chirality of inorganic crystals to those of organic compounds with very high ee. In addition, L-leucine [55] and P-hexahelicene [56] with very low (<2%) ee, the ees of which are induced by circularly polarized light [57], act as a chiral initiator of asymmetric autocatalysis to afford pyrimidyl alkanols with very high ee. These results again correlate the chirality of circularly polarized light to those of organic compounds with very high ee. In addition, asymmetric autocatalysis is a powerful tool of chiral recognition. Asymmetric autocatalysis can also be initiated by using 2-butanol with 0.5% ee [55], chiral primary alkanol-α-d [58], chiral cobalt complex due to the coordination manner of achiral ligand [59], 1,1′-binaphthyl [60], and monosubstituted [2.2]paracyclophanes [61].

References

1. Soai K, Niwa S (1992) Chem Rev 92:833
2. Noyori R, Kitamura M (1991) Angew Chem Int Ed Engl 30:49
3. Pu L, Yu HB (2001) Chem Rev 101:757
4. Soai K, Shibata T (1999) In: Jacobsen EN, Pfaltz A, Yamamoto H (eds) Comprehensive asymmetric catalysis, vol II. Springer, Berlin Heidelberg New York, p 911
5. Soai K, Shibata T, Sato I (2000) Acc Chem Res 33:382
6. Dosa PI, Ruble, JC, Fu GC (1997) J Org Chem 62:444
7. Priego J, Mancheño OG, Cabrera S, Carretero JC (2001) Chem Commun 2026
8. Dangel BD, Polt R (2000) Org Lett 2:3003
9. Schinnerl M, Seitz M, Kaiser A, Reiser O (2001) Org Lett 3:4259
10. Uozumi Y, Mizutani K, Nagai S (2001) Tetrahedron Lett 42:407
11. Liu DX, Zhang LC, Wang Q, Da CS, Xin ZQ, Wang R, Choi MCK, Chan ASC (2001) Org Lett 3:2733
12. Lawrence, CF, Nayak SK, Thijs L, Zwanenburg B (1999) Synlett 1571
13. Soai K, Ookawa A, Kaba T, Ogawa K (1987) J Am Chem Soc 109:7111
14. Shiina I, Konishi K, Kuramoto Y (2002) Chem Lett 164
15. Shibata T, Tabira H, Soai K (1998) J Chem Soc Perkin Trans 1 177
16. Shibata T, Nakatsui K, Soai K (1999) Inorg Chim Acta 296:33
17. Yang XW, Shen JH, Da CS, Wang HS, Su W, Liu DX, Wang R, Choi MCK, Chan ASC (2001) Tetrahedron Lett 42:6573
18. Bauer T, Tarasiuk J (2002) Tetrahedron Lett 43:687
19. You JS, Hsieh SH, Gau HM (2001) Chem Commun 1546
20. Ueki M, Matsumoto Y, Jodry JJ, Mikami K (2001) Synlett 1889
21. Seebach D, Marti RE, Hintermann T (1996) Helv Chim Acta 79:1710
22. Sato I, Shibata T, Ohtake K, Kodaka R, Hirokawa Y, Shirai N, Soai K (2000) Tetrahedron Lett 41:3123
23. Sato I, Kodaka R, Hosoi K, Soai K (2002) Tetrahedron Asymmetry 13:805
24. Huang WS, Hu QS, Zheng XF, Anderson J, Pu L (1997) J Am Chem Soc 119:4313; Hu QS, Huang WS, Pu L (1998) J Org Chem 63:2798
25. Soai K, Yokoyama S, Hayasaka T (1991) J Org Chem 56:4264
26. Niwa S, Soai K (1990) J Chem Soc Perkin Trans 1 937
27. Corey EJ, Cimprich KA (1994) J Am Chem Soc 116:3151
28. Thompson AS, Corley EG, Huntington MF, Grabowski EJJ, Remenar JF, Collum DB (1998) J Am Chem Soc 120:2028
29. Soai K, Yokoyama S, Hayasaka T (1991) J Org Chem 56:424; Soai K, Konishi T, Shibata T (1999) Heterocycles 59:1421
30. Anand NK, Carreira EM (2001) J Am Chem Soc 123:9687
31. Lu G, Li X, Chan WL, Chan ASC (2002) Chem Commun 172
32. Soai K, Kawase Y, Oshio A (1991) J Chem Soc Perkin Trans 1 1613
33. Dosa PI, Fu GC (1997) J Org Chem 62:444
34. Huang WS, Pu L (1999) J Org Chem 64:4222
35. Zhao G, Li XG, Wang XR (2001) Tetrahedron Asymmetry 12:399
36. Bolm C, Hermanns N, Hildebrand JP, Muñiz K (2000) Angew Chem Int Ed 39:3465
37. Bolm C, Kesselgruber M, Hermanns N, Hildebrand JP, Raabe G (2001) Angew Chem Int Ed 40:1488
38. Dosa PI, Fu GC (1998) J Am Chem Soc 120:445
39. Sakai M, Ueda M, Miyaura N (1998) Angew Chem Int Ed 37:3279
40. Soai K, Sato I, Shibata T (2001) Chem Rec 1:321; Soai K (1999) Enantiomer 4:591; Soai K, Shibata T (1999) In:Pályi G, Zucchi C, Caglioti L (eds) Advances in biochirality. Elsevier, Amsterdam, p 125; Soai K, Sato I, Shibata T (2002) Yuki Gosei Kagaku Kyokaishi (J Synth Org Chem Jpn) 60:668; Soai K (2002) In: Pályi G, Zucchi C, Caglioti L (eds) Fundamentals of life. Elsevier, Paris, p 427; Bolm C, Bienewald F, Seger A (1966) Angew Chem Int Ed Engl

35:1657; Buschmann H, Thede R, Heller D (2000) Angew Chem Int Ed 39:4033; Avalos M, Babiano R, Cintas P, Jiménez JL, Palacios JC (2000) Chem Commun 887
41. Soai K, Niwa S, Hori H (1990) J Chem Soc Chem Commun 982
42. Soai K, Hayase T, Shimada C, Isobe K (1994) Tetrahedron Asymmetry 5:789
43. Soai K, Hayase T, Takai K (1995) Tetrahedron Asymmetry 6:637
44. Shibata T, Morioka H, Hayase T, Choji K, Soai K (1996) J Am Chem Soc 118:471
45. Shibata T, Choji K, Morioka H, Hayase T, Soai K (1996) Chem Commun 751
46. Shibata T, Morioka H, Tanji S, Hayase T, Kodaka Y, Soai K (1996) Tetrahedron Lett 37:8783
47. Shibata T, Yonekubo S, Soai K (1999) Angew Chem Int Ed 38:659
48. Soai K, Shibata T, Morioka H, Choji K (1995) Nature (London) 378:767
49. Shibata T, Choji K, Hayase T, Aizu Y, Soai K (1996) Chem Commun 1235
50. Tanji S, Kodaka Y, Ohno A, Shibata T, Sato I, Soai K (2000) Tetrahedron Asymmetry 11:4249
51. Sato I, Yanagi T, Soai K (2002) Chirality 14:166
52. Feringa BL, van Delden RA (1999) Angew Chem Int Ed 38:3418; Eschenmoser A (1999) Science 284:2118; Kondepudi DK, Nelson GW (1985) Nature 314:438; Mason S (1988) Chem Soc Rev 17:347; Podlech J (2001) Cell Mol Life Sci 58:44
53. Soai K, Osanai S, Kadowaki K, Yonekubo S, Shibata T, Sato I (1999) J Am Chem Soc 121:11235
54. Sato I, Kadowaki K, Soai K (2000) Angew Chem Int Ed 39:1510
55. Shibata T, Yamamoto J, Matsumoto N, Yonekubo S, Osanai S, Soai K (1998) J Am Chem Soc 120:12157
56. Sato I, Yamashima R, Kadowaki K, Yamamoto J, Shibata T, Soai K (2001) Angew Chem Int Ed 40:1096
57. Flores JJ, Bonner WA, Massey GA (1977) J Am Chem Soc 99:3622; Nishino H, Kosaka A, Hembury GA, Shitomi H, Onuki H, Inoue Y (2001) Org Lett 3:921; Moradpour A, Nicoud JF, Balavoine G, Kagan HB, Tsoucaris G (1971) J Am Chem Soc 93:2353; Bernstein WJ, Calvin M, Buchardt O (1972) J Am Chem Soc 94:494; Inoue Y (1992) Chem Rev 92:741
58. Sato I, Omiya D, Saito T, Soai K (2000) J Am Chem Soc 122:11739
59. Sato I, Kadowaki K, Ohgo Y, Soai K, Ogino H (2001) Chem Commun 1022
60. Sato I, Osanai S, Kadowaki K, Sugiyama T, Shibata T, Soai K (2002) Chem Lett 168
61. Tanji S, Ohno A, Sato I, Soai K (2001) Org Lett 3:287

Supplement to Chapter 26.2
Alkylation of C=N

Keiko Hatanaka, Hisashi Yamamoto

Department of Chemistry, University of Chicago, 5735 South Ellis Avenue, Chicago, IL 60637, USA
e-mail: yamamoto@uchicago.edu

Keywords: Catalyst, Alkylation, Allylation, Arylation, Mannich reaction, Carbon–nitrogen double bond, Imine, Nitrone, Aldimine, Organozinc reagents, Silyl ketene acetal, Silyl enol ether, Amine, β-Amino acid

1
Introduction

The alkylation of the carbon–nitrogen double bond of imines provides an attractive route to amines [1–4]. Compared to the catalytic asymmetric alkylation of carbonyl compounds, the corresponding study of alkylation of the azomethine group has remained relatively unexplored [1, 2, 5]. This disparity is primarily due to the low reactivity of imines toward nucleophilic additions. The low reactivities of imines are explained by the difference in electronegativity between oxygen and nitrogen, the steric hindrance of imines, and the preference of enolizable imines to undergo α-deprotonation rather than addition. Recently, however, Kobayashi and co-workers have discovered the remarkable ability of lanthanide salts to reverse the normal reactivity trends for aldehydes and their derived aldimines [6]. They demonstrated that the aldimine generated in situ reacted preferentially over the aldehyde in the reaction mixture using a catalytic amount of lanthanide triflate as a Lewis acid. This discovery had a great influence on the following research.

The last five years have witnessed explosive advances in the development of catalytic enantioselective alkylation of imines. This chapter collects some of the important advances that have taken place during the past few years in catalytic enantioselective alkylation of imines.

2
Catalytic Enantioselective Alkylation of Imines with Organozinc Reagents

The amino alcohol-mediated enantioselective addition of dialkylzinc reagents to aldehydes is a very effective and general method for the synthesis of chiral alcohols. In contrast, only very few examples are known for their aza analogues [1]. Imines are considerably less reactive than aldehydes toward nucleophilic additions, and activation of the imino group is required. The use of activated *N*-acyl- and *N*-phosphinoylimines has been crucial to the alkylation of imines using dialkylzinc reagents in the presence of the chiral amino alcohol. Due to the poor electrophilicity of imines, excess dialkylzinc reagent and stoichiometric amounts of amino alcohol ligand are normally required to establish high conversion and enantioselectivity.

To solve this problem, Pericás and co-workers have introduced a dual catalytic system consisting of a chiral amino alcohol 2, to control the enantioselectivity of the addition process, and a bulky silylating agent, to further activate the imine substrate (Scheme 1) [7]. When the 2/TIPSCl system was used to promote the addition to imines derived from aromatic aldehydes, the addition reactions took place with good yield (63–75%) and high enantioselectivities (72–91%). Even in this case, a substoichiometric amount of chiral amino alcohol is required for a satisfactory result.

Quite recently, there has been significant expansion and development in the alkylation of imines with organozinc reagents using chiral Lewis acid catalysts. In 2000, Tomioka and co-workers reported a copper(II)–chiral amidophosphine 4-catalyzed asymmetric process for the addition of diethylzinc to *N*-sulfonylimines (Scheme 2) [8]. Excellent enantioselectivities (90–94%) and yields (83–99%) were obtained in reactions of *N*-sulfonylimines derived from arylaldehydes.

In 2001, Hoveyda and co-workers disclosed Zr-catalyzed imine alkylations promoted by peptide-based chiral ligands 7 and 8 that afford arylimines 9 in 82–

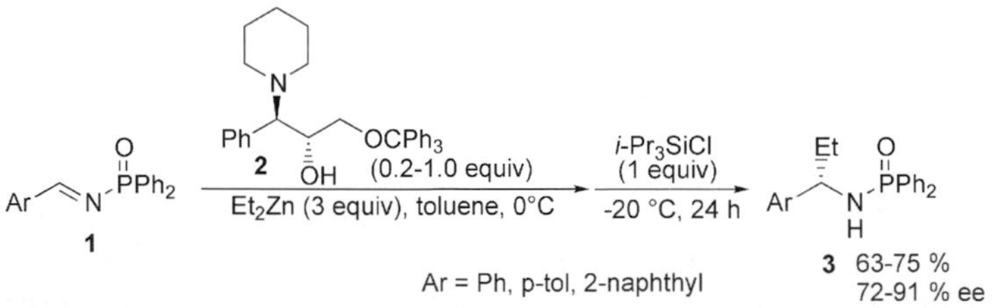

Ar = Ph, p-tol, 2-naphthyl

Scheme 1

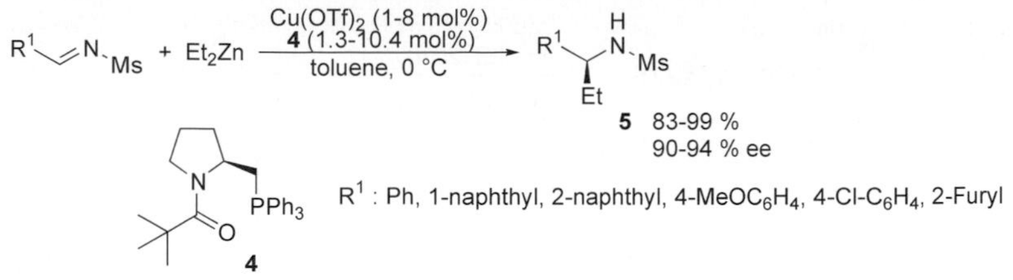

$$R^1 \diagdown = N_{\diagdown Ms} + Et_2Zn \xrightarrow[\text{toluene, 0 °C}]{\substack{\text{Cu(OTf)}_2 \text{ (1-8 mol\%)} \\ \textbf{4} \text{ (1.3-10.4 mol\%)}}} R^1 \diagdown \underset{Et}{\overset{H}{\diagup}} N_{\diagdown Ms}$$

5 83-99 %
        90-94 % ee

4

R^1 : Ph, 1-naphthyl, 2-naphthyl, 4-MeOC$_6$H$_4$, 4-Cl-C$_6$H$_4$, 2-Furyl

Scheme 2

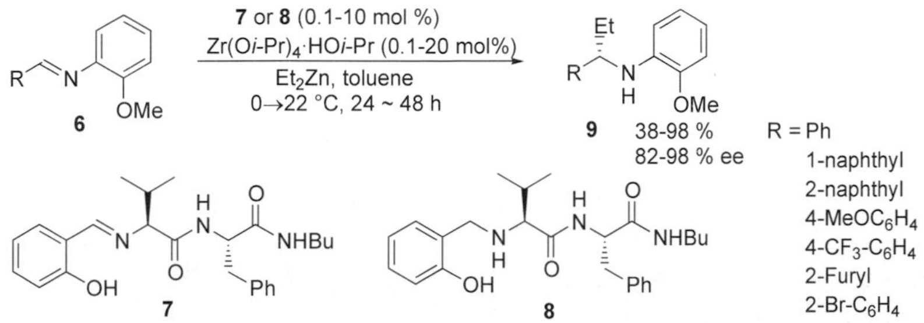

$$R \diagdown = N \xrightarrow[\substack{Et_2Zn, \text{ toluene} \\ 0 \to 22\text{ °C, 24 ~ 48 h}}]{\substack{\textbf{7 or 8} \text{ (0.1-10 mol \%)} \\ Zr(Oi\text{-}Pr)_4 \cdot HOi\text{-}Pr \text{ (0.1-20 mol\%)}}} R \diagdown \underset{H}{\overset{Et}{\diagup}} N$$

6   OMe

9   38-98 %   R = Ph
        82-98 % ee   1-naphthyl
                       2-naphthyl
                       4-MeOC$_6$H$_4$
                       4-CF$_3$-C$_6$H$_4$
                       2-Furyl
                       2-Br-C$_6$H$_4$

7   **8**

Scheme 3

$$R^1CHO + H_2N \diagdown \xrightarrow[\substack{Et_2Zn \text{ (6 eq), toluene} \\ 0 \to 22\text{ °C, 24 ~ 48 h}}]{\substack{\textbf{11} \text{ (10 mol\%)} \\ Zr(Oi\text{-}Pr)_4 \cdot HOi\text{-}Pr \text{ (10 mol\%)}}} R^1 \diagdown \underset{H}{\overset{Et}{\diagup}} N$$

                                                                            OMe

10   48->98 %
         95 ->98 %ee

$R^1 = C_4H_9, C_6H_{13}, PhCH_2CH_2, (CH_3)_2CH, Br(CH_2)_5 , HOCH_2$

11

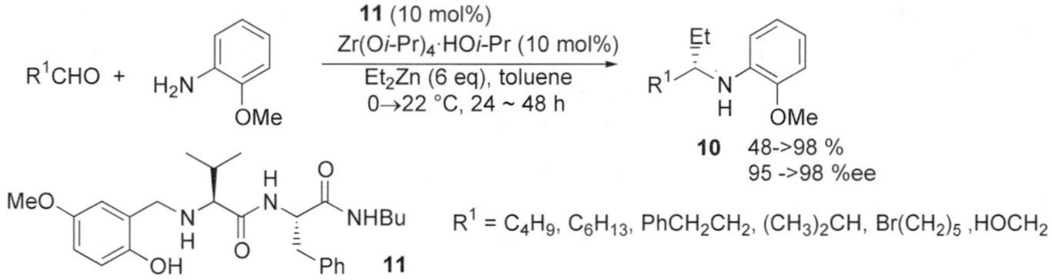

Scheme 4

95% ee and 32–87% isolated yield, respectively (Scheme 3) [9]. The attempts to examine the catalytic alkylations of aliphatic o-anisidyl imines **6** were thwarted by their lack of stability upon isolation. To circumvent this situation, they examined the three-component asymmetric amine synthesis involving imine formation from an aldehyde and o-anisidine, followed by in situ catalytic alkylation (Scheme 4) [10]. Optically enriched aliphatic amines **10** were obtained with excellent enantioselectivity (94 to >98%ee) from three-component catalytic alkylations. Prior isolation of unstable imines is not necessary with this method.

Carreira and co-workers reported the addition of terminal alkynes to nitrones **12** in the presence of both catalytic Zn(OTf)$_2$ (10 mol%) and a tertiary amine base (25 mol%) under mild conditions (23 °C) (Scheme 5) [11]. The addition reactions are quite general for a broad range of nitrones and terminal acetylenes. They postulate that the process proceeds through the intermediary

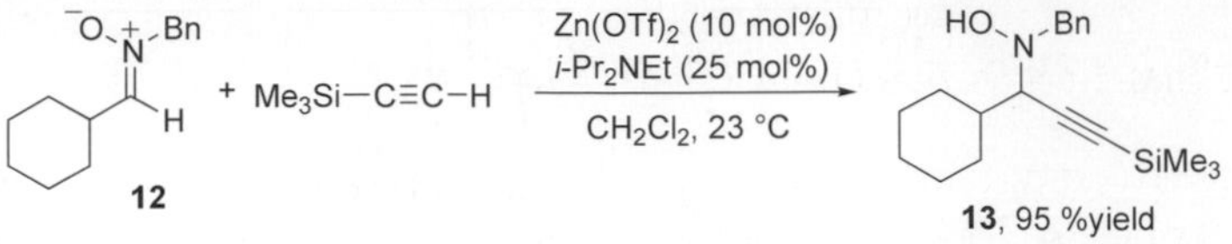

Scheme 5

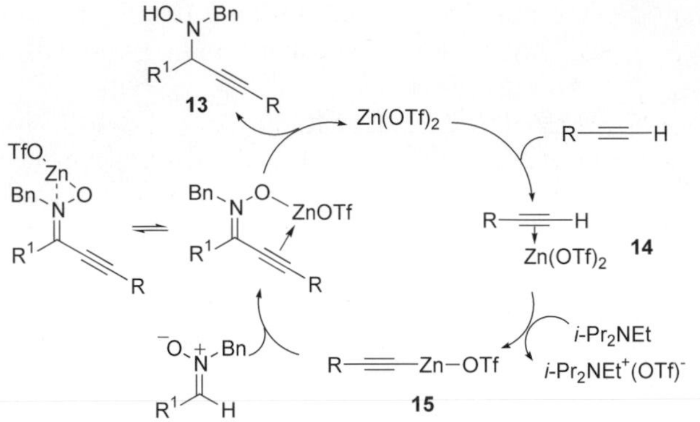

Fig. 1

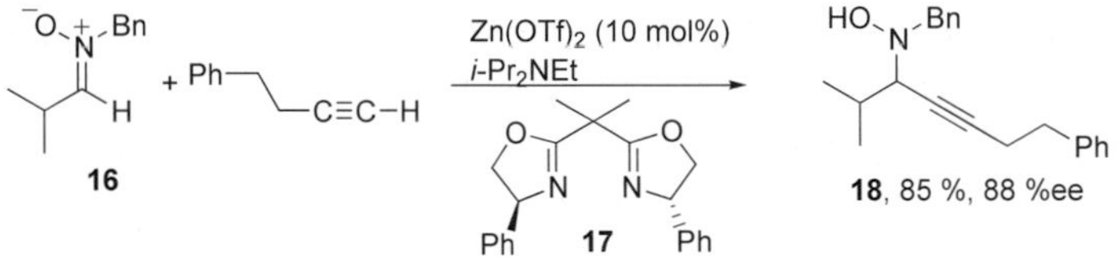

Scheme 6

of a Zn(II)-alkynilide. Their working model is illustrated in Fig. 1 [12]. Analogous with the Ag(I) and Cu(I) chemistry they hypothesize that Zn(II) forms a π-complex **14** with the terminal acetylene, thereby acidifying the terminal C(sp)–H bond. The amine base subsequently participates in a proton abstraction to deliver the corresponding zinc acetylide. Following addition of the metalated alkynilide **15**, the adduct undergoes protonation by either the trialkylammonium hydrotriflate or the starting terminal acetylene to provide Zn(II). The potential of this methodology in asymmetric catalytic synthesis is also mentioned. For example, the addition of 4-phenylbutyne to *C*-isopropyl *N*-benzyl nitrone **16** in the presence of catalytic amounts of Zn(OTf)$_2$ and chiral bisoxazoline **17** as a ligand for Zn(II) furnished adduct **18** in 88% ee and 85% yield (Scheme 6).

3
Catalytic Enantioselective Allylation of Imines

Enantiomerically pure homoallylic amines are very important chiral building blocks for the synthesis of natural products. However, enantioselective methods for homoallylamine are quite undeveloped. In 1995, Itsuno and co-workers reported the first example of enantioselective allylation of an imine (Scheme 7) [13]. The reaction of N-trimethylsilylbenzaldimine **19** with a chiral allylboron reagent **20** in ether at -78 °C afforded the corresponding homoallylamine **22** in 73% ee.

A more detailed study has been made by Brown et al. [14]. They found the critical importance of water in the asymmetric allylboration of N-trimethylsilylaldimines, and concluded that the reaction takes place during the aqueous workup. The allylboration of **19** with **20** proceeded only in the presence of one molar equivalent of water to give **22** in 92% ee and 90% yield (Scheme 8). They suggested that the reactive aldimines could be generated in situ from N-trimethylsilylimines upon addition of one equivalent of water and captured by the allylborating agent.

In 1998, Yamamoto et al. reported the first catalytic enantioselective allylation of imines with allyltributylstannane in the presence of a chiral π-allylpalladium complex **23** (Scheme 9) [15]. The imines derived from aromatic aldehydes underwent the allylation with high ee values. Unfortunately, the allylation reaction of aliphatic imines resulted in modest enantioselectivities. They proposed that a bis-π-allylpalladium complex is a reactive intermediate for the allylation and reacts with imines as a nucleophile. The bis-π-allylpalladium complex seemed the most likely candidate for the Stille coupling [16]. Indeed, the Stille coupling reaction takes place in the presence of triphenylphosphine even if imines are present, whereas the allylation of imines occurs in the absence of the phosphine [17]. They suggested the phosphine ligand played a key role in controlling the

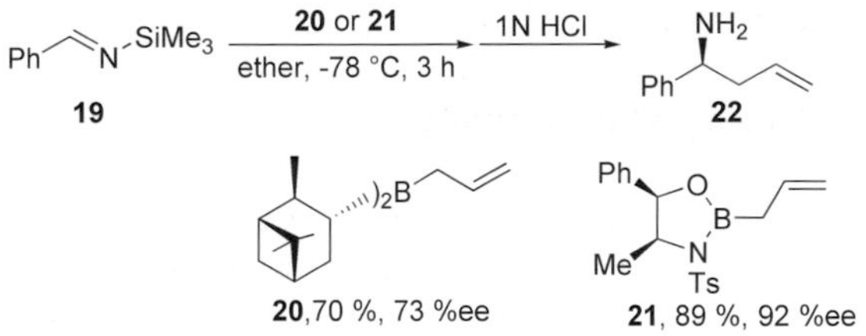

Scheme 7

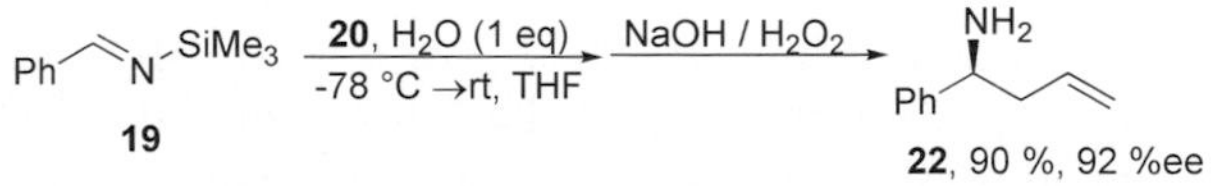

Scheme 8

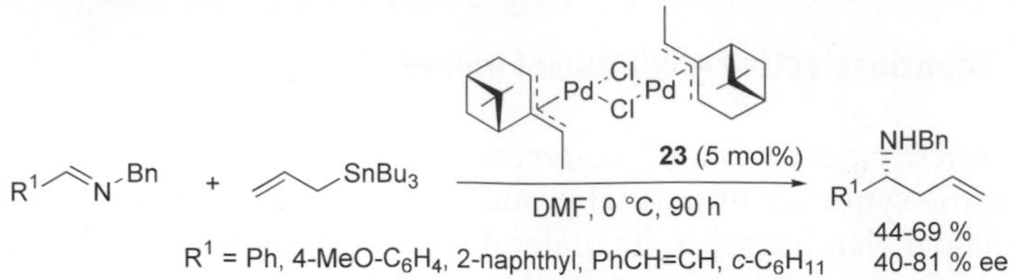

Scheme 9

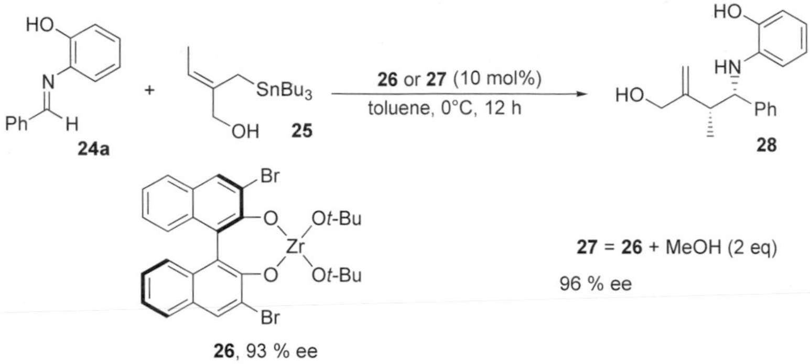

Scheme 10

chemoselectivity. The use of allyltrimethylsilane instead of allyltributylstannane was also reported [18].

Kobayashi and colleagues developed a catalytic enantioselective method for the allylation of imines 24 by substituted allylstannanes 25 with chiral zirconium catalysts 26 and 27 prepared from zirconium alkoxides and 1,1′-bi-2-naphthol derivatives (Scheme 10) [19]. The allylation of aromatic imines 24 with 25 afforded the corresponding homoallylic amines 28 in good yields (71–85%) with high stereoselectivities (87–99% ee).

4
Catalytic Asymmetric Arylation of Imines

The nucleophilic addition of organometallic reagents to imines provides an attractive route to amines [4]. Recently, however, some completely different approaches to the synthesis of α-aryl amine were reported. Hayashi and Ishigedani found a new catalytic system for the asymmetric addition of arylstannanes to imines derived from aromatic aldehydes (Scheme 11) [20].

The chiral phosphine 31 or 32–rhodium complex catalyzed the addition of arystannanes 30 to N-sulfonylimines 29 to give diarylmethylamines 33 with high enantioselectivity (75–96% ee) [21]. The choice of the chiral monodentate phosphine ligand is essential for their catalytic asymmetric arylation. With chelating bisphosphine ligands the arylation was very slow. The authors hypoth-

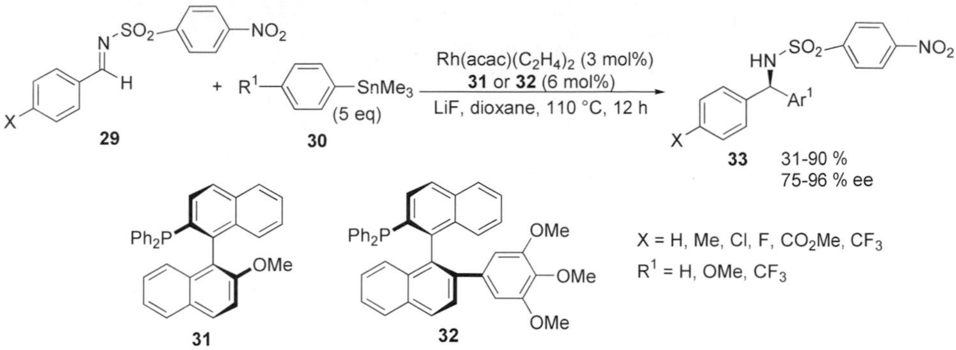

Scheme 11

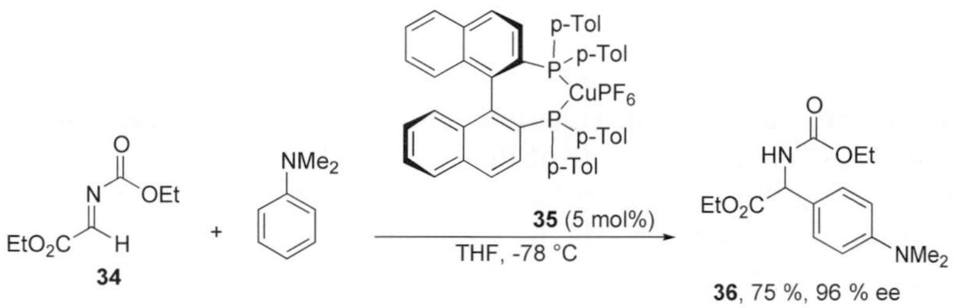

Scheme 12

esized that the catalytic cycle of reaction involved a rhodium–aryl species generated from arylstannane and its enantioselective addition to the carbon–nitrogen double bond of the imine.

Jørgensen and co-workers reported a catalytic enantioselective addition reaction of α-imino esters to electron-rich aromatic compounds leading to protected optically active aromatic α-amino acids (Scheme 12) [22]. The reaction of the α-imino ester **34** with an N,N-dimethylaniline derivative in the presence of (R)-Tol-BINAP/CuPF$_6$ **35** gave the aromatic α-amino ester **36** in good yield (44–88%) and with high ee (52–98%). The reaction is highly regioselective, as only the *para*-substituted product relative to the N,N-dimethylamino substituent is formed.

5
Catalytic Enantioselective Mannich-Type Reactions

The aldol reaction of an enolate or enolate equivalent with an imine is referred to as the Mannich-type reaction. Asymmetric Mannich-type reactions provide useful routes for the synthesis of enantiomerically enriched β-amino acid derivatives, which are versatile chiral building blocks for the synthesis of nitrogen-containing biologically important compounds [23]. Despite the enormous progress made in asymmetric aldol reactions [24], the corresponding asymmet-

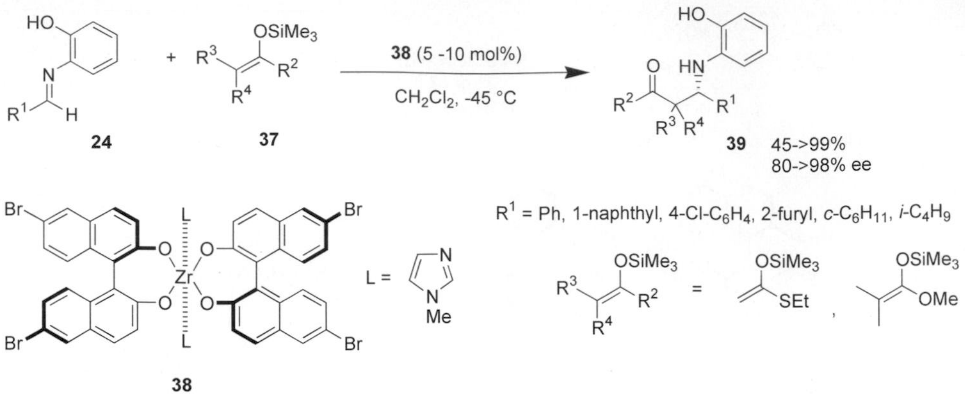

Scheme 13

ric Mannich reactions remain a relatively undeveloped area of research. In 1991, Corey reported the first example of the enantioselective synthesis of β-amino acid esters using chiral diazaborolidine [25]. However, a stoichiometric amount of chiral source was needed. The first example of a catalytic enantioselective Mannich-type reaction was presented by Tomioka and co-workers [26].

In 1997, Kobayashi and colleagues reported the first truly catalytic enantioselective Mannich-type reactions of aldimines **24** with silyl enolates **37** using a novel chiral zirconium catalyst **38** prepared from zirconium (IV) *tert*-butoxide, 2 equivalents of (*R*)-6,6′-dibromo-1,1′-bi-2-naphthol, and *N*-methylimidazole (Scheme 13) [27, 28]. In addition to imines derived from aromatic aldehydes, those derived from heterocyclic aldehydes also worked well in this reaction, and good to high yields and enantiomeric excess were obtained. The hydroxy group of the 2-hydroxyphenylimine moiety, which coordinates to the zirconium as a bidentate ligand, is essential to obtain high selectivity in this method.

Since then, efficient catalytic asymmetric methods have been developed for the addition of silyl enol ethers or silyl ketene acetals to imines with chiral metal catalysts [29–34]. Recently, direct catalytic asymmetric Mannich reactions which do not require preformation of enolate equivalents have appeared.

List gave the first examples of the proline-catalyzed direct asymmetric three-component Mannich reactions of ketones, aldehydes, and amines (Scheme 14) [35]. This was the first organocatalytic asymmetric Mannich reaction. These reactions do not require enolate equivalents or preformed imine equivalent. Both α-substituted and α-unsubstituted aldehydes gave the corresponding β-amino ketones **40** in good to excellent yield and with enantiomeric excesses up to 91%. The aldol addition and condensation products were observed as side products in this reaction. The application of their reaction to the highly enantioselective synthesis of 1,2-amino alcohols was also presented [36]. A plausible mechanism of the proline-catalyzed three-component Mannich reaction is shown in Fig. 2. The ketone reacts with proline to give an enamine **41**. In a second pre-equilib-

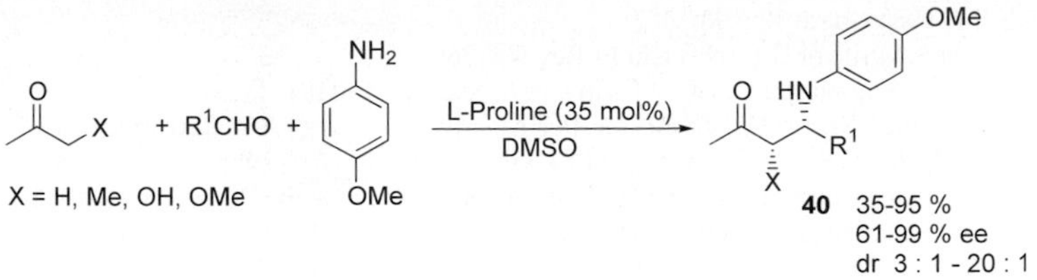

Scheme 14

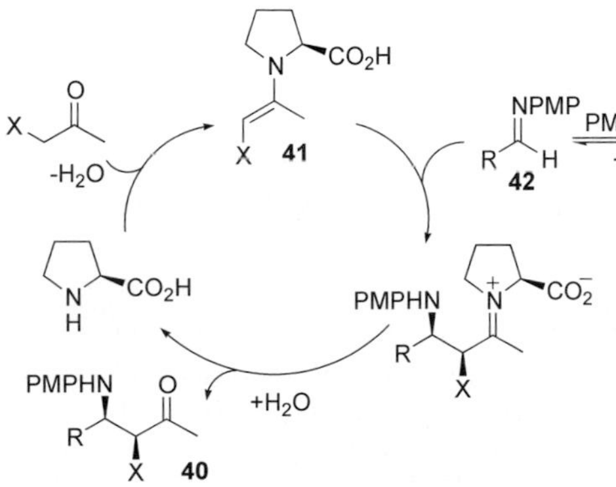

Fig. 2

rium between the aldehyde and *p*-anisidine, an imine **42** is formed. Enamine **41** and imine **42** then react to give **40** after hydrolysis.

Catalytic asymmetric nitro-Mannich-type reactions have also been introduced very recently by the groups of Shibasaki [37] and Jørgensen [38, 39].

6
Conclusions

Although several excellent examples of the catalytic asymmetric alkylation of imines have been reported, especially in the past few years, the scope of the reactions is still limited with regard to substrate generality, experimental simplicity, catalyst loading, and the enantiomeric purity of the isolated products. Research in this field has just started and further development can be expected in the near future.

References

1. Denmark SE, Nicaise OJ-C (1999) Alkylation of imino group. In: Jacobsen EN, Pfaltz A, Yamamoto H (eds) Comprehensive asymmetric catalysis, vol II. Springer, Berlin Heidelberg New York, p 923
2. Denmark SE, Nicaise OJ-C (1996) Chem Commun 999
3. Risch N, Arend M (1996) In: Helmchen G, Hoffmann RW, Mulzer J, Schaumann E (eds) Stereoselective synthesis (Houben-Weyl) E21b. Thieme, Stuttgart, p 1833

4. Bloch R (1998) Chem Rev 98:1407
5. Kobayashi S, Ishitani H (1999) Chem Rev 99:1069
6. Kobayashi S, Nagayama S (1997) J Am Chem Soc 119:10049
7. Jimeno C, Vidal-Ferran A, Moyano A, Pericás MA, Riera A (1999) Tetrahedron Lett 40:777
8. Fujihara H, Nagai K, Tomioka K (2000) J Am Chem Soc 122:12055
9. Porter JR, Traverse JF, Hoveyda AH, Snapper ML (2001) J Am Chem Soc 123:984
10. Porter JR, Traverse JF, Hoveyda AH, Snapper ML (2001) J Am Chem Soc 123:10409
11. Frants DE, Fässler R, Carreira EM (1999) J Am Chem Soc 121:11245
12. Frants DE, Fässler R, Tomooka CS, Carreira EM (2000) Acc Chem Res 33:373
13. Itsuno S, Watanabe K, Ito K, El-Shehawy AA, Sarhan AA (1997) Angew Chem Int Ed Engl 36:109
14. Chen G-M, Ramachandran V, Brown HC (1999) Angew Chem Int Ed Engl 38:825
15. Nakamura H, Nakamura K, Yamamoto Y (1998) J Am Chem Soc 120:4242
16. Farina V, Krishnamurthy B, Scott WJ (1997) Org React 50:1
17. Nakamura H, Bao M, Yamamoto Y (2001) Angew Chem Int Ed Engl 40:3208
18. Nakamura K, Nakamura H, Yamamoto Y (1999) J Org Chem 64:2614
19. Gastner T, Ishitani H, Akiyama R, Kobayashi S (2001) Angew Chem Int Ed Engl 40:1896
20. Hayashi T, Ishigedani M (2000) J Am Chem Soc 122:976
21. Ishiyama and Hartwig disclosed a set of rhodium(I)-catalyzed intermolecular Heck-type reactions between aryl iodides and N-heterocyclic aldimines to form the corresponding ketimines. Ishiyama T, Hartwig J (2000) 122:12043
22. Saaby S, Fang X, Gathergood N, Jørgensen KA (2000) Angew Chem Int Ed Engl 39:4114
23. Juraisti E (ed) (1997) Enantioselective synthesis of β-amino acids. Wiley-VHC, New York
24. Carreira EM (1999) Mukaiyama aldol reaction. In: Jacobsen EN, Pfaltz A, Yamamoto H (eds) Comprehensive asymmetric catalysis, vol III. Springer, Berlin Hidelberg New York, p 997
25. Corey EJ, Decicco CP, Newbold RC (1991) Tetrahedron Lett 32:5287
26. Fujieda H, Kanai M, Kambara T, Iida A, Tomioka K (1997) J Am Chem Soc 119:2060
27. Ishitani H, Ueno M, Kobayashi S (1997) J Am Chem Soc 119:7153
28. Ishitani H, Ueno M, Kobayashi S (2000) J Am Chem Soc 122:8180
29. Hagiwara E, Fujii A, Sodeoka M (1998) J Am Chem Soc 120:2474
30. Fujii A, Hagiwara E, Sodeoka M (1999) J Am Chem Soc 121:5450
31. Ferraris D, Young B, Dudding T, Lectka T (1998) J Am Chem Soc 120:4548
32. Ferraris D, Dudding T, Young B, Drury WJ III, Lectka T (1998) J Org Chem 64:2168
33. Xue S, Yu S, Deng Y, Wulff WD (2001) Angew Chem Int Ed Engl 40:2271
34. Murahashi S, Imada Y, Kawakami T, Harada K, Yonemushi Y, Tomita N (2002) J Am Chem Soc 124:2888
35. List B (2000) J Am Chem Soc 122:9336
36. List B, Pojarliv P, Biller WT, Martin HJ (2002) J Am Chem Soc 124:827
37. Yamada K, Harwood SJ, Gröger H, Shibasaki M (1999) Angew Chem Int Ed Engl 38:3504
38. Nishiwaki N, Knudsen KR, Gothelf KV, Jørgensen KA (2001) Angew Chem Int Ed Engl 40:2992
39. Juhl K, Gathergood N, Jørgensen KA (2001) Angew Chem Int Ed Engl 40:2995

Supplement to Chapter 28
Cyanation of Carbonyl and Imino Groups

Petr Vachal, Eric N. Jacobsen

Department of Chemistry and Chemical Biology, Harvard University, 12 Oxford St., Cambridge, MA, USA
e-mail: jacobsen@chemistry.harvard.edu

Keywords: Cyanation, α-Cyanohydrin, α-Aminonitrile, Cyanide, HCN, TMSCN, Lewis acid, Metal-free, Organocatalyst, C=O bond, C=N bond, Strecker, Reissert, Aldehydes, Ketones, Imines, Aldimines, Ketoimines

1
Asymmetric Cyanation of C=O Bonds

Cyanation of carbonyl compounds has one of the richest histories of any transformation in the field of asymmetric catalysis, and intensive research efforts have continued unabated since the editorial deadline for the first edition of *Comprehensive Asymmetric Catalysis* in 1998. This chapter will summarize all efforts in this area from 1998 to date, highlighting the most important catalytic systems from a synthetic and/or mechanistic standpoint. Significant advances in both the cyanation of aldehydes (formation of secondary cyanohydrins; Section 28.2.1) and the cyanation of ketones (formation of tertiary cyanohydrins; Section 28.2.2) will be addressed [1, 2].

1.1
Cyanation of Aldehydes

Asymmetric methodology should always be measured against the best existing alternatives for accessing the targeted compounds. In the case of hydrocyanation of aromatic aldehydes, the bar is set very high, as both enantiomers of

madelonitrile and of several of its substituted analogues are available easily and
cost-effectively via enzymatic methods [1, 3, 4].

That said, a significant number of methods have been developed for the cyanation of aromatic aldehydes with synthetic chiral catalysts. Ti(IV)-based chiral salen-derived catalysts developed by Jiang [5, 6] and Che [7] as well as chiral phosphine ligands developed by Fang [8] and Bueno [9] allow for the preparation of several electron-rich secondary aromatic cyanohydrins in good enantioselectivity. Walsh and Somanathan modified catalysts previously reported by
Oguni [10] and achieved high enantioselectivity for the cyanation of 4-meth-
oxybenzaldehyde [11, 12]. The pybox/AlCl$_3$ system developed by Iovel and Lukevics is an efficient catalyst for the cyanation of electron-rich aromatic and heterocyclic aldehydes [13, 14]. Unfortunately, similarly high levels of asymmetric
induction were unattainable for aliphatic and electronically-deficient aromatic
aldehydes using the aforementioned methodologies.

Very few methods developed recently are applicable to the cyanation of not
only electron-rich but also electron-deficient aromatic, as well as aliphatic aldehydes. Titanium(IV)-derived complex 1 (Fig. 1), reported by Uang, catalyzes the hydrocyanation of aromatic, α,β-unsaturated, and aliphatic aldehydes
with high enantioselectivity (>88% ee for all substrates reported) [15]. Similarly, Ti(IV)-catalyst 2 developed by Choi has proved to be highly enantioselective for the cyanation of various classes of aldehydes (>90% ee for most substrates) [16]. Belokon and co-workers reported two chiral salen-based systems:
(salen)VO catalyst 3a and [(salen)TiO]$_2$ 3b, both of which provided moderate
levels of enantioselection when applied to the asymmetric cyanation of diverse

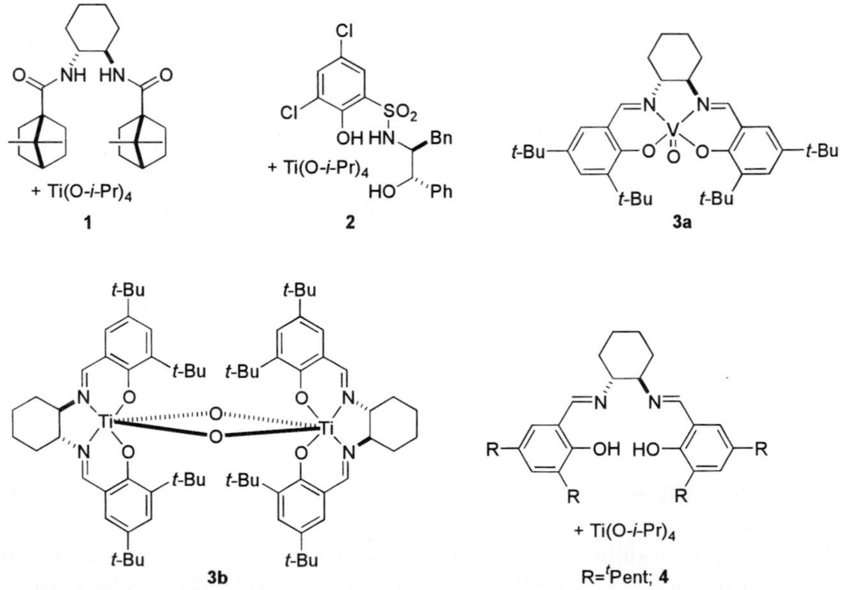

Fig. 1. Catalysts 1–4

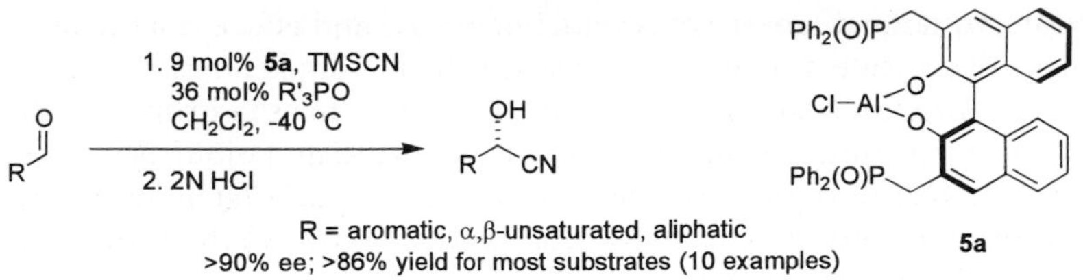

R = aromatic, α,β-unsaturated, aliphatic
>90% ee; >86% yield for most substrates (10 examples)

Scheme 1. (BINOL)Al(III)-based methodology developed by Shibasaki and co-workers

aldehyde classes [17–24]. Belokon recently disclosed the use of **3b** in combination with acetic anhydride and KCN as a superior cyanide source. This modification proved beneficial to both the level of enantioselectivity and the generality of the substrate scope; aromatic as well as some aliphatic aldehydes were converted into secondary cyanohydrins with good to excellent enantioselectivity [25].

Recently, Bu and Liang reported that the level of enantioinduction provided by chiral (salen)Ti(IV) catalysts was considerably dependent on the substitution of the chiral ligand, prompting them to prepare catalyst **4**. As a result, they observed excellent enantioselectivity for the cyanation of all aromatic aldehydes investigated (>92% ee for all 10 examples; no example of an aliphatic aldehyde was reported) [26].

Undoubtedly, the most significant development within the past few years for the asymmetric cyanation of aldehydes has been reported by Shibasaki and co-workers [27–30]. They demonstrated that (BINOL)Al(III) complex **5a** is a general catalyst for the cyanation of aldehydes: aromatic, α,β-unsaturated, and aliphatic aldehydes were converted to their corresponding cyanohydrins in nearly quantitative yield with high enantioselectivity (>90% ee for most substrates; Scheme 1). Mechanistic investigations suggest that catalyst **5a** serves as both a Lewis acid (Al center, for activation of the carbonyl functionality) and a Lewis base (phosphine oxide moiety, for cyanide delivery). Najera and Saa recently reported a close analogue of this system, wherein the phosphine oxide functionality is replaced by an amino group [31].

Finally, several examples of only moderately effective (≤75% ee) catalytic enantioselective cyanation of benzaldehyde derivatives have been reported recently [32–37].

1.2
Cyanation of Ketones

Relatively few of the enzymatic methods applicable to the preparation of secondary cyanohydrins have been adapted successfully to the synthesis of optically pure tertiary cyanohydrins [1, 3, 4]. Similarly, progress in asymmetric hydrocyanation of ketones with synthetic catalysts lagged far behind advances in aldehyde cyanation. This situation has changed fairly dramatically over the past few years, with the discovery of several interesting and effective systems for the synthesis of optically active tertiary cyanohydrins from ketones.

In 1999, Belokon reported the use of catalyst **3b** for the hydrocyanation of several acetophenone derivatives in up to 70% ee (Fig. 1) [38]. More recently, three research groups have provided even more impressive solutions for the cyanation of a wide variety of ketones. Discussion of these methods follows.

Based on the principle of dual activation proposed for catalyst **5a** (Scheme 1), Shibasaki and co-workers devised a modified, carbohydrate-derived, catalyst bearing the same functional groups as the parent catalyst **5a** (i.e., Al(III) Lewis acid and phosphine oxide Lewis base centers) [29]. Although less enantioselective for the cyanation of aldehydes than **5a**, the new system was found to display increased catalytic activity relative to **5a**. Systematic optimization of the ligand structure and the identity of the metal led to the discovery of the novel and highly effective Ti complex **6**. It was found that **6** catalyzed the cyanation of both aromatic and α,β-unsaturated ketones with high levels of enantioselectivity (>90% ee for most substrates reported, Scheme 2) [39–41]. Aliphatic ketones also underwent transformation into tertiary cyanohydrins with synthetically useful levels of enantioselectivity (82–93% ee). Significant effort has been directed toward broadening the utility of the methodology [42]. For example, since only one enantiomer of the carbohydrate-derived ligand is readily accessible, Shibasaki provided a solution to access both enantiomers of tertiary cyanohydrins using the same enantiomer of the chiral ligand by changing the metal center from Ti(IV) or Al(III) to Gd(III) [43–45].

In a key contribution to the renaissance of cinchona alkaloids in asymmetric catalysis, Deng and Tian reported a highly enantioselective cyanation of ketones catalyzed by **7** [46]. The methodology is especially effective for α-substituted and α,α-disubstituted ketones (>85% ee for most substrates reported; Scheme 3). Even simple cyclic ketones were converted into the corresponding cyanohydrins with high levels of enantioselectivity. Either enantiomer of the product can be accessed using either dihydroquinidine- or dihydroquinine-derived catalysts. Since both pseudoenantiomers of catalysts **7a** and **7b** are commercially available or accessible in one step from commercial materials, and all catalysts are fully recyclable upon completion of the reaction, this methodology represents a practical route to a wide range of α-substituted tertiary cyanohydrins.

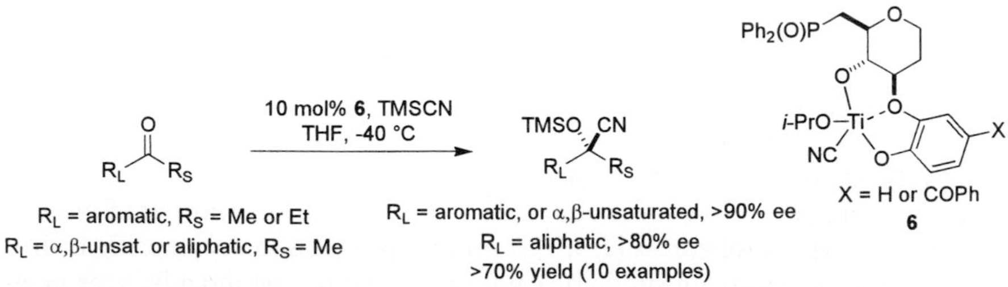

Scheme 2. Methodology for cyanation of ketones developed by Shibasaki and co-workers

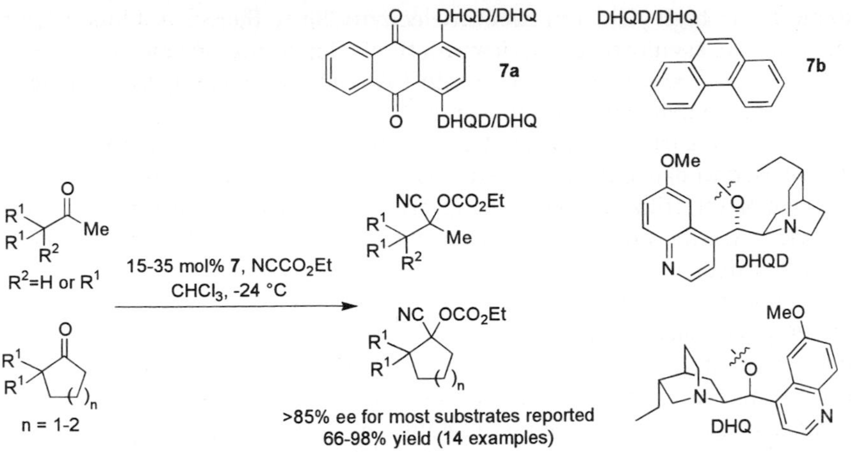

Scheme 3. Cinchona alkaloid-based methodology for cyanation of ketones developed by Deng and Tian

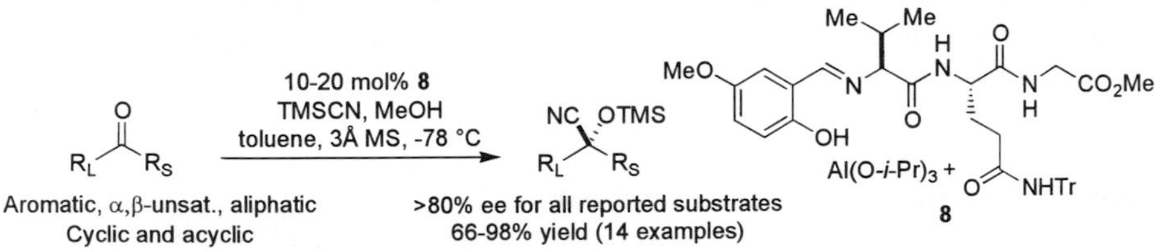

Scheme 4. Methodology for cyanation of ketones developed by Hoveyda, Snapper, and co-workers

Hoveyda, Snapper, and co-workers have reported a range of interesting asymmetric transformations catalyzed by Lewis acids bearing peptide-like Schiff base ligands [47]. During the course of optimizing this system for the asymmetric cyanation of ketones, they discovered that Al(III)-based **8** is a remarkably general and highly enantioselective catalyst (Scheme 4) [47]. Aromatic, α,β-unsaturated, and aliphatic cyclic and acyclic ketones were converted into the corresponding cyanohydrins in good yields and high enantioselectivities (≥ 80% ee for all reported substrates). Thus, this methodology provides efficient access to all major classes of ketone-derived cyanohydrins. Furthermore, the ligand was found to be recyclable, further underlining the practical potential of this system.

2
Asymmetric Cyanation of C=N Bonds (Strecker-Type Reaction)

The cyanation of imines, generally known as the Strecker reaction, has been one of the most aggressively studied transformations of asymmetric catalysis over the past several years. Very recent efforts in this area have resulted in the discovery of several highly efficient catalytic systems capable of providing α-amino nitriles in high yield and enantioselectivity. Since there was a lone example of a catalytic asymmetric Strecker reaction prior to the editorial deadline for *Comprehensive Asymmetric Catalysis* in 1998 [48, 49], the following update will serve as a complete review of this topic [50].

The wide assortment of catalytic asymmetric Strecker reaction methodologies devised to date can be divided into two major categories based on the nature of catalyst utilized: 1) Lewis acid-promoted and 2) metal-free (or organocatalytic) systems. Both classes of catalysis will be discussed and key results will be highlighted.

The potential substrates for the Strecker reaction fall into two categories: aldimines (derived from aldehydes, for which cyanide addition results in formation of a tertiary stereocenter) and ketoimines (derived from ketones, for which addition results in a quaternary stereocenter). As in the case of carbonyl cyanation, significant differences are observed between the substrate subclasses. To date, while a few catalyst systems have been found to display broad substrate scope with respect to aldimine substrates, successful Strecker reactions of ketoimines have been reported in only two cases. As is the case for all asymmetric catalytic methodologies, the breadth of the substrate scope constitutes a crucial criterion for the application of the Strecker reaction to a previously unexplored substrate.

2.1
Lewis Acid-Catalyzed Cyanation of C=N Bonds

The first highly effective Lewis acid-based asymmetric catalytic Strecker reaction was developed by Sigman and Jacobsen [51, 52]. Chiral (salen)AlCl complex **9** was found to catalyze the hydrocyanation of a wide range of aromatic aldimines with high levels of enantioselectivity (>90% ee for the majority of aromatic substrates) (Scheme 5). Since catalyst **9** is commercially available, this methodology represents a convenient means of accessing precursors to arylglycine derivatives. On the other hand, aliphatic imines represented a limitation of this catalyst system, with the corresponding aminonitriles generated in only 30–50% ee.

Hoveyda, Snapper, and co-workers identified a series of Ti(IV)-based catalysts **10** through systematic screening [53] of modular Schiff base ligands [54, 55]. Their methodology proved to be very general for hydrocyanation of ald-

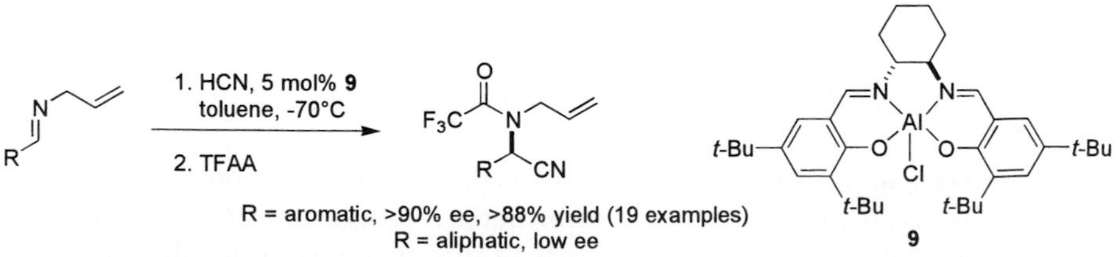

R = aromatic, >90% ee, >88% yield (19 examples)
R = aliphatic, low ee

Scheme 5. Chiral (salen)AlCl-based methodology developed by Sigman and Jacobsen

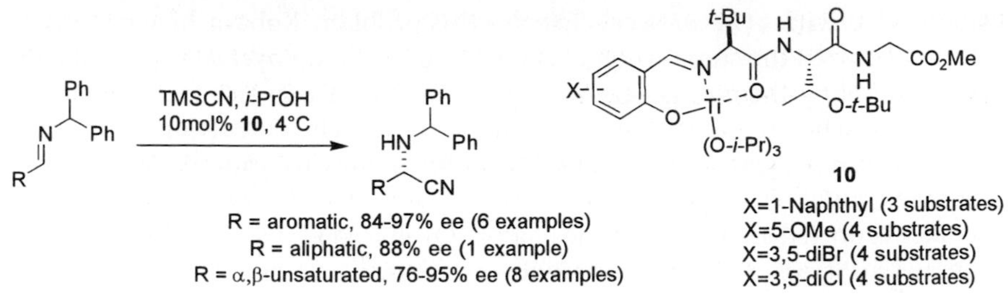

Scheme 6. Chiral Ti(IV)-based methodology developed by Hoveyda, Snapper, and co-workers

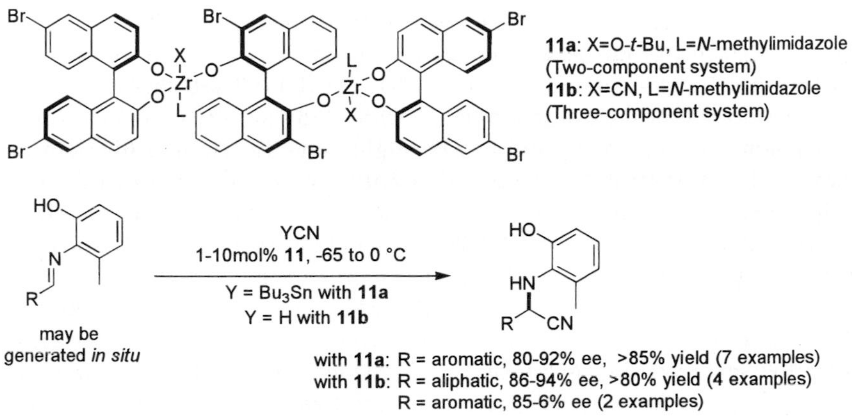

Scheme 7. (BINOL)Zr(IV)-based methodology developed by Kobayashi and co-workers

imine substrates: aromatic, α,β-unsaturated, and aliphatic substrates were converted into the Strecker adducts in high yields and enantioselectivities (Scheme 6). The necessity to match up each substrate with a specific (and often different) catalyst remains the only significant drawback of this methodology, at least as far as application to previously unexplored substrates is concerned. Recent investigations of this catalyst system provide a mechanistic basis for these reactions, and may therefore serve as a useful tool to assist in the identification of the optimal catalyst for a particular substrate of interest [56].

Nakai reported that (BINOL)Ti(Oi-Pr)$_2$ catalyzes the Strecker reaction with limited success (30% ee) [57]. Kobayashi and co-workers reported the first example of a highly enantioselective (BINOL)Zr(IV)-based catalyst. Aside from changing the identity of the metal, the key modifications to the system were replacement of one of the achiral ligands on the metal center by a BINOL-derived ligand and incorporation of a chelating group into the imine substrate to induce two-point binding (Scheme 7). The oligomeric system 11a was reported to be an efficient catalyst for a range of aromatic substrates using Bu$_3$SnCN as the cyanide source (80–91% ee) [58, 59]. Although general for aromatic imines, catalyst 11a yielded Strecker adducts derived from aliphatic aldimines with reduced

enantioselectivities (74–83% ee). To solve this problem, Kobayashi and collabo-
rators optimized the counter-ion of catalyst **11a** and employed a new set of conditions enabling them to generate the imine substrates in situ from aldehydes
and amines. The three-component (amine, aldehyde, HCN) reaction setup using
catalyst **11b** represents a practical solution for the hydrocyanation of aliphatic
aldimines [60].

Vallee reported another example of a BINOL-based Lewis acid catalyst for the
asymmetric Strecker reaction of ketoimines. While a traditional (BINOL)Ti(IV)-
based system provided poor enantioselectivity [61], Sc(BINOL)$_2$Li proved to be
highly enantioselective for the cyanation of N-benzyl acetophenonimine (95%
ee at 50% conversion, 91% ee at 80% conversion) [62]. Unfortunately, results
were provided only for a single ketoimine and a single aromatic aldimine, leaving the generality of the methodology in question.

Shibasaki and co-workers applied (BINOL)Al(III)-derived catalyst **5a**, previously developed for the cyanation of aldehydes [28], to the asymmetric Strecker reaction. This catalyst proved to be highly enantioselective for both aromatic and α,β-unsaturated acyclic aldimines (>86% ee for most substrates)
(Scheme 8) [63–65]. Aliphatic aldimines underwent cyanide addition with lower levels of enantioselectivity (70–80% ee). A significant distinction of **5** relative
to other catalysts is, undoubtedly, its successful application to the hydrocyanation of quinolines and isoquinolines, followed by in situ protection of the sensitive α-amino nitrile formed (this variant of the Strecker reaction is also known
as the Reissert reaction [66]). Thus, Shibasaki has shown that high enantioselectivities (>80% ee for most substrates) and good yields are generally obtainable
in the Reissert reaction catalyzed by **5b** [67, 68]. When applied to 1-substituted

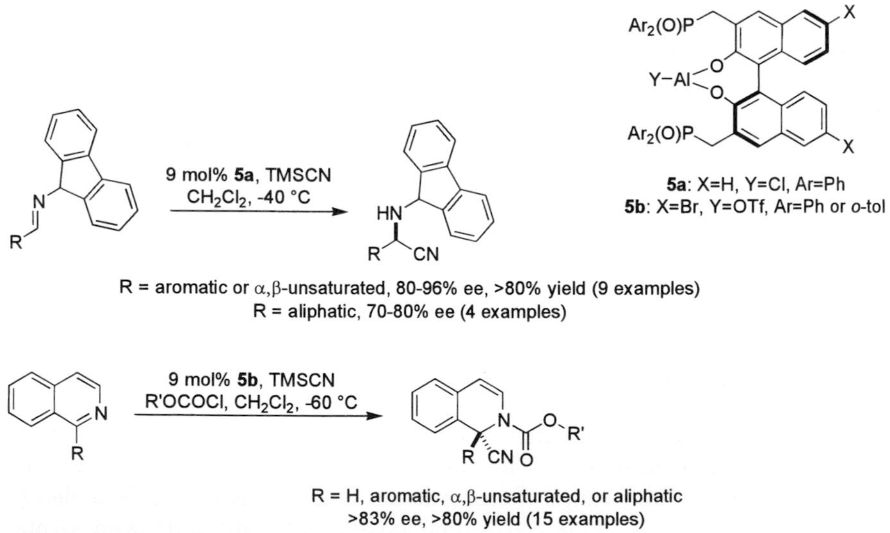

R = aromatic or α,β-unsaturated, 80–96% ee, >80% yield (9 examples)
R = aliphatic, 70–80% ee (4 examples)

R = H, aromatic, α,β-unsaturated, or aliphatic
>83% ee, >80% yield (15 examples)

Scheme 8. (BINOL)Al(III)-based methodology developed by Shibasaki and co-workers

isoquinolines, the asymmetric Reissert reaction provides efficient access to chiral α,α-disubstituted cyclic α-amino nitriles [69]. The high enantioselectivities provided by **5a** for cyanation of both C=O and C=N represent an exception to the general observation that the same catalysts rarely afford favorable results in asymmetric additions to both carbonyl compounds and imines.

2.2
Metal-free Catalysts for the Cyanation of C=N Bonds

Organocatalysts have been identified as highly effective in asymmetric catalytic Strecker reactions. Jacobsen and co-workers reported a highly active and enantioselective catalyst **12** that was identified and optimized by a parallel, combinatorial approach. Both resin-bound and soluble versions of catalyst **12** were prepared and investigated, with the former proving highly practical and completely recyclable, although the latter afforded slightly (2–5%) higher enantioselectivity. Easily prepared soluble catalyst **12a** [70] proved to be extremely general in substrate scope: a wide range of diverse aromatic, α,β-unsaturated, as well as aliphatic aldimine substrates were converted into the corresponding Strecker adducts in high yield and enantioselectivity (generally >90% ee; Scheme 9) [71–73]. Moderate enantioselectivity was observed for unbranched aliphatic aldimines using catalyst **12a**. A detailed mechanistic investigation elucidated the basis of imine activation and revealed the key structural elements controlling the enantioselectivity of the catalysts [74]. Based on the mechanistic model, Vachal and Jacobsen introduced rational modifications to the structure of catalyst **12**. The improved system **12b** proved to be an effective catalyst for all aldimine substrates investigated, with especially striking improvements observed for substrates displaying limited success with catalyst **12a**. In addition, **12a** and **12b** were shown to be enantioselective catalysts for the Strecker reaction of acetophenone imine derivatives (>89% ee for essentially all substrates), providing

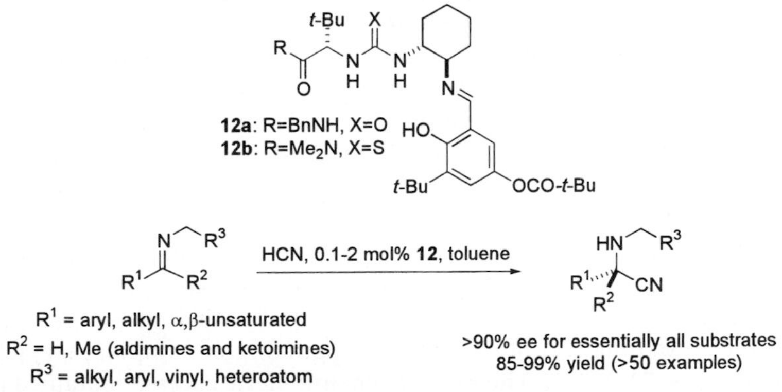

12a: R=BnNH, X=O
12b: R=Me₂N, X=S

HCN, 0.1-2 mol% **12**, toluene

R¹ = aryl, alkyl, α,β-unsaturated
R² = H, Me (aldimines and ketoimines)
R³ = alkyl, aryl, vinyl, heteroatom

>90% ee for essentially all substrates
85-99% yield (>50 examples)

Scheme 9. Strecker methodology using metal-free catalyst **12** as developed by Jacobsen and co-workers

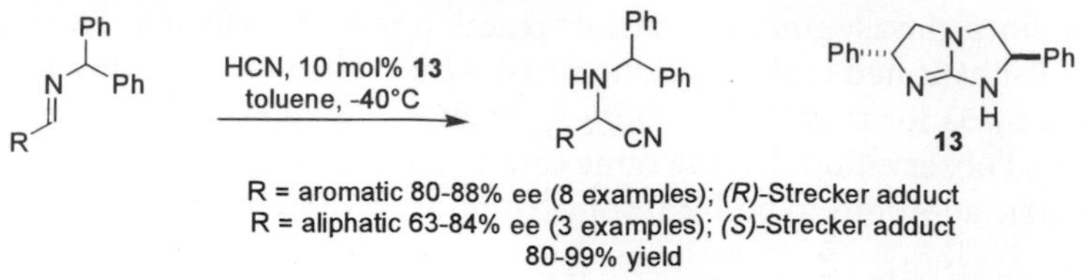

R = aromatic 80-88% ee (8 examples); *(R)*-Strecker adduct
R = aliphatic 63-84% ee (3 examples); *(S)*-Strecker adduct
80-99% yield

Scheme 10. Methodology using metal-free quanidine 13 developed by Corey and Grogan

efficient access to chiral α,α-disubstituted α-amino nitriles that, in turn, were readily converted into the corresponding quaternary α-amino acids [75].

Corey and Grogan reported another example of a metal-free catalytic system for the asymmetric Strecker reaction, the C2-symmetric guanidine derivative 13 [76]. This catalyst displayed effectiveness for aromatic aldimine substrates, affording the corresponding enantioenriched Strecker adducts in 80–88% ee for most substrates in almost quantitative yield (Scheme 10). Although quite general for aromatic substrates, 13 proved less effective for aliphatic aldimines (63–84% ee). This difference was postulated to arise from the absence of π-π stacking interaction crucial for the high enantioselectivity of the aromatic substrates. Consistent with this notion, the Strecker adducts derived from aliphatic aldimines had an opposite sense of absolute stereochemistry from those derived from aromatic aldimines.

2.3
Summary of the Asymmetric Cyanation of C=N Bonds

Several highly efficient and enantioselective methods have been developed for the asymmetric Strecker reaction within the past four years. Several of these methods are general for various types of aldimine substrates, and a few are even applicable to the hydrocyanation of ketoimines. As is generally the case for all asymmetric catalytic methodologies, the breadth of the substrate scope constitutes a crucial criterion for the application of the Strecker reaction to a novel, previously unexplored substrate. To help the reader evaluate which of the methods may be optimal for a particular imine of interest, we have summarized the reported enantioselectivity achieved with all available methodologies in Table 1. The results are divided into several substrate classes: aromatic, aliphatic, α,β-unsaturated aldimines, and ketoimines. Since most of the methods provide the Strecker adducts in essentially quantitative yield, only enantioselectivities are listed in Table 1. Indeed, the enantiomeric excesses of the aminonitrile products are tied closely to the theoretical yield of optically pure compounds that can be obtained after crystallization (Strecker adducts and/or their derivatives are usually highly crystalline and often undergo upgrading of ee upon recrystallization) [54–56, 68, 72, 73, 75]. The range of enantiomeric excesses obtained for all reported imines is provided in first column of every substrate class, and the ee range for the majority (best 80%) of substrates is given in second column, since

Table 1. Overview of the methods for the catalytic asymmetric Strecker reaction

Method (catalyst)[a]	Enantioselectivity (% ee) Aromatic aldimines All[b]	Most[c]	Aliphatic aldimines All[b]	Most[c]	α,β-Unsaturated aldimines All[b]	Most[c]	Ketoimines All[b]	Most[c]
Lipton	10–99 (10 examples)	64–99	10–17 (2 examples)	10–17	Not reported		Not reported	
Jacobsen (salen)AlCl (9)	79–96 (21 examples)	91–96	37–57 (2 examples)	37–57	Not reported		Not reported	
Jacobsen metal-free (12)	89–99 (34 examples)	92–99	86–99 (16 examples)	90–99	91–96 (3 examples)[d]	91–96	45–98 (18 examples)	90–98
Hoveyda (10)	90–97 (6 examples)	93–97	85 (1 example)	85	76–97 (8 examples)[d]	84–97	Not reported	
Corey (13)	50–88 (10 examples)	80–88	63–84 (3 examples)	76–84	Not reported		Not reported	
Kobayashi (11)[e]	76–92 (8 examples)	87–92	91–94 (4 examples)	91–94	Not reported		Not reported	
Vallee Sc(BINOL)$_2$Li	86 (1 example)	86	Not reported		Not reported		95[f] (1 example)	95[f]
Shibasaki (5)	54–96 (17 examples)[g]	75–96	70–81 (4 examples)	72–81	86–96 (2 examples)	86–96	73–95 (10 examples)[h]	87–95

[a] Catalysts listed in the order in which they were first reported
[b] Range of ee values for all reported substrates for a given method
[c] Range of ee values for the majority of substrates (best 80% of reported substrates for a given method)
[d] Exclusive 1,2-addition reported
[e] Bu$_3$SnCN or HCN used as the cyanide source
[f] 95% ee at 50% conversion, 91% ee at 80% conversion
[g] Including Reissert-type reaction of quinolines and isoquinolines
[h] Reissert-type reaction of 1-substituted isoquinolines

some reports include poor results in order to help define the limits of the methodology. The total number of imines investigated is also included to help evaluate the extent to which the different methodologies have been examined. These methodologies certainly appear to be well poised for synthetic application in the immediate future.

References

1. For an excellent review on cyanohydrins, see: Gregory RJH (1991) Chem Rev 99:3649
2. Oguni N (1998) Transition metals for organic synthesis. Wiley-VCH, Weinheim, Germany, chap 2.20, 1:325
3. Schmidt M, Griengl H (1999) Top Curr Chem 200:193
4. Effenberger F (1999) Chimia 53:3
5. Feng X, Gong L-Z, Hu W-H, Li Z, Pan W-D, Mi A-Q, Jiang Y-Z (1998) Gaodeng Xuexiao Huaxue Xuebao 19:1416
6. Jiang Y, Gong L, Feng X, Hu W, Pan W, Li Z, Mi A (1997) Tetrahedron 53:14327
7. Zhou X-G, Huang J-S, Ko P-H, Cheung K-K, Che C-M (1999) J Chem Soc, Dalton Trans 18: 3303
8. Yang WB, Fang JM (1998) J Org Chem 63:1356
9. Brunel J-M, Legrand O, Buono G (1999) Tetrahedron Asymmetry 10:1979
10. Hayashi M, Inoue T, Miyamoto Y, Oguni N (1994) Tetrahedron 50:4385
11. Gama A, Flores-Lopez LZ, Aguirre G, Parra-Hake M, Somanathan R, Walsh PJ (2002) Tetrahedron Asymmetry 13:149
12. Flores-Lopez LZ, Parra-Hake M, Somanathan R, Walsh PJ (2000) Organometallics 19: 2153
13. Iovel I, Popelis Y, Fleisher M, Lukevics E (1997) Tetrahedron Asymmetry 8:1279
14. For an earlier report on the use of box-ligands for cyanation of aldehydes, see: Corey EJ, Wang Z (1993) Tetrahedron Lett 34:4001
15. Hwang C-D, Hwang D-R, Uang B-J (1998) J Org Chem 63:6762
16. You J-S, Gau H-M, Choi MCK (2000) Chem Commun 19:1963
17. Belokon YN, North M, Parsons T (2000) Org Lett 2:1617
18. Belokon YN, Caveda-Cepas S, Green B, Ikonnikov NS, Khrustalev VN, Larichev VS, Moscalenko MA, North M, Orizu C, Tararov VI, Tasinazzo M, Timofeeva GI, Yashkina LV (1999) J Am Chem Soc 121:3968
19. Belokon YN, Green B, Ikonnikov NS, Larichev VS, Lokshin BV, Moscalenko MA, North M, Orizu C, Peregudov AS, Timofeeva GI (2000) Eur J Org Chem 14:2655
20. Belokon YN, Green B, Ikonnikov NS, North M, Parsons T, Tararov VI (2001) Tetrahedron 57:771
21. Tararov VI, Orizu C, Ikonnikov NS, Larichev VS, Moscalenko MA, Yashkina LV, North M, Belokon YN (1999) Russ Chem Bull 48:1128
22. Tararov VI, Hibbs DE, Hursthouse MB, Ikonnikov NS, Abdul MKM, North M, Orizu C, Belokon YN (1998) Chem Commun 3:387
23. Belokon YN, Flego M, Ikonnikov N, Moscalenko M, Orizu C, Tararov V, Tasinazzo MJ (1997) Chem Soc Perkin Trans1 9:1293
24. Belokon YN, Ikonnikov N, Moscalenko M, North M, Orlova S, Tararov V, Yashkina L (1996) Tetrahedron Asymmetry 7:851
25. Belokon YN, Gutnov AV, Moskalenko MA, Yashkina LV, Lesovoy DE, Ikonnikov NS, Larichev VS, North M (2002) Chem Commun 3:244
26. Liang S, Bu XR (2002) J Org Chem 67:2702
27. Hamashima Y, Sawada D, Nogami H, Kanai M, Shibasaki M (2001) Tetrahedron 57:805
28. Hamashima Y, Sawada D, Kanai M, Shibasaki M (1999) J Am Chem Soc 121:2641
29. For catalyst derived from **5a,** see (more details provided in Sect. 28.2.2): Kanai M, Hamashima Y, Shibasaki M (2000) Tetrahedron Lett 41:2405

30. For a review, see: Groeger H (2001) Chem Eur J 7:5247
31. Casas J, Najera C, Sansano JM, Saa JM (2002) Org Lett
32. Yang Z, Zhou Z, Tang C (2001) Synth Commun 31:3031
33. Aspinall HC, Greeves N, Smith PM (1999) Tetrahedron Lett 40:1763
34. Qian C, Zhu C, Huang TJ (1998) Chem Soc, Perkin Trans 1 14:2131
35. Zi G-F, Yin C-L (1998) J Mol Catal A Chem 132:L1
36. Wada M, Takahashi T, Domae T, Fukuma T, Miyoshi N, Smith K (1997) Tetrahedron Asymmetry 8:3939
37. Kim J-H, Kim G-J (2001) Stud Surf Sci Catal 135:3646
38. Belokon YN, Green B, Ikonnikov NS, North M, Tararov VI (1999) Tetrahedron Lett 40:8147
39. Hamashima Y, Kanai M, Shibasaki M (2001) Tetrahedron Lett 42:691
40. Hamashima Y, Kanai M, Shibasaki M (2000) J Am Chem Soc 122:7412
41. Manickam G, Nogami H, Kanai M, Groger H, Shibasaki M (2001) Synlett 617
42. Masumoto S, Yabu K, Kanai M, Shibasaki M (2002) Tetrahedron Lett 43:2919
43. Yabu K, Masumoto S, Yamasaki S, Hamashima Y, Kanai M, Du W, Curran DP, Shibasaki M (2001) J Am Chem Soc 123:9908
44. Yabu K, Masumoto S, Kanai M, Curran DP, Shibasaki M (2002) Tetrahedron Letters 43:2923
45. Shibasaki M, Yoshikawa N (2002) Chem Rev 102:2187
46. Tian SK, Deng L (2001) J Am Chem Soc 123:6195
47. Deng H, Isler MP, Snapper ML, Hoveyda AH (2002) Angew Chem Int Ed 41:1009; see cited references for a complete list of asymmetric transformations catalyzed by peptide-like Schiff base derived Lewis acids
48. Iyer MS, Gigstad KM, Namdev,ND, Lipton M (1996) J Am Chem Soc 118:4910
49. Iyer MS, Gigstad KM, Namdev ND, Lipton M (1996) Amino Acids 11:259
50. For a brief review covering papers published prior to 2001, see: Yet L (2001) Angew Chem Int Ed 40:875
51. Sigman MS, Jacobsen EN (1998) J Am Chem Soc 120:5315
52. Sigman MS, Espino CG, Jacobsen EN (1998–2000) Unpublished results
53. Kuntz KW, Snapper ML, Hoveyda AH (2001) High-throughput synthesis. Marcel Dekker, New York, p 283
54. Krueger CA, Kuntz KW, Dzierba CD, Wirschun WG, Gleason JD, Snapper ML, Hoveyda AH (1999) J Am Chem Soc 121:4284
55. Porter JR, Wirschun WG, Kuntz KW, Snapper ML, Hoveyda AH (2000) J Am Chem Soc 122:2657
56. Josephsohn NS, Kuntz KW, Snapper ML, Hoveyda AH (2001) J Am Chem Soc 123:11594
57. Mori M, Imma H, Nakai T (1997) Tetrahedron Lett 38:6229
58. Ishitani H, Komyiama S, Kobayashi S (1998) Angew Chem Int Ed 37:3186
59. Kobayashi S, Ishitani H (2000) Chirality 12:540
60. Ishitani H, Komyiama S, Hasegawa Y, Kobayashi S (2000) J Am Chem Soc 122:762
61. Byrne JJ, Chavarot M, Chavant P-Y, Vallee Y (2000) Tetrahedron Lett 41:873
62. Chavarot M, Byrne JJ, Chavant P-Y, Vallee Y (2001) Tetrahedron Asymmetry 12:1147
63. Takamura M, Hamashima Y, Yoshitaka U, Usuda H, Kanai M, Shibisaki M (2000) Angew Chem Int Ed 39:1650
64. Takamura M, Hamashima Y, Usuda H, Kanai M, Shibasaki M (2000) Chem Pharm Bull 48:1586
65. For a report on the solid supported version of 5, see: Nogami H, Matsunaga S, Kanai M, Shibasaki M (2001) Tetrahedron Lett 42:279
66. Reissert A (1905) Chem Ber 38:1603
67. Takamura M, Funabashi K, Kanai M, Shibasaki M (2000) J Am Chem Soc 122:6327
68. Takamura M, Funabashi K, Kanai M, Shibasaki M (2001) J Am Chem Soc 123:6801
69. Funabashi K, Ratni H, Kanai M, Shibasaki M (2001) J Am Chem Soc 123:10784
70. Su JT, Vachal P, Jacobsen EN (2001) Adv Synth Catal 343:197

71. Sigman MS, Jacobsen EN (1998) J Am Chem Soc 120:4901
72. Sigman MS, Vachal P, Jacobsen EN (2000) Angew Chem Int Ed 39:1279
73. Sigman MS, Vachal P, Jacobsen EN (1998–2001) Unpublished results
74. Vachal P, Jacobsen EN (2002) J Am Chem Soc (in press)
75. Vachal P, Jacobsen EN (2000) Org Lett 2:867
76. Corey EJ, Grogan MJ (1999) Org Lett 1:157

Supplement to Chapter 29.3
Nitroaldol Reaction

Masakatsu Shibasaki, Harald Gröger, Motomu Kanai

Graduate School of Pharmaceutical Sciences, The University of Tokyo, 7–3–1 Hongo,
Bunkyo-ku, 113–0033 Tokyo, Japan
e-mail: mshibasa@mol.f.u-tokyo.ac.jp

Keywords: copper complex, nitroaldol, quaternary ammonium fluoride, zinc complex

1
Supplement

Because of the utility of the products and the potentially atom-economic nature of the reaction, the catalytic asymmetric nitroaldol (Henry) reaction [49] continues to be under intensive investigation. Very recently, Trost et al. reported an application of their dinuclear zinc complex with chiral semi-azacrown ligand (catalyst **58**) to a catalytic asymmetric nitroaldol reaction [50]. With 5 mol % of the catalyst and 10 equivalents of nitromethane, high enantioselectivity was obtained from a wide range of aromatic and aliphatic aldehydes [Scheme 14, Eq. (1)]. The authors proposed a reaction mechanism in which one of the zinc alkoxides acts as a Brönsted base to generate a zinc nitronate, while the other zinc acts as a Lewis acid to activate an aldehyde. This mechanism is just like a homo-bimetallic version of Shibasaki's hetero-bimetallic complex (Scheme 2). Jørgensen et al. reported a catalytic asymmetric nitroaldol reaction to pyruvate derivatives using t-Bu-BOXCu(OTf)$_2$ complex (20 mol %) in the presence of Et$_3$N (20 mol %) [Scheme 14, Eq. (2)] [51]. The product contains a chiral tetrasubstituted carbon. Interestingly, a 1:1 ratio of the Lewis acid (Cu) and the Brönsted base (Et$_3$N), as well as Et$_3$N as a base were essential for high enantioselectivity. Other amines such as Hünigs base gave lower enantioselectivity. Considering that the reaction proceeded only in the presence of Et$_3$N (in the absence of the Lewis acid), the observed high enantioselectivity might be due in part to a narrow balance between the activities of the Lewis acid and the Brönsted base. There appears to be a need for a stricter balance between these two functions in Jørgensen's case, in which the Lewis acid and the Brönsted base exist independently, than in Shibasaki's and Trost's cases, in which the two functions exist in one enantioselective catalyst. Corey et al. demonstrated that a chiral quaternary ammonium fluoride promotes the catalytic diastereoselective nitroaldol reaction [Scheme 14, Eq. (3) and (4)] [52]. With 10 mol % of the cinchonidine-derived conformationally rigid catalyst **60**, *syn*-(2*R*,3*S*)-nitroaldol

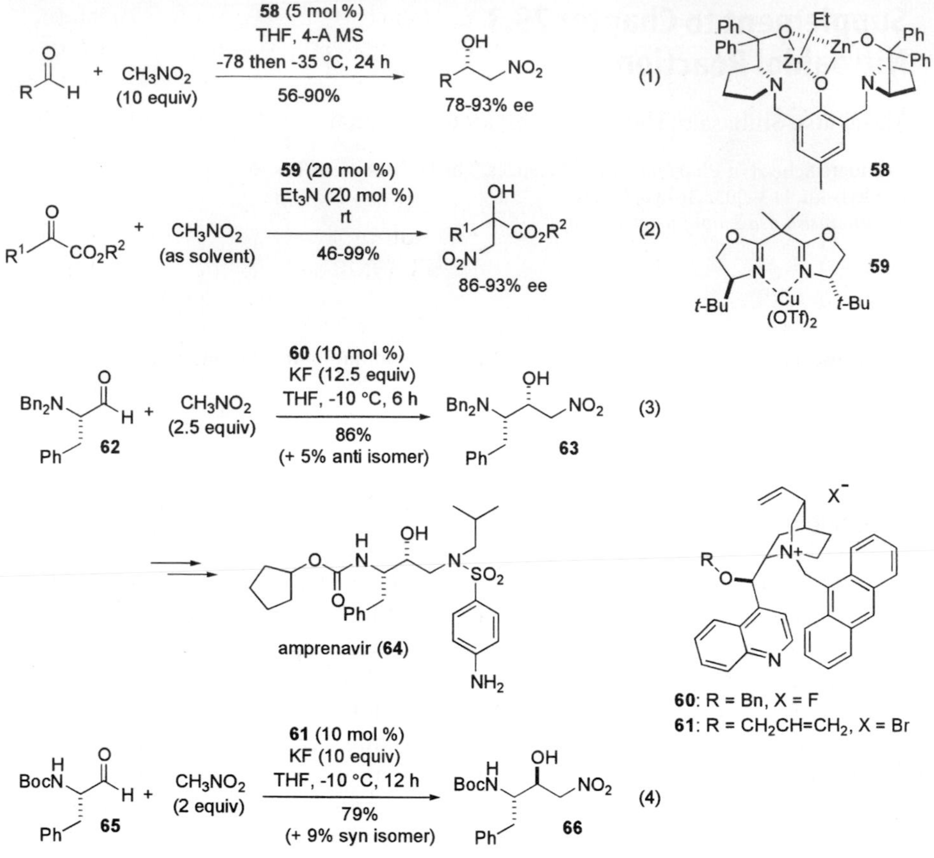

Scheme 1

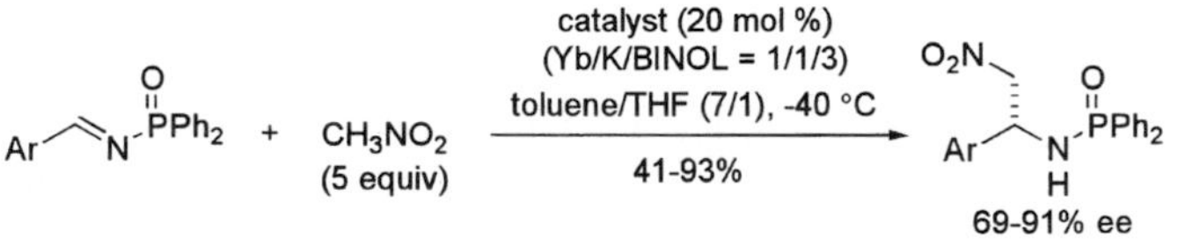

Scheme 2

adduct **63** was obtained in a 17:1 ratio from N,N-dibenzyl-(S)-phenylalaninal (**62**) (Scheme 1). Adduct **63** was converted to an HIV protease inhibitor, amprenavir (**64**), in high overall yield. When the Boc-protected (S)-phenylalaninal **65** was used as a substrate, however, the diastereoselectivity was reversed and *anti*-(2S,3S)-nitroaldol adduct **66** was obtained as a major isomer (9:1), by using the quaternary ammonium salt **61** containing the same absolute configuration as **60**. The authors rationalized these results from three-dimensional models of the aldehyde–chiral quaternary ammonium assemblies. The interaction between the formyl oxygen and the ammonium nitrogen atoms, as well as the van der Waals contact between the aromatic groups (such as between the anthracene or the quinoline of the catalyst and the *N*-benzyl of the aldehyde), have a key role in defining the structure of the assemblies. The nitronate should react with the aldehyde in the assembly from the open space.

The catalytic asymmetric nitroaldol reaction was extended to a direct catalytic asymmetric nitro-Mannich-type reaction promoted by hetero-bimetallic catalysts (Scheme 2) [53–55] or by Et₃NBOX–Cu complexes [56]. These topics are reviewed in Chap. 28.2.

References

49. Luzzio FA (2001) Tetrahedron 57:915
50. Trost BM, Yeh VSC (2002) Angew Chem Int Ed Engl 41:861
51. Christensen C, Juhl K, Jørgensen KA (2001) Chem Commun 2222
52. Corey EJ, Zhang F. (1999) Angew Chem Int Ed Engl 38:1931
53. Yamada K-i, Harwood SJ, Göger H, Shibasaki M (1999) Angew Chem Int Ed Engl 38:3504
54. Yamada K-i, Moll G, Shibasaki M (2001) Synlett SI:980
55. Tsuritani N, Yamada K-i, Yoshikawa N, Shibasaki M, (2002) Chem Lett 276
56. Nishiwaki N, Knudsen KR, Gothelf KV, Jørgensen KA (2001) Angew Chem Int Ed Engl 40: 2992

Chapter 29.4
Direct Catalytic Asymmetric Aldol Reaction

Masakatsu Shibasaki, Naoki Yoshikawa, Shigeki Matsunaga

Graduate School of Pharmaceutical Sciences, The University of Tokyo, 7–3–1 Hongo,
Bunkyo-ku, Tokyo 113–0033, Japan
e-mail: mshibasa@mol.f.u-tokyo.ac.jp

Keywords: Aldol, Direct, Ketone, Asymmetric catalysis, Enantioselective reaction, Diastereoselectivity, 1,2-Diol, Aldehyde, Lewis acid, Brønsted base, Organocatalysis, Bimetallic catalyst, Proline

1
Introduction

The aldol reaction is well established in organic chemistry as a remarkably useful synthetic tool, providing access to β-hydroxycarbonyl compounds and related building blocks. Intensive efforts have raised this classic process to a highly enantioselective transformation employing only catalytic amounts of chiral
promoters, as reviewed in the previous section (Chap. 29.1). While some effective applications have been reported, most of the methodologies necessarily in-
volve the preformation of latent enolates **2**, such as ketene silyl acetals, using

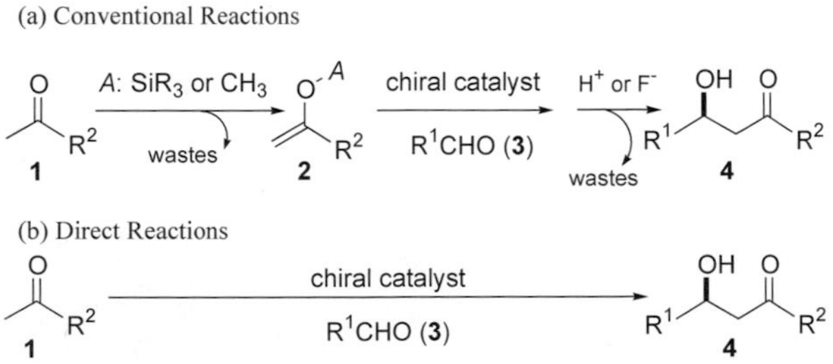

(a) Conventional Reactions

(b) Direct Reactions

Scheme 1

stoichiometric amounts of silylating agents (Scheme 1, top). Due to the growing demand for an atom-economic process, the development of a *direct* catalytic asymmetric aldol reaction (Scheme 1, bottom), which eliminates the necessity of preforming latent enolates, is an exciting and challenging subject. This chapter focuses on the notable advances which have been achieved over the last few years.

2
Catalysis by Metal Complexes

Studies of catalytic asymmetric Mukaiyama aldol reactions were initiated in the early 1990s. Until recently, however, there have been few reports of direct catalytic asymmetric aldol reactions [1]. Several groups have reported metallic and non-metallic catalysts for direct aldol reactions. In general, a metallic catalysis involves a synergistic function of the Brønsted basic and the Lewis acidic moieties in the catalyst (Scheme 2). The Brønsted basic moiety abstracts an α-proton of the ketone to generate an enolate (6), and the Lewis acidic moiety activates the aldehyde (3).

In 1997, Shibasaki et al. discovered the promotion of enantioselective aldol reactions of unmodified ketones 1 with a broad range of applicable substrates [2]. The reactions were catalyzed by the $Li_3[La(BINOL)_3]$ complex (LLB, 9) (Scheme 3, right) [3] to afford the desired aldol products (4) with up to 94% ee. In 1999, a large acceleration of this reaction was achieved using a heteropolymetallic catalyst that was prepared from LLB (9), KOH, and H_2O, allowing for a reduction in the amount of catalyst from 20 mol % to 3–8 mol % with a significantly shorter reaction time (Scheme 3) [4]. Kinetic studies indicated that the rate-determining step was the deprotonation of the ketone. The heteropolymetallic catalyst (LLB+KOH+H_2O) was successfully applied to the direct aldol reaction of 2-hydroxyacetophenones (10) without the need to protect the hydroxyl group, providing a valuable method for enantioselective synthesis of *anti*-1,2-diols (*anti*-11) (Scheme 4, top) [5,6]. A catalyst that was prepared from

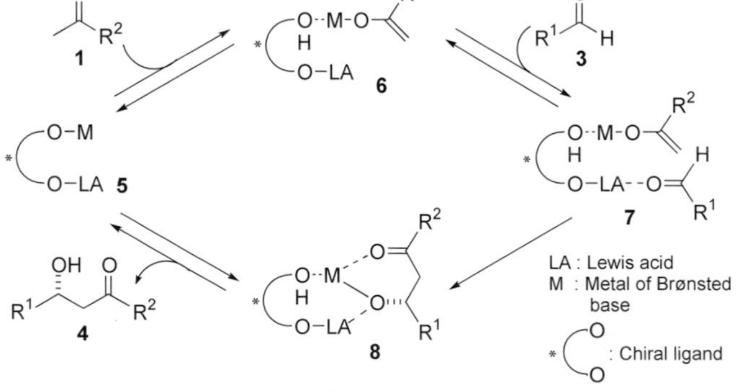

Scheme 2

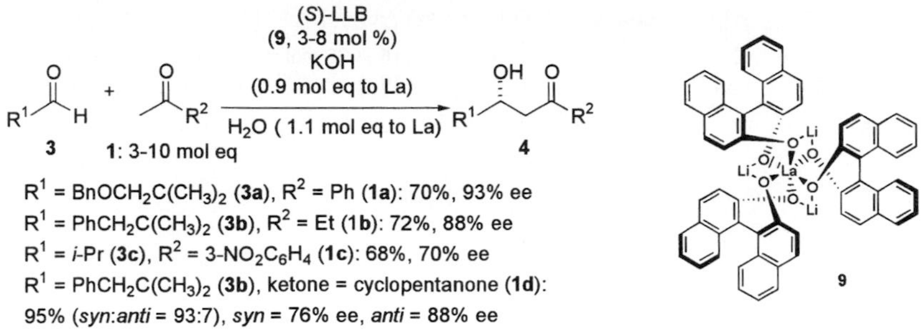
(S)-LLB
(9, 3-8 mol %)
KOH
(0.9 mol eq to La)
H2O (1.1 mol eq to La)
3
1: 3-10 mol eq
4
R1 = BnOCH2C(CH3)2 (3a), R2 = Ph (1a): 70%, 93% ee
R1 = PhCH2C(CH3)2 (3b), R2 = Et (1b): 72%, 88% ee
R1 = i-Pr (3c), R2 = 3-NO2C6H4 (1c): 68%, 70% ee
R1 = PhCH2C(CH3)2 (3b), ketone = cyclopentanone (1d):
95% (syn:anti = 93:7), syn = 76% ee, anti = 88% ee
9

Scheme 3

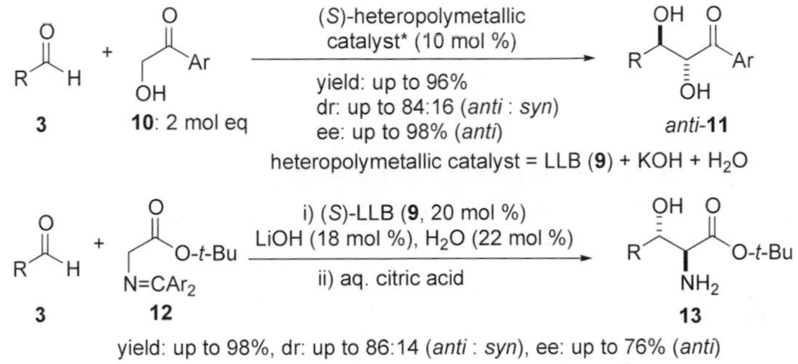
(S)-heteropolymetallic
catalyst* (10 mol %)
yield: up to 96%
dr: up to 84:16 (anti : syn)
ee: up to 98% (anti)
3
10: 2 mol eq
anti-11
heteropolymetallic catalyst = LLB (9) + KOH + H2O
i) (S)-LLB (9, 20 mol %)
LiOH (18 mol %), H2O (22 mol %)
ii) aq. citric acid
3
12
13
yield: up to 98%, dr: up to 86:14 (anti : syn), ee: up to 76% (anti)

Scheme 4

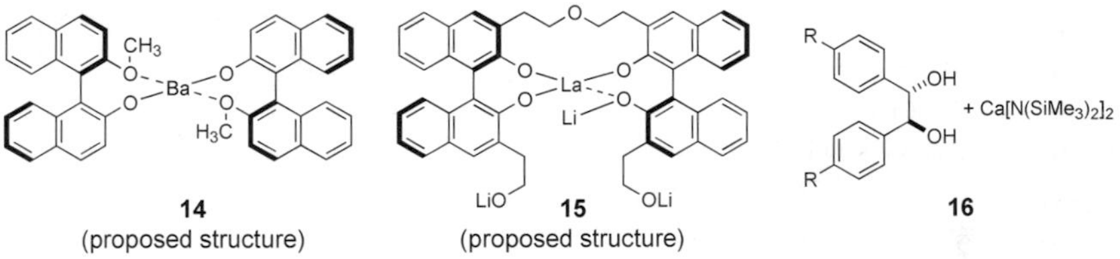
14
(proposed structure)
15
(proposed structure)
+ Ca[N(SiMe3)2]2
16

Scheme 5

LLB, LiOH, and H_2O promoted the direct aldol reaction of glycinate Schiff bases **12** with aldehydes **3**, providing access to β-hydroxy-α-amino acid esters **13** (Scheme 4, bottom) [7].

Shibasaki et al. also developed a barium complex (BaB-M, **14**, Scheme 5) for the aldol reaction of acetophenone (**1a**), making use of the strongly basic characteristic of barium alkoxide. The catalyst was prepared from $Ba(O\text{-}i\text{-}Pr)_2$ and BINOL monomethyl ether, and the products were obtained in excellent yield with up to 70% ee (Scheme 6) [8]. Shibasaki et al. attempted to incorporate a strong Brønsted base into the catalyst and developed a lanthanide heterobimetallic catalyst (**15**) possessing lithium alkoxide moieties, which promoted the aldol reaction with up to 74% ee (Scheme 6) [9]. Noyori and Shibasaki et al. reported a calcium alkoxide catalyst (**16**) that was prepared from $Ca[N(SiMe_3)_2]_2$,

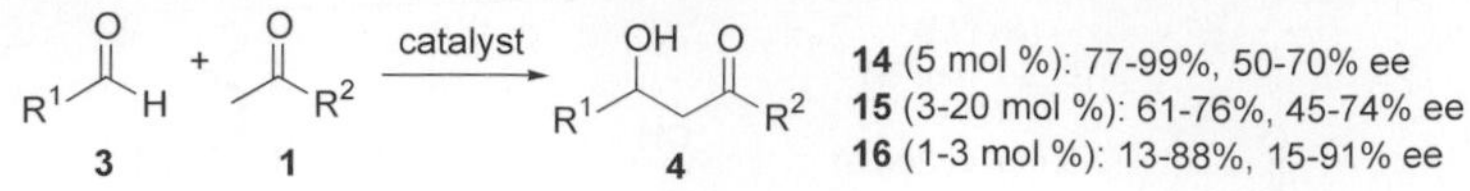
catalyst
3
1
4
14 (5 mol %): 77-99%, 50-70% ee
15 (3-20 mol %): 61-76%, 45-74% ee
16 (1-3 mol %): 13-88%, 15-91% ee

Scheme 6

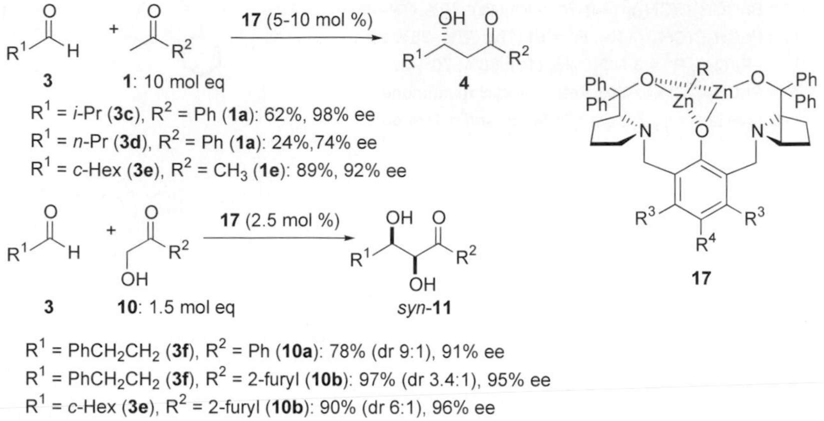
17 (5-10 mol %)
3
1: 10 mol eq
4
R1 = i-Pr (3c), R2 = Ph (1a): 62%, 98% ee
R1 = n-Pr (3d), R2 = Ph (1a): 24%,74% ee
R1 = c-Hex (3e), R2 = CH3 (1e): 89%, 92% ee
17 (2.5 mol %)
3
10: 1.5 mol eq
syn-11
17
R1 = PhCH2CH2 (3f), R2 = Ph (10a): 78% (dr 9:1), 91% ee
R1 = PhCH2CH2 (3f), R2 = 2-furyl (10b): 97% (dr 3.4:1), 95% ee
R1 = c-Hex (3e), R2 = 2-furyl (10b): 90% (dr 6:1), 96% ee

Scheme 7

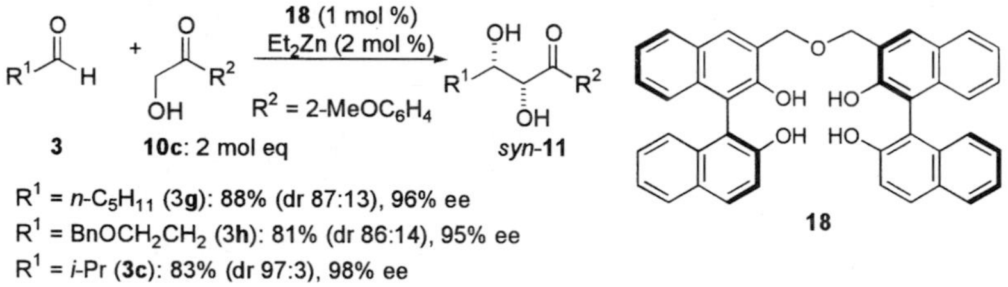
18 (1 mol %)
Et2Zn (2 mol %)
R2 = 2-MeOC6H4
3
10c: 2 mol eq
syn-11
18
R1 = n-C5H11 (3g): 88% (dr 87:13), 96% ee
R1 = BnOCH2CH2 (3h): 81% (dr 86:14), 95% ee
R1 = i-Pr (3c): 83% (dr 97:3), 98% ee

Scheme 8

hydrobenzoin, and KSCN, with sufficient activity, even with only 1 mol % loading (Scheme 6) [10].

Trost et al. [11] reported another impressive example of bimetallic catalysts in which a Zn–Zn homobimetallic complex (**17**, Scheme 7) serves as an effective catalyst for direct aldol reactions [11–13]. The proposed structure of the catalyst was verified by mass spectrometry and the best ratio of Et_2Zn and the ligand. The chemical yield was moderate in the reaction of methyl ketones (**1**) (Scheme 7, top) [11, 12], but a highly atom-economic system was achieved when α-hydroxylated ketones (**10**) were used as a substrate (Scheme 7, bottom) [13]. Excellent diastereo- and enantioselectivity were obtained under mild conditions. In contrast to the case of Shibasaki's heteropolymetallic catalyst, *syn*-1,2-diols (*syn*-**11**) were obtained as the major diastereomers.

Shibasaki and colleagues achieved further efficient catalysis using a zinc complex as a catalyst [5, 14], which was prepared from Et_2Zn and linked-BINOL

(18) [15] (Scheme 8). The system is applicable to a wide variety of aldehydes, and a series of *syn*-α,β-dihydroxyketones (*syn*-11) can be synthesized in excellent yield. Moreover, the diastereo- and enantioselectivity are almost perfect in some cases using as little as 1 mol % of the catalyst. The introduction of a methoxy group at the *ortho* position of 2-hydroxyacetophenone has crucial effects on the reactivity and selectivity. The structure of the real catalytic species is of great interest and is currently under investigation [16].

3
Metal-free Organocatalysis

Aminocatalysis is a biomimetic strategy used by enzymes such as class I aldolases. Application of aminocatalysis in an asymmetric aldol reaction was reported in the early 1970s. Proline (19) efficiently promoted an intramolecular direct aldol reaction to afford Wieland–Miescher ketone in 93% ee [17, 18]. More than 25 years later, in 2000, List, Barbas, and co-workers reported that proline (19) is also effective for intermolecular direct aldol reactions of acetone (1e) and various aldehydes 3. Notably, the reaction proceeded smoothly in anhydrous DMSO at an ambient temperature to afford aldol adducts in good yield and in modest to excellent enantioselectivity (up to >99% ee, Scheme 9) [19–22]. The chemical yields and selectivity of proline catalysis are comparable to the best metallic catalysts, although high catalyst loading (30 mol %) is required. Proline (19)

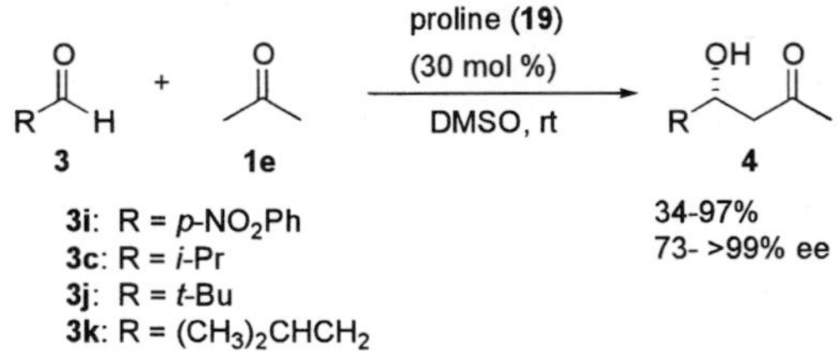

Scheme 9

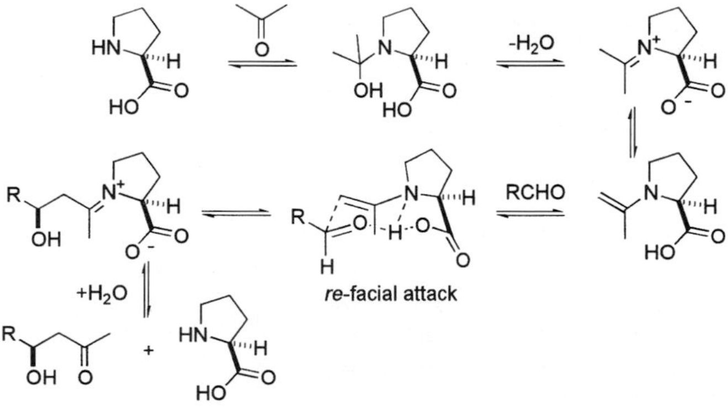

Scheme 10

also promoted direct aldol reactions of cyclic ketones such as cyclopentanone (**20**) and cyclohexanone (**21**) to give *anti*-aldol adducts selectively [22]. Based on mechanistic investigations, the enamine mechanism is postulated, in which one proline molecule might be involved and functions as a class I aldolase mimic (Scheme 10) [20, 22]. Screening of organocatalysts suitable for direct aldol reactions of **1e** and aldehyde **3i** revealed two other effective catalysts, 5,5-dimethylthiazolidinium-4-carboxylate (**22**) [20] and diamine-protonic acid catalyst **23** [23], which were reported by Barbas and Yamamoto, respectively (Fig. 1). Both **22** and **23** promoted direct aldol reactions to afford good results comparable with proline.

Important extensions of proline catalysis in direct aldol reactions were also reported. Pioneering work by List and co-workers demonstrated that hydroxyacetone (**24**) effectively serves as a donor substrate to afford *anti*-1,2-diol **25** with excellent enantioselectivity (Scheme 11) [24]. The method represents the first catalytic asymmetric synthesis of *anti*-1,2-diols and complements the asymmetric dihydroxylation developed by Sharpless and other researchers (described in Chap. 20). Barbas utilized proline to catalyze asymmetric self-aldolization of acetaldehyde [25]. Jørgensen reported the cross aldol reaction of aldehydes and activated ketones like diethyl ketomalonate, in which the aldehyde

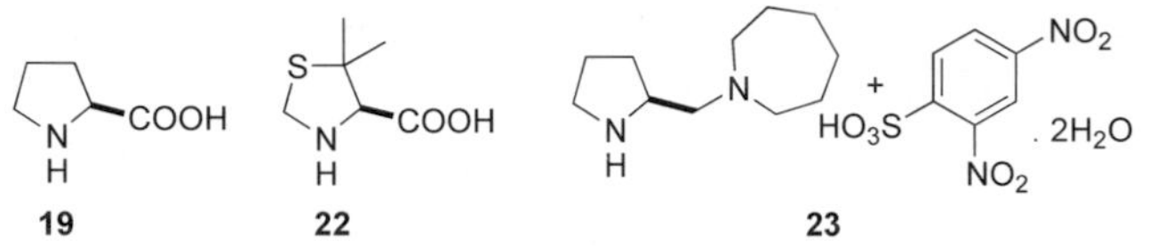
COOH
COOH
HO_3S
NO_2
NO_2
. $2H_2O$
19
22
23

Fig. 1

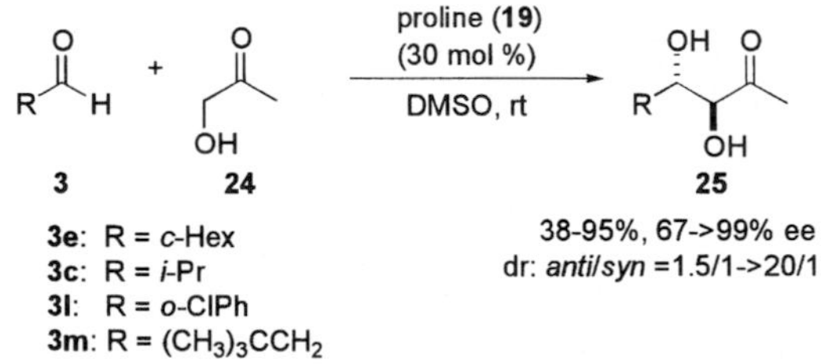
proline (19)
(30 mol %)
DMSO, rt
3
24
25
3e: R = c-Hex
3c: R = i-Pr
3l: R = o-ClPh
3m: R = $(CH_3)_3CCH_2$
38-95%, 67->99% ee
dr: anti/syn =1.5/1->20/1

Scheme 11

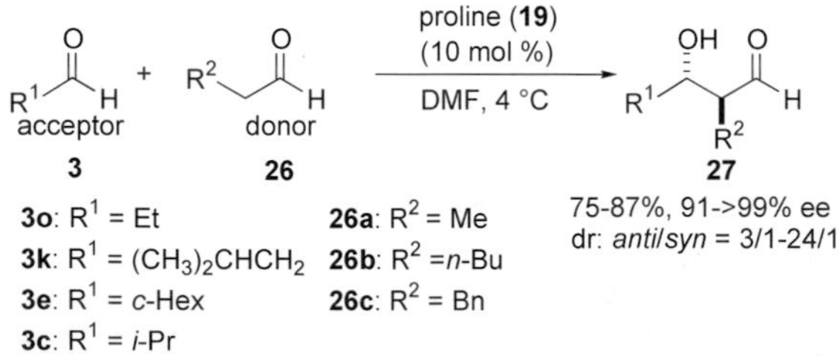
proline (19)
(10 mol %)
DMF, 4 °C
acceptor
donor
3
26
27
3o: R^1 = Et
3k: R^1 = $(CH_3)_2CHCH_2$
3e: R^1 = c-Hex
3c: R^1 = i-Pr
26a: R^2 = Me
26b: R^2 = n-Bu
26c: R^2 = Bn
75-87%, 91->99% ee
dr: anti/syn = 3/1-24/1

Scheme 12

functions as a donor and the ketone as an acceptor [26]. MacMillan succeeded in the first cross-aldol reaction of aldehydes, which has not yet been achieved by metallic catalysis. With 10 mol % proline (**19**), the reaction of aldehyde **26a** with aldehyde **3c** proceeded smoothly in DMF at 4 °C to afford cross-aldol adducts *anti*-selectively (24/1) in 82% yield and greater than 99% ee (Scheme 12) [27].

4
Future Perspective

Two different strategies to achieve direct catalytic asymmetric aldol reactions are summarized. The field has grown rapidly over the past five years and many researchers are now investigating this "classical" yet new paradigm using metallic catalysts and organocatalysts. Application of these systems to direct Mannich reactions and Michael reactions are also being studied intensively. Future investigation is necessary to overcome problems in substrate generalities, reaction time, catalyst loading, volumetric productivity, etc.

References

1. Nakagawa M, Nakao H, Watanabe K-I (1985) Chem Lett 391
2. Yamada YMA, Yoshikawa N, Sasai H, Shibasaki M (1997) Angew Chem Int Ed Engl 36: 1871
3. Shibasaki M, Yoshikawa N (2002) Chem Rev 102:2187
4. Yoshikawa N, Yamada YMA, Das J, Sasai H, Shibasaki M (1999) J Am Chem Soc 121:4168
5. Yoshikawa N, Kumagai N, Matsunaga S, Moll G, Ohshima T, Suzuki T, Shibasaki M (2001) J Am Chem Soc 123:2466
6. Yoshikawa N, Suzuki T, Shibasaki M (2002) J Org Chem 67:2556
7. Yoshikawa N, Shibasaki M (2002) Tetrahedron 58:8989
8. Yamada YMA, Shibasaki M (1998) Tetrahedron Lett 39:5561
9. Yoshikawa N, Shibasaki M (2001) Tetrahedron 57:2569
10. Suzuki T, Yamagiwa N, Matsuo Y, Sakamoto S, Yamaguchi K, Shibasaki M, Noyori R (2001) Tetrahedron Lett 42:4669
11. Trost BM, Ito H (2000) J Am Chem Soc 122:12003
12. Trost BM, Silcoff ER, Ito H (2001) Org Lett 3:2497
13. Trost BM, Ito H, Silcoff ER (2001) J Am Chem Soc 123:3367
14. Kumagai N, Matsunaga S, Yoshikawa N, Ohshima T, Shibasaki M (2001) Org Lett 3:1539
15. Matsunaga S, Ohshima T, Shibasaki M (2002) Adv Synth Catal 344:3
16. Kumagai N, Matsunaga S, Kinoshita T, Harada S, Okada S, Sakamoto S, Yamaguchi K, Shibasaki M (2003) J Am Chem Soc 125:2169
17. Eder U, Sauer G, Wiechert R (1971) Angew Chem Int Ed Engl 10:496
18. Hajos ZG, Parrish DR (1974) J Org Chem 39:1615
19. List B, Lerner RA, Barbas CFIII (2000) J Am Chem Soc 122:2395
20. Sakthivel K, Notz W, Bui T, Barbas CF III (2001) J Am Chem Soc 123:5260
21. List B, Pojarliev P, Castello C (2001) Org Lett 3:573
22. List B (2001) Synlett 1675
23. Saito S, Nakadai M, Yamamoto H (2001) Synlett 1245
24. Notz W, List B (2000) J Am Chem Soc 122:7386
25. Córdova A, Notz W, Barbas CF III (2002) J Org Chem 67:301
26. Bøgevig A, Kumaragurubaran N, Jørgensen KA (2002) Chem Commun 620
27. Northrup AB, MacMillan DWC (2002) J Am Chem Soc 124:6798

Chapter 29.5
Mannich Reaction

Shū Kobayashi, Masaharu Ueno

Graduate School of Pharmaceutical Sciences, The University of Tokyo, Hongo, Bunkyo-ku, Tokyo 113–0033, Japan
e-mail: skobayas@mol.f.u-tokyo.ac.jp

Keywords: Mannich reaction, β-Amino carbonyl compounds, Imine, Enolate

1
Introduction

Asymmetric Mannich reactions provide useful routes for the synthesis of optically active β-amino ketones or esters, which are versatile chiral building blocks for the preparation of many nitrogen-containing biologically important compounds [1–6]. While several diastereoselective Mannich reactions with chiral auxiliaries have been reported, very little is known about enantioselective versions. In 1991, Corey et al. reported the first example of the enantioselective synthesis of β-amino acid esters using chiral boron enolates [7]. Yamamoto et al. disclosed enantioselective reactions of imines with ketene silyl acetals using a Brønsted acid-assisted chiral Lewis acid [8]. In all cases, however, stoichiometric amounts of chiral sources were needed. Asymmetric Mannich reactions using small amounts of chiral sources were not reported before 1997. This chapter presents an overview of catalytic asymmetric Mannich reactions.

2
Catalytic Asymmetric Mannich Reactions Using Lewis Acid Catalysts

Although asymmetric reactions using chiral Lewis acids are of great current interest as one of the most efficient methods for the preparation of chiral compounds, examples using imines as electrophiles are rare compared to those using aldehydes. This is due to two main difficulties; first, many Lewis acids are de-activated or sometimes decomposed by the nitrogen atoms of starting materials or products, and even when the desired reactions proceed, more than stoichio-metric amounts of the Lewis acids are needed because the acids are trapped by the nitrogen atoms. Second, imine–chiral Lewis acid complexes are rather flexi-ble and often have several stable conformers (including E/Z-isomers of imines), while aldehyde–chiral Lewis acid complexes are believed to be rigid. Therefore, in the additions to imines activated by chiral Lewis acids, plural transition states would exist and the selectivities are decreased.

In 1997, the first truly catalytic enantioselective Mannich reactions of imines with silicon enolates using a novel zirconium catalyst was reported [9, 10]. To solve the above problems, various metal salts were first screened in achiral reac-tions of imines with silylated nucleophiles, and then, a chiral Lewis acid based on Zr(IV) was designed. On the other hand, as for the problem of the confor-mation of the imine–Lewis acid complex, utilization of a bidentate chelation was planned; imines prepared from 2-aminophenol were used [(Eq. (1)]. This moiety was readily removed after reactions under oxidative conditions. Imi-nes derived from heterocyclic aldehydes worked well in this reaction, and good to high yields and enantiomeric excesses were attained. As for aliphatic alde-hydes, similarly high levels of enantiomeric excesses were also obtained by us-ing the imines prepared from the aldehydes and 2-amino-3-methylphenol. The present Mannich reactions were applied to the synthesis of chiral β-amino alco-hols from α-alkoxy enolates and imines [11], and *anti*-α-methyl-β-amino acid derivatives from propionate enolates and imines [12] via diastereo- and enan-tioselective processes [(Eq. (2)]. Moreover, this catalyst system can be utilized in Mannich reactions using hydrazone derivatives [13] [(Eq. (3)] as well as the aza-Diels–Alder reaction [14–16], Strecker reaction [17–19], allylation of imi-nes [20], etc.

(1)

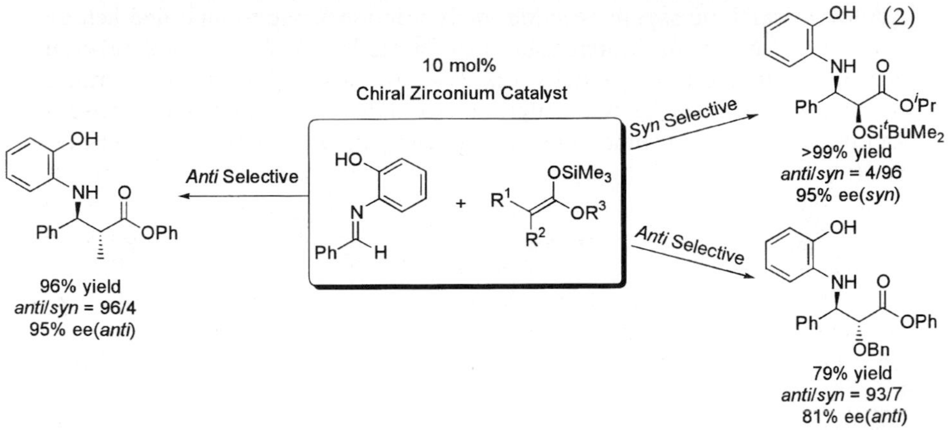

10 mol%
Chiral Zirconium Catalyst

Anti Selective

Syn Selective

Anti Selective

(2)

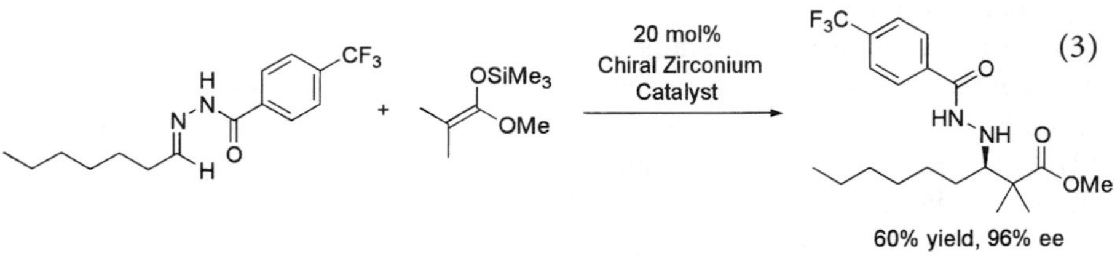

Ph

OH
NH O
O^{i}Pr
OSi^{t}BuMe$_2$
>99% yield
anti/syn = 4/96
95% ee(syn)

Ph

OH
NH O
OPh
OBn
79% yield
anti/syn = 93/7
81% ee(anti)

Ph

OH
NH O
OPh
96% yield
anti/syn = 96/4
95% ee(anti)

HO

N
Ph H

+ R^1

OSiMe$_3$
OR^3
R^2

20 mol%
Chiral Zirconium
Catalyst

(3)

60% yield, 96% ee

Use of 2,2′-diphenyl-(3,3′-biphenanthrene)-4,4′-diol(VAPOL) instead of BINOL in a similar zirconium system was reported [21].

The vinylogous Mannich reaction of triisopyloxyfurans with aldimines prepared from aldehydes and 2-aminophenol proceeded with moderate selectivity in the presence of a catalytic amount of a Ti(IV)–BINOL complex [22].

In the presence of water-free late transition metalphosphine cation complexes as Lewis acids, glyoxylatetosylamine imine reacted with silicon enolates stereoselectivity [23–26]. It was proposed that imine coordinated to the metal such as Ag(I), Pd(II), and Cu(I) in a bidentate manner [23]. The copper-based catalyst was the most effective, and the desired product was obtained in high yields with high enantioselectivities [(Eq. (4)].

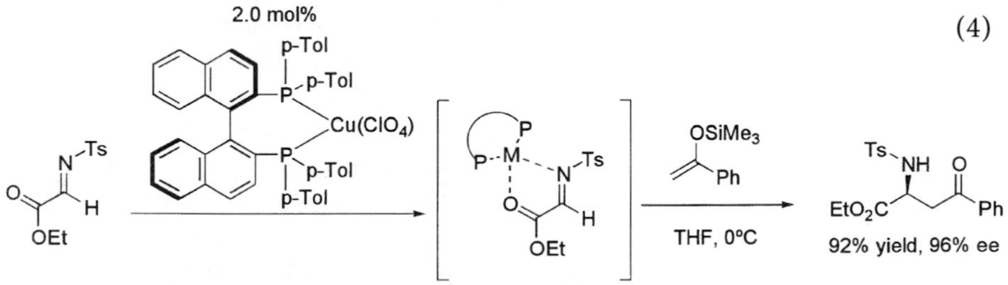

(4)

2.0 mol%
p-Tol
p-Tol
P
Cu(ClO$_4$)
p-Tol
p-Tol

OSiMe$_3$
Ph

THF, 0°C

Ts
NH O
EtO$_2$C Ph
92% yield, 96% ee

A direct catalytic asymmetric Mannich reaction using unmodified ketones was reported using cooperative catalysis of a AlLibis((R)-binaphthoxide) complex ((R)-ALB) and La(OTf)$_3$·nH$_2$O [27, 28]. It was also reported that enantioselective and diastereoselective catalytic nitro-Mannich reactions of N-phosphinoylimines proceeded smoothly using the complex of ALB and tert-BuOK [29, 30] [(Eq. (5)].

$$(5)$$

It was also reported that diastereo- and enantioselective Mannich reactions of activated carbonyl compounds with α-imino esters were catalyzed by a chiral Lewis acid derived from Cu(OTf)$_2$ and a bisoxazoline (BOX) ligand [31] [(Eq. (6)]. Catalytic enantioselective addition of nitro compounds to imines [32], and aza-Henry reactions of nitronates with imines [33] also proceeded under similar reaction conditions.

$$(6)$$

A convenient method for the preparation of N-acylamino esters using a polymer-supported amine has been developed, and catalytic asymmetric Mannich reactions have been successfully performed using a chiral copper catalyst prepared from copper triflate and a chiral diamine ligand [34]. This reaction proceeded smoothly in high yield and excellent ee, not only with silicon enolates derived from ketones, esters, and thioesters as for enolate components, but also with alkyl vinyl ethers. In the reactions with alkyl vinyl ethers, the initial adducts were vinyl ethers, which were converted to the corresponding amino acid derivative under acidic conditions. A possible mechanism for the formation of vinyl ethers is the [4+2]-cycloaddition of N-acylimino esters to vinyl ethers followed by proton transfer. It was also reported that a new inhibitor of ceramide trafficking from endoplasmic reticulum to the site of sphingomyelin (SM) synthesis, (1R,3R)-N-(3-hydroxy-1-hydroxymethyl-3-phenylpropyl)dodecanamide (HPA-12), was efficiently synthesized using this method (Scheme 1).

In recent years, organic reactions in aqueous media have attracted a great deal of attention, not only because these reactions eliminate the necessity for vigorous drying of solvents and substrates, but also because unique reactiv-

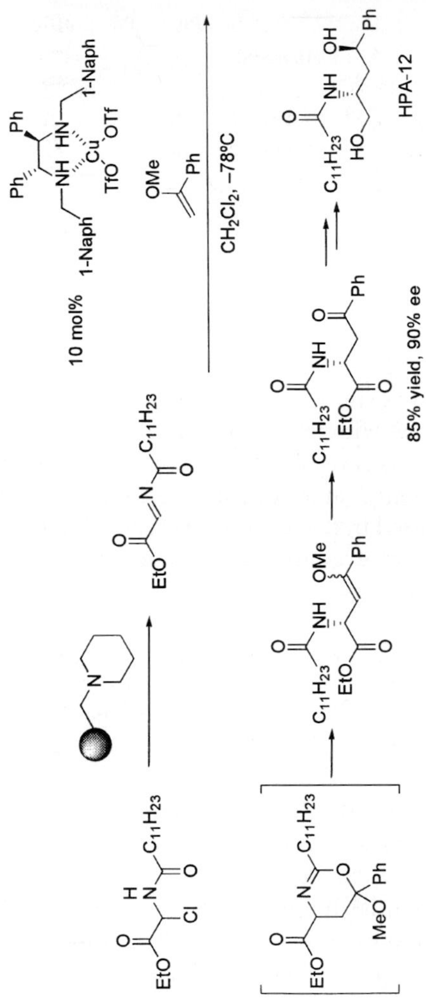

Scheme 1

ity and selectivity are often observed in aqueous reactions. The first reported example of a Mannich reaction in aqueous media proceeded smoothly using a combination of ZnF_2 and a chiral diamine [(Eq. (7)] [35]. In this reaction, a catalytic amount of TfOH dramatically increaced the yield. It is assumed that this reaction proceeds with double activation where Zn^{2+} acts as a Lewis acid to activate the hydrazono ester and the fluoride anion acts as a Lewis base to attack the silicon atom of the enolate. The N–N bond of the hydrazine was easily cleaved by SmI_2.

$$\text{(7)}$$

The reaction scheme for Eq. (7) is shown with reactants imine (EtO-C(=O)-CH=N-NHBz) and silyl enol ether (OSiMe₃, Ph), with 10 mol% chiral diamine ligand (Ph-NH HN-Ph with Ph substituents), 50 mol% ZnF₂, 1 mol% TfOH, in H₂O/THF = 1/9, giving product BzHN-NH-CH(CO₂Et)-CH₂-C(=O)-Ph, **89% yield, 92% ee**.

3
Catalytic Asymmetric Mannich Reactions via Metal Enolates

The catalytic asymmetric Mannich reaction of lithium enolates with imines was reported in 1997 using an external chiral ligand [36]. First, it was found that reactions of lithium enolates with imines were accelerated by addition of external chiral ligands. Then, it was revealed that reactions were in most cases accelerated in the presence of excess amounts of lithium amides. A small amount of a chiral source was then used in the asymmetric version [(Eq. (8)], and chiral ligands were optimized to achieve suitable catalytic turnover [37].

$$\text{(8)}$$

Reaction of imine (Ph-CH=N-(p-MeO)Ph) with lithium enolate (OLi, 2.0 equiv), 20 mol% chiral ligand (Ph-CH-CH-Ph with MeO and OMe, Li), 2.4 equiv ^{i}Pr-N(^{c}Hex)-Li, Toluene, -78°C, 20 h, giving β-lactam product ((p-MeO)Ph-N, Ph), **80% yield, 75% ee**.

In 1998, a new type of Pd(II) binuclear complex was reported which was effective for Mannich reactions of an imine derived from glyoxylate and anisidine with silicon enolates [38, 39]. In these reactions, use of solvents including a small amount of water was essential. It was shown that water played an important role in this system: water not only activated the Pd(II) complex to generate a cation complex, but also cleaved the N–Pd bond of the intermediate to regenerate the chiral catalyst. This reaction reportedly proceeded via an optically active palladium enolate on the basis of NMR and ESIMS analyses. A unique binuclear palladium-sandwiched enolate was obtained in the reaction of the μ-hydroxo palladium complex with the silyl enol ether [(Eq. (9)].

$$\text{(9)}$$

Reaction of imine (^{i}PrO₂C-CH=N-(p-MeO)Ph) with silyl enol ether (OSiMe₃, Ph), 5 mol% Chiral Pd-catalyst, DMF, 25°C, giving product ((p-MeO)Ph-NH-CH(^{i}PrO₂C)-CH₂-C(=O)-Ph), **95% yield, 90% ee**.

Chiral Pd-catalyst = binuclear palladium complex with p-Tol substituents on phosphorus, Pd⁺ centers bridged by OH groups, (BF₄⁻)₂.

4
Miscellaneous

It was reported that proline catalyzed the direct catalytic asymmetric Mannich reactions of hydroxyacetone, aldehydes, and aniline derivatives [(Eq. (10)] [40–44]. Not only aromatic aldehydes but also aliphatic aldehydes worked well in this reaction, and good to excellent enantioselectivity and moderate to excellent yields were observed. Mannich reactions of glyoxylate imines with aldehydes or ketones were also successfully performed [45, 46].

$$(10)$$

References

1. Hart DJ, Ha D-C (1989) Chem Rev 89:1447
2. Kleinmann EF (1991) In: Trost BM, Fleming I (eds) Comprehensive organic synthesis, vol 2. Pergamon, Oxford, p 893
3. Juaristi E (ed) (1997) Enantioselective synthesis of β-amino acids. VCH, Weinheim
4. Arend M, Westermann B, Risch N (1998) Angew Chem Int Ed Engl 37:1044
5. Kobayashi S, Ishitani H (1999) Chem Rev 99:1069
6. Denmark SE, Nicaise OJ-C (1999) In: Jacobsen EN, Pfaltz A, Yamamoto H (eds) Comprehensive asymmetric catalysis, vol 2. Springer, Berlin Heidelberg New York, p 923
7. Corey EJ, Decicco CP, Newbold RC (1991) Tetrahedron Lett 39:5287
8. Ishihara K, Miyata M, Hattori K, Tada T, Yamamoto H (1994) J Am Chem Soc 116:7153
9. Ishitani H, Ueno M, Kobayashi S (1997) J Am Chem Soc 119:7153
10. Ishitani H, Ueno M, Kobayashi S (2000) J Am Chem Soc 122:8180
11. Kobayashi S, Ishitani H, Ueno M (1998) J Am Chem Soc 120:431
12. Kobayashi S, Kobayashi J, Ishitani H, Ueno M (2002) Chem Eur J 8:4185
13. Kobayashi S, Hasegawa Y, Ishitani H (1998) Chem Lett 1131
14. Kobayashi S, Komiyama S, Ishitani H (1998) Angew Chem Int Ed Engl 37:979
15. Kobayashi S, Kusakabe K, Komiyama S, Ishitani H (1999) J Org Chem 64:4220
16. Kobayashi S, Kusakabe K, Ishitani H (2000) Org Lett 2:1225
17. Ishitani H, Komiyama S, Kobayashi S (1998) Angew Chem Int Ed Engl 37:3186
18. Ishitani H, Komiyama S, Hasegawa Y, Kobayashi S (2000) J Am Chem Soc 122:762
19. Kobayashi S, Ishitani H (2000) Chirality 12:540
20. Gastner T, Ishitani H, Akiyama R, Kobayashi S (2001) Angew Chem Int Ed Engl 40:1896
21. Xue S, Yu S, Deng Y, Wulff WD (2001) Angew Chem Int Ed Engl 40:2271
22. Martin SF, Lopez OD (1999) Tetrahedron Lett 40:8949
23. Ferraris D, Young B, Dudding T, Lectka T (1998) J Am Chem Soc 120:4548
24. Ferraris D, Young B, Cox C, Drury WJ III, Dudding T, Lectka T (1998) J Org Chem 63:6090
25. Ferraris D, Dudding T, Young B, Drury WJ III, Lectka T (1999) J Org Chem 64:2168
26. Ferraris D, Young B, Cox C, Dudding T, Drury WJ III, Ryzhkov L, Taggi AE, Lectka T (2002) J Am Chem Soc 124:67
27. Yamasaki S, Iida T, Shibasaki M (1999) Tetrahedron Lett 40:307

28. Yamasaki S, Iida T, Shibasaki M (1999) Tetrahedron 55:8857
29. Yamada K-I, Harwood SJ, Gröger H, Shibasaki M (1999) Angew Chem Int Ed Engl 38: 3504
30. Yamada K-I, Moll G, Shibasaki M (2001) Synlett 980
31. Juhl K, Gathergood N, Jørgensen KA (2001) Angew Chem Int Ed Engl 40:2995
32. Nishiwaki N, Knudsen KR, Gothelf KV, Jørgensen KA (2001) Angew Chem Int Ed Engl 40: 2992
33. Knudsen KR, Risgaard T, Nishiwaki N, Gothelf KV, Jørgensen KA (2001) J Am Chem Soc 123:5843
34. Kobayashi S, Matsubara R, Kitagawa H (2002) Org Lett 4:143
35. Kobayashi S, Hamada T, Manabe K (2002) J Am Chem Soc 124:5640
36. Fujieda H, Kanai M, Kambara T, Iida A, Tomioka K (1997) J Am Chem Soc 119:2060
37. Tomioka K, Fujieda H, Hayashi S, Hussein MA, Kambara T, Nomura Y, Kanai M, Koga K (1999) Chem Commun 715
38. Hagiwara E, Fujii A, Sodeoka M (1998) J Am Chem Soc 120:2474
39. Fujii A, Hagiwara E, Sodeoka M (1999) J Am Chem Soc 121:5450
40. List B (2000) J Am Chem Soc 122:9336
41. List B (2001) Synlett 1675
42. List B, Pojarliev P, Biller WT, Martin HJ (2002) J Am Chem Soc 122:827
43. Notz W, Sakthivel K, Bui T, Zhong G, Barbas CF III (2001) Tetrahedron Lett 42:199
44. Sakthivel K, Notz W, Bui T, Barbas CF III (2001) J Am Chem Soc 123:5260
45. Córdova A, Notz W, Zhong G, Betancort JM, Barbas CF III (2002) J Am Chem Soc 124: 1842
46. Córdova A, Watanabe S-I, Tanaka F, Notz W, Barbas CF III (2002) J Am Chem Soc 124: 1866

Supplement to Chapter 31.2
Catalytic Conjugate Addition of Stabilized Carbanions

Masahiko Yamaguchi

Department of Organic Chemistry, Graduate School of Pharmaceutical Sciences, Tohoku University, Aoba, Sendai 980–8578, Japan
e-mail: yama@mail.pharm.tohoku.ac.jp

Keywords: Michael addition, Amine, Phase transfer catalyst, Phenoxide, Crown ether, Transition metal, Lewis acid

1
Introduction

Since the previous review in this comprehensive book [1], the asymmetric catalytic conjugate additions of stabilized carbanions have made considerable advances. Publications between 1996 and early 2002 are summarized in this supplementary according to the same categorization as before. Several reviews have also appeared on this subject [2–4].

2
Amine Catalysts

Use of proline as a catalyst has become an important methodology in the catalytic asymmetric addition of stabilized carbanions to conjugated carbonyl compounds. Hannessian employed *L*-proline (*S*)-**1** in the addition of nitroalkanes to enones (Scheme 1) [5]. In the presence of 3–7 mol % of (*S*)-**1** and an excess of *trans*-2,5-dimethylpiperazine in chloroform, comparable or higher enantioselectivities were attained compared to the Yamaguchi's method using *L*-proline rubidium salt. An unusual nonlinear effect was observed in the catalysis. List and Enders used (S)-1 (15–20 mol %) in the addition of ketones to nitroalkenes, and obtained optically active *syn*-isomers predominantly (Scheme 2) [6, 7]. Enders obtained higher selectivity by using methanol as solvent. Barbas employed (S)-(N-morpholinomethyl)pyrrolidine [(S)-2] (20 mol %) derived from L-proline for the addition of aldehydes to nitroalkenes (Scheme 3) [8, 9]. The diamine (S)-2 exhibits higher catalytic activity and enantioselectivity compared to proline (S)-1, and an enamine was suggested as the reactive intermediate. Brunner re-examined the reaction of 1-indanone-2-carboxylate and nitroalkenes in the presence of cinchona alkaloids [10]. The related asymmetric addition of N-alkylpyrroles to unsaturated aldehydes is catalyzed by benzylimidazolidone trifluoroacetic acid salt (S)-3 (20 mol %) and proceeds with high enantioselectivity. MacMillan suggested that this process involves an iminium intermediate (Scheme 4) [11].

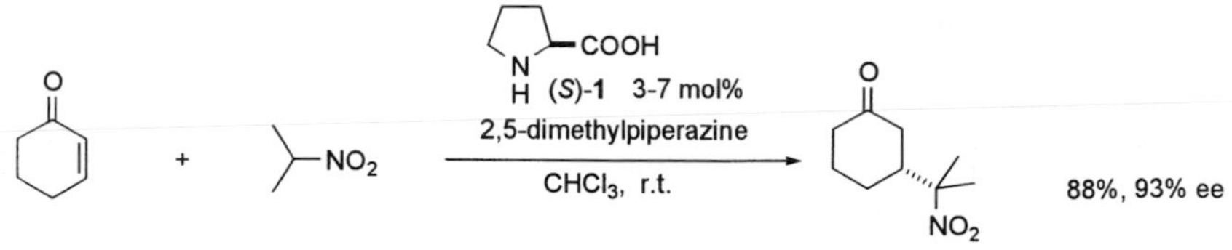

Scheme 1

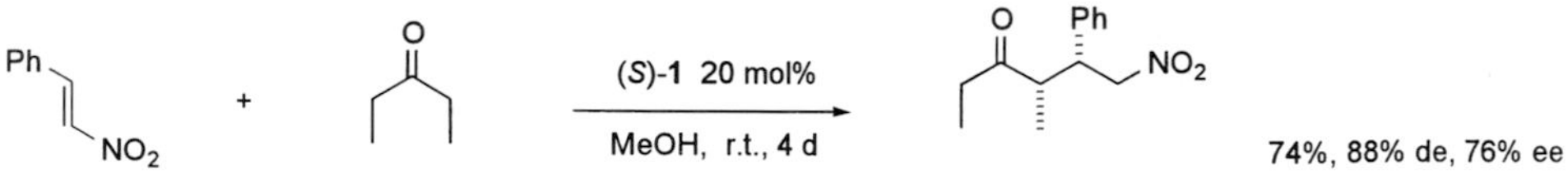

Scheme 2

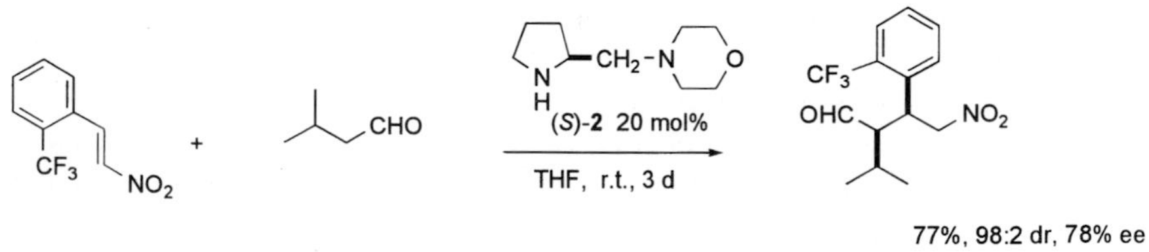

Scheme 3

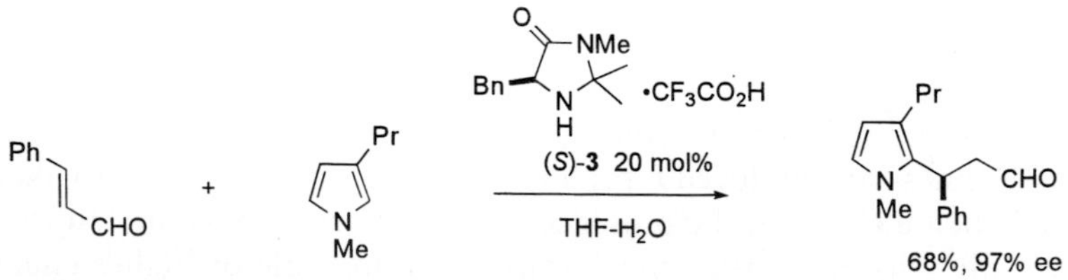

Scheme 4

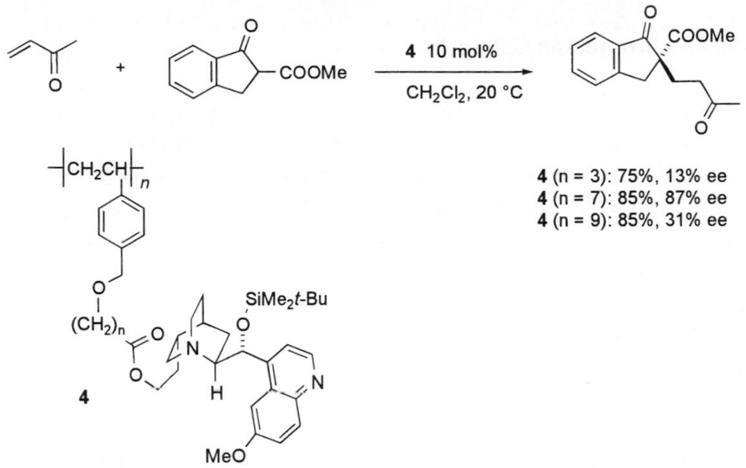

Scheme 5

Polymer catalysts containing cinchona alkaloids were re-examined by d'Angelo for the reaction of 1-indanone-2-carboxylate and methyl vinyl ketone, in which the structure of the spacer connecting the base moiety to the Merrifield resin dramatically influenced the enantioselectivity (Scheme 5) [12]. Catalyst 4 (*n*=7) with a 7-atom-length spacer to quinine exhibits 87% ee, while 4 (*n*=3) with a 5-atom spacer and 4 (*n*=9) with an 11-atom spacer gave only 13% and 31% ee, respectively.

3
Phase Transfer Catalysts

Corey employed a cinchona alkaloid-derived ammonium salt 5 for the solid–liquid phase transfer catalyst, and attained 99% ee in the addition of a glycine-derived imine to 2-cyclohexenone (Scheme 6) [13, 14].

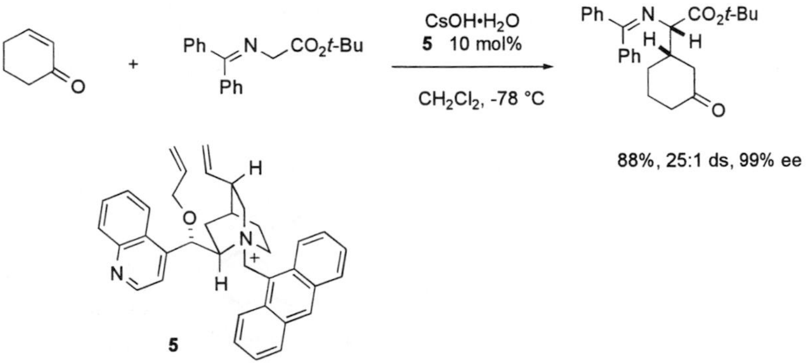

Scheme 6

4
Alkoxide and Phenoxide Catalysts

Shibasaki made several improvements in the asymmetric Michael addition re-action using the previously developed BINOL-based (R)-ALB, (R)-6, and (R)-LPB, (R)-7 [1]. The former is prepared from (R)-BINOL, diisobutylaluminum hydride, and butyllithium, while the latter is from (R)-BINOL, La(Oi-Pr)₃, and potassium t-butoxide. Only 0.1 mol % of (R)-6 and 0.09 mol % of potassium t-butoxide were needed to catalyze the addition of dimethyl malonate to 2-cy-clohexenone on a kilogram scale in >99% ee, when 4-Å molecular sieves were added [15,16]. (R)-6 in the presence of sodium t-butoxide catalyzes the asym-metric 1,4-addition of the Horner–Wadsworth–Emmons reagent [17]. (R)-7 cat-alyzes the addition of nitromethane to chalcone [18]. Feringa prepared another aluminum complex from BINOL and lithium aluminum hydride and used this in the addition of nitroacetate to methyl vinyl ketone [19]. Later, Shibasaki de-veloped a linked lanthanum reagent (R,R)-8 for the same asymmetric addition, in which two BINOLs were connected at the 3-positions with a 2-oxapropylene

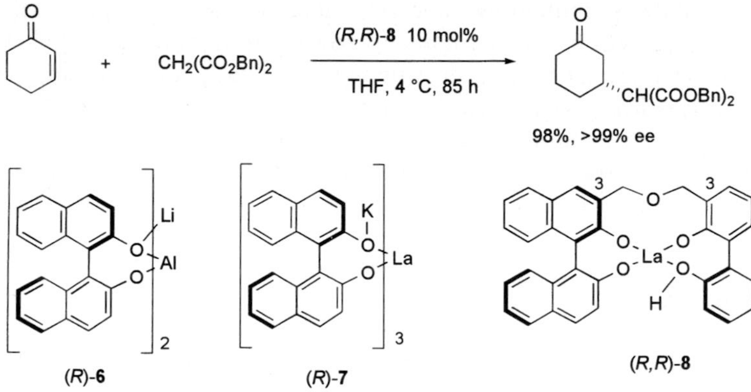

Scheme 7

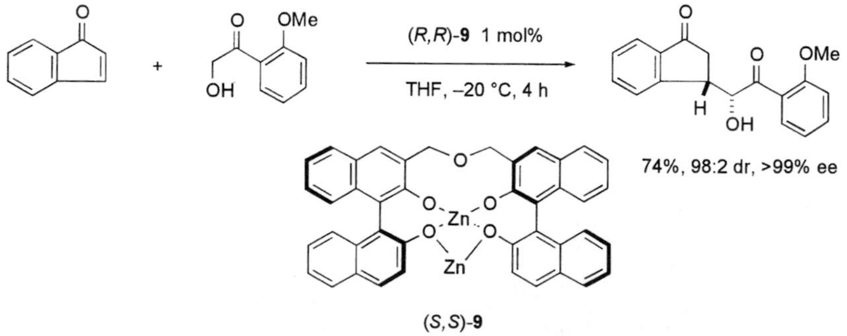

Scheme 8

group (Scheme 7) [20]. This storable and reusable catalyst can be immobilized on an insoluble polymer [21]. In the presence of 1,1,1,3,3,3-hexafluoro-2-propanol, the catalyst (R,R)-**8** is effective even for the asymmetric addition of α-alkyl-malonates [22]. Narasimhan developed an aminophenol ligand for the Michael addition [23]. A zinc complex (S,S)-**9** prepared by treating the linked BINOL with diethylzinc was reported by Shibasaki to catalyze the addition of hydroxy-acetophenone to enones (Scheme 8) [24].

5
Crown Ether/Alkali Metal Base Catalysts

Toke continued to study sugar-derived crown ethers for the catalysis of asymmetric Michael addition reactions. Enantiomeric excesses exceeding 80% were attained for the addition of 2-nitropropane to chalcone by using glucose-derived crown ethers **10** or **11** in the presence of potassium t-butoxide (Scheme 9) [25–27]. The substituents on the nitrogen atom play an important role both on the chemical yield and optical yield, although they are located at a remote position from the chiral centers. In the case of **10**, the phenylethyl derivative exhibited much higher ee than the benzyl derivative; in case of **11**, the unsubstituted derivative gave satisfactory results compared to the phenylethyl derivative. Deracemization of α-phenylglutarate in the presence of potassium t-butoxide and a chiral crown ether was also examined [28].

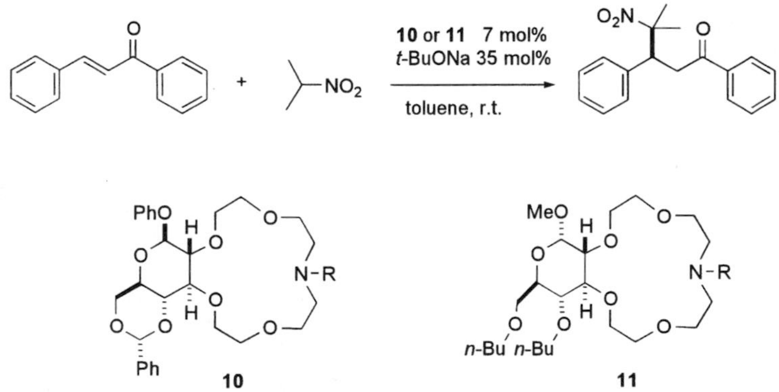

R = PhCH$_2$CH$_2$: 78%, 84% ee
R = PhCH$_2$: 56%, 6% ee

R = H: 82%, 90% ee
R = PhCH$_2$CH$_2$: 43%, 42% ee

Scheme 9

6
Transition Metal Catalysts

Transition metal catalysts containing cobalt or nickel metal exhibit high ees. Cristoffers and Brunner examined the transition metal catalysts for the addition of 2-oxocycloalkanecarboxylates to methyl vinyl ketone using various combinations of metal acetates and ligands including diamines, amino alcohols, and mercapto phosphines [29–33]. Cristoffers found (R,R)-1,2-diaminocyclohexane [(R,R)-12] in the presence of Ni(OAc)$_2$ to be effective (Scheme 10) [32]. Pfaltz attained a high ee in the addition of malonate to chalcone by using a combination of (S,S)-bisoxazoline (S,S)-13 and Co(OAc)$_2$, although the chemical yield was low (Scheme 11) [34]. Kozlowski employed a salen BINOL ligand complexed to Ni for the asymmetric addition of dibenzylmalonate to 2-cyclohexenone in 71% ee [35]. Nozaki and Takaya developed a rhodium complex derived from a novel binaphthylbiphosphine for the enantioselective addition of cyanoacetate to methyl vinyl ketone [36]. A platinum complex derived from oxazolinophosphine was used by Williams in the same reaction [37].

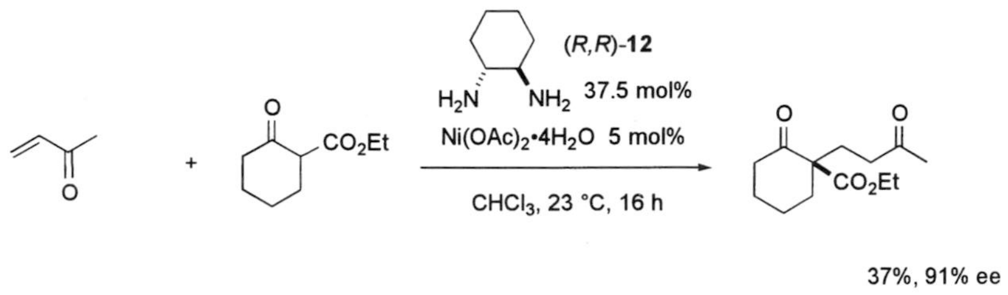

Scheme 10

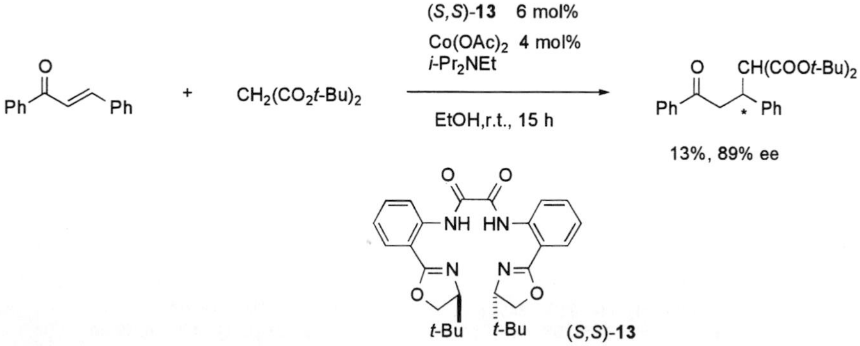

Scheme 11

7
Lewis Acid Catalysts

Bernadi and Scolastico, and later Evans in a more effective manner, indicated that the enantioselective addition reaction using silyl enol ethers can be catalyzed by Lewis acidic copper(II) cation complexes derived from bisoxazolines [38–40]. In the presence of the copper complex (*S,S*)-**14** (10 mol %), silyl enol ethers derived from thioesters add to alkylidenemalonates or 2-alkenoyloxazolidone in high ees (Scheme 12). Bernadi, Scolastico, and Seebach employed a titanium complex derived from TADDOL for the addition of silyl enol ethers to nitroalkenes or 2-cyclopentenone [41–43], although these are stoichiometric reactions.

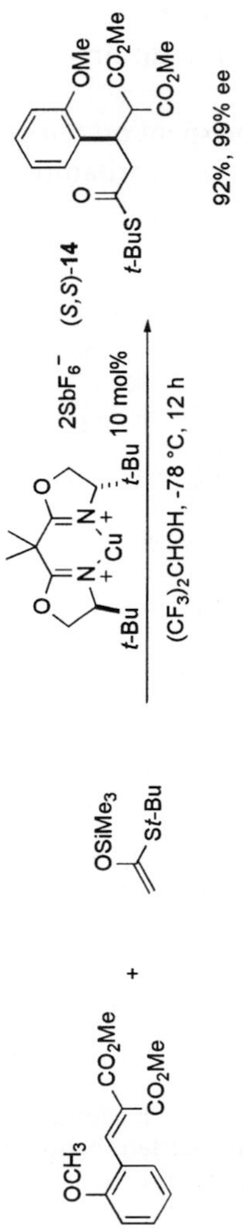

Scheme 12

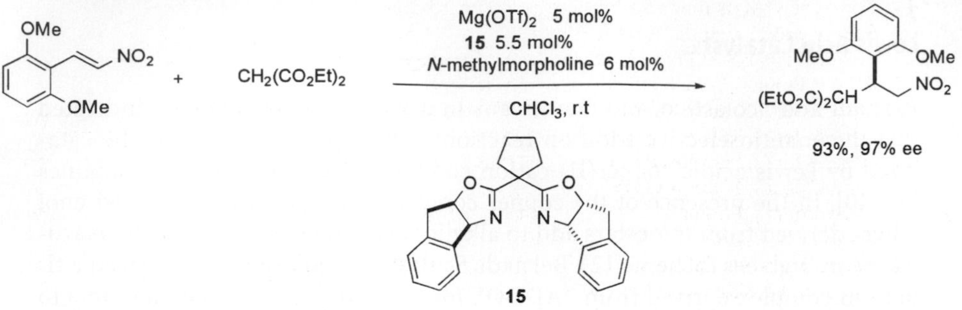

Scheme 13

A combination of chiral oxazoline **15** and magnesium triflate was employed by Ji and Barnes in the asymmetric addition of keto esters and malonates to nitroalkenes (Scheme 13) [44].

More efforts will continue to be spent on the development of effective asymmetric addition reactions of stabilized carbanions employing various catalysts.

References

1. Yamaguchi M (1999) In: Jacobsen EN, Pfaltz A, Yamamoto H (eds) Comprehensive asymmetric catalysis, vol III. Springer, Berlin Heidelberg New York, p 1121
2. Sibi MP, Manyem S (2000) Tetrahedron 56:8033
3. Krause N, Hoffmann-Röder A (2001) Synthesis 171
4. Christoffers J (1998) Eur J Org Chem 1259
5. Hanessian S, Pham V (2000) Org Lett 2:2975
6. Enders D, Seki A (2002) Synlett 26
7. List B, Pojarliev P, Martin HJ (2001) Org Lett 3:2423
8. Betancort JM, Barbas CF III (2001) Org Lett 3:3737
9. Betancort JM, Sakthivel K, Thayumanavan R, Barbas CF III (2001) Tetrahedron Lett 42: 4441
10. Brunner H, Kimel B (1996) Monat Chem 127:1063
11. Paras NA, MacMillan DWC (2001) J Am Chem Soc 123:4370
12. Alvarez R, Hourdin M-A, Cavé C, d'Angelo J, Chaminade P (1999) Tetrahedron Lett 40: 7091
13. Corey EJ, Noe MC, Xu F (1998) Tetrahedron Lett 39:5347
14. Zhang F-Y, Corey EJ (2000) Org Lett 2:1097
15. Shimizu S, Ohori K, Arai T, Sasai H, Shibasaki M (1998) J Org Chem 63:7547
16. Xu Y, Ohori K, Ohshima T, Shibasaki M (2002) Tetrahedron 58:2585
17. Arai T, Sasai H, Yamaguchi K, Shibasaki M (1998) J Am Chem Soc 120:441
18. Funabashi K, Saida Y, Kanai M, Arai T, Sasai H, Shibasaki M (1998) Tetrahedron Lett 39: 7557
19. Keller E, Veldman N, Spek AL, Feringa BL (1997) Tetrahedron: Asymmetry 8:3403
20. Kim YS, Matsunaga S, Das J, Sekine A, Ohshima T, Shibasaki M (2000) J Am Chem Soc 122: 6506
21. Matsunaga S, Ohshima T, Shibasaki M (2000) Tetrahedron Lett 41:8473
22. Takita R, Ohshima T, Shibasaki M (2002) Tetrahedron Lett 43:4661
23. Narasimhan S, Velmathi S, Balakumar R, Radhakrishnan V (2001) Tetrahedron Lett 42: 719

24. Kumagai N, Matsunaga S, Shibasaki M (2001) Org Lett 3:4251
25. Bakó P, Szöllosy A, Bombicz P, Toke L (1997) Synlett 291
26. Bakó P, Vizvárdi K, Bajor Z, Toke L (1998) Chem Commun 1193
27. Bakó P, Bajor Z, Toke L (1999) Perkin Trans 1 3651
28. Toke L, Bakó P, Keseru GM, Albert M, Fenichel L (1998) Tetrahedron 54:213
29. Christoffers J, Rößler U (1999) Tetrahedron: Asymmetry 10:1207
30. Christoffers J, Mann A, Pickardt J (1999) Tetrahedron 55:5377
31. Christoffers J, Mann A (1999) Eur J Org Chem 1475
32. Christoffers J, Rößler U, Werner T (2000) Eur J Org Chem 701
33. Brunner H, Krumey C (1999) J Mol Cat A 142:7
34. End N, Macko L, Zehnder M, Pfaltz A (1998) Chem Eur J 4:818
35. DiMauro EF, Kozlowski MC (2001) Org Lett 3:1641
36. Inagaki K, Nozaki K, Takaya H (1997) Synlett 119
37. Blacker AJ, Clarke ML, Loft MS, Mahon MF, Williams JMJ (1999) Organometallics 18:
    2867
38. Bernardi A, Colombo G, Scolastico C (1995) Tetrahedron Lett 37:8921
39. Evans DA, Rovis T, Kozlowski MC, Tedrow JS (1999) J Am Chem Soc 121:1994
40. Evans DA, Willis MC, Johnston JN (1999) Org Lett 1:865
41. Seebach D, Lyapkalo IM, Dahinden R (1999) Helv Chim Acta 82:1829
42. Bernardi A, Karamfilova K, Boschin G, Scolastico C (1995) Tetrahedron Lett 36:1363
43. Bernardi A, Karamfilova K, Sanguinetti S, Scolastico C (1997) Tetrahedron 53:13009
44. Ji J, Barnes DM, Zhang J, King SA, Wittenberger SJ, Morton HE (1999) J Am Chem Soc 121:
    10215

Supplement to Chapter 34.1
Alkylation of Enolates

David L. Hughes

Merck and Co., Inc., PO BOX 2000, Mail Drop R800-B275, Rahway, NJ 07065, USA
e-mail: Dave_Hughes@Merck.com

Keywords: Alkylation, enolate, phase transfer, Cinchona alkaloid, arylation, chiral amide base, desymmetrization, allylic alkylation, NOBIN, binaphthyl

1
Phase Transfer Catalysis

In 1997 the Corey [1] and Lygo [2] groups disclosed the use of *N*-(anthracenyl)methyl-modified Cinchona alkaloids (e.g., **1**) as catalysts in phase transfer alkylations, which afforded remarkable enantiomeric excesses of up to 99%. During the ensuing years, these groups have expanded the scope and limitations of these catalysts, as summarized below.

The Lygo group has shown that alkylation using *o*, *m*, and *p*-bis-(bromomethyl)benzene gives the desired bis-alkylated products in 95% ee [3]. Enantioselective alkylation of alanine-derived imines provide α,α-disubstituted amino acids in up to 87% ee [4]. A survey of solvents revealed that similar enantiomeric excesses were obtained in dichloromethane, toluene, *t*-BuOMe, and diethyl ether. The catalyst could be readily recovered from toluene by simply filtering the reaction mixture through a short column of magnesium sulfate, and eluting with a more polar solvent such as chloroform [5]. A survey of alkylating agents showed that alkyl, allyl, and benzyl-type bromides and iodides gave good ee, but *t*-butyl bromoacetate gave a lower ee, primarily due to an increase in the uncatalyzed reaction [5]. The Lygo group also carried out a structure-selectivity study by synthesizing a series of catalysts with one major portion altered. The conclusion was that the *N*-anthracenylmethyl group is essential in providing high ee in the alkylation reaction. In addition, the 1-quinolyl group also plays a key role in enantioselectivity [6].

The Corey group has used the anthracenyl-modified catalyst to prepare a number of novel α-amino acids [7] and has applied the technology to a non-imine enolate **2** with excellent results [8]. Increasing electron-donation in the substituents of **2** leads to higher ee, in line with the hypothesis that tight ion pairs are responsible for the enantioselectivity [8].

While the majority of work with these catalysts has been carried out by the Lygo and Corey groups, a few other groups have made important contributions. O'Donnell has shown that the Corey/Lygo catalysts can be used under homogeneous conditions along with the strong neutral phosphazene bases developed by Schwesinger [9]. Use of homogeneous solvent conditions has allowed implementation of this technology to the solid-phase synthesis of α-amino acid derivatives [10]. Takemoto has developed an all-water solvent system with these catalysts using Triton X–100 to form micelles, which has allowed catalyst loadings to be reduced to 0.1% in the best case [11].

Perhaps spurred by the outstanding ee and versatility obtained with the anthracenylmethyl Cinchona alkaloids developed by Lygo and Corey [12], numerous groups have designed and developed new phase transfer catalysts over the past four years. Tables 1–3 list the new catalysts that have been designed between 1998 and 2001, along with the substrates employed and the best enantioselectivities obtained. The catalysts in Table 1 are based on the binaphthyl template, those in Table 2 on Chinchona alkaloid templates, and those in Table 3 are miscellaneous structures.

Two groups have developed effective immobilized Chinchona alkaloid phase transfer catalysts, with a connection to a polymer support either through the *N*-benzyl group or an *O*-benzyl group [13–15].

Table 1. Chiral phase transfer catalysts based on the binaphthyl template developed between 1998 and 2001

Catalyst	Substrate	Best ee's	Reference
(NOBIN)		68%	16
Same		97%	17
		68%	18
		29%	19
R = Ph, beta-naphthyl, C_6F_5		96%	20

Table 2. Chiral phase transfer catalysts derived from Cinchona alkaloids developed between 1998 and 2001

Catalyst	Substrate	Best ee's	Reference
		91%	21
		32%	22
	$\underset{Ph}{\overset{Ph}{>}}=N\diagdown CO_2t\text{-}Bu$	95%	23
	$\underset{Ph}{\overset{Ph}{>}}=N\diagdown CO_2t\text{-}Bu$	97%	24

Table 3. Other chiral phase transfer catalysts developed between 1998 and 2001

Catalyst	Substrate	Best ee's	Reference
		50%	25
(TADDOL)		93%	26
		90%	27
		81%	28

2
Transition Metal-Catalyzed Asymmetric Allylic Alkylations of Enolates

Significant progress has been made on the asymmetric Pd-catalyzed allylic alkylation of prochiral enolates, with a number of ligands now available that provide products with high ee. Trost was the first to demonstrate that high enantiomeric excesses were capable with ketoester substrates [29]; now asymmetric allylic alkylation of ketoesters and simple ketone substrates has been achieved in several more cases. Table 4 summarizes the ligands, substrates, and ee for recent examples.

Table 4. Catalysts for asymmetric allylic alkylations developed between 1998 and 2001

Catalyst	Substrates	Best ee's	Reference
BINAP		88-95%	30-31
		95%	32
(c-C$_6$H$_{11}$)$_2$N—...—S—p-Tol		50%	33
		72-99%	34-36
		95%	37
		64%	38
N-Anthracenylmethyl Cinchona alkaloids		61 - 96%	39-40

3
Transition Metal-Catalyzed Asymmetric Arylation of Enolates

Buchwald has designed a hindered dialkylphosphino–binaphthyl ligand (3) that is much more active than the original ligand for asymmetric arylation of ketone enolates. Reactions occur at room temperature using only 2 mol % catalyst with enantioselectivities up to 94% [41]. Additionally, the Buchwald group has developed an electron-rich monodentate ligand (4) capable of vinylation of ketone enolates with up to 92% ee [42].

Hartwig has reported the asymmetric intramolecular arylation of amides using a chiral carbene ligand (5) with up to 76% ee [43].

4
Asymmetric Alkylation of Enolates Using Chiral Ligands

Koga has continued his research program in the enantioselective alkylations of achiral lithium enolates using chiral ligands. Previous examples had provided excellent ee and yields for tertiary carbon centers; ligand 6 has now been designed that allows formation of quaternary carbon centers with high ee [44].

5
Deprotonation by Chiral Amide Bases

Deprotonation of carbonyl compounds by chiral amide bases followed by trapping with silylating agents or aldehydes has become a common method for desymmetrizing prochiral and conformationally locked 4-substituted cyclohexanones and bicyclic ketones. The literature through 1997 has been reviewed [45].

Recent advances include the use of new chiral bases, extention to substrates other than ketones, and trapping with electrophiles other than silylating reagents and aldehydes. Regarding alternate substrates and electrophiles, the Simpkins group reported alkylation of a prochiral diester with common alkyl halides with >98% ee [46]. Simpkins and coworkers have also demonstrated desymmetrization of cyclic imides, in this case with trapping by silyl groups [47].

Novel chiral bases introduced recently include the magnesium amide base **7** and analogs reported by Henderson [48] and Knochel's urea base **8** (mono and di-anions) [49].

References

1. Corey EJ, Xu F, Noe MC (1997) J Am Chem Soc 119:12414
2. Lygo B, Crosby J, Lowdon TR, Wainwright PG (1997) Tetrahedron Lett 38:2343
3. Lygo B, Crosby J, Peterson JA (1999) Tetrahedron Lett 40:1385; Lygo B (1999) Tetrahedron Lett 40:1389; Lygo B, Crosby J, Peterson JA (2001) Tetrahedron 57:6447
4. Lygo B, Crosby J, Peterson JA (1999) Tetrahedron Lett 40:8671
5. Lygo B, Crosby J, Lowdon TR, Peterson JA, Wainwright PG (2001) Tetrahedron 57:2403
6. Lygo B, Crosby J, Lowdon TR, Wainwright PG (2001) Tetrahedron 57:2391
7. Corey EJ, Noe MC, Xu F (1998) Tetrahedron Lett 39:5347
8. Corey EJ, Bo Y, Busch-Petersen (1998) J Am Chem Soc 120:13000
9. O'Donnell MJ, Delgado F, Hostettler C, Schwesinger R (1998) Tetrahedron Lett 39:8775
10. O'Donnell MJ, Delgado F, Pottorf RS (1999) Tetrahedron 55:6347
11. Okino T, Takemoto Y (2001) Org Lett 3:1515
12. As of the end of 2001, the Corey [1] and Lygo [2] papers have been cited by nearly 200 authors
13. Thierry B, Perrard T, Audouard C, Plaquevent JC, Cahard D (2001) Synthesis 1742
14. Thierry B, Plaquevent JC, Cahard D (2001) Tetrahedron: Asymmetry 12:983
15. Chinchilla R, Mazon P, Najera C (2000) Tetrahedron: Asymmetry 11:3277
16. Belokon YN, Kochetkov KA, Churkina TD, Ikonnikov NS, Vyskocil S, Kagan HB (1999) Tetrahedron: Asymmetry 10:1723
17. Belokon YN, Kochetkov KA, Churkina TD, Ikonnikov NS, Larionov OV, Harutyunyan SR, Vyskocil S, North M, Kagan HB (2001) Angew Chem Int Ed 40:1948
18. Casas J, Najera C, Sansano JM, Gonzalez J, Saa JM, Vega M (2001) Tetrahedron: Asymmetry 12:699
19. Ooi T, Kameda M, Maruoka K (1999) J Am Chem Soc 121:6519
20. Ooi T, Takeuchi M, Kameda M, Maruoka K (2000) J Am Chem Soc 122:5228
21. Arai S, Oku M, Ishida T, Shiori T (1999) Tetrahedron Lett 40:6785
22. Ducry L, Diederich F (1999) Helv Chim Acta 82:981
23. Jew S, Jeong B, Yoo M, Huh H, Park H (2001) Chem Comm 1244
24. Park H, Jeong B, Yoo M, Park M, Huh H, Jew S (2001) Tetrahedron Lett 42:4645
25. Manabe K (1998) Tetrahedron 54:14465; Manabe K (1998) Tetrahedron Lett 39:5807
26. Belokon YN, Kochetkov KA, Churkina TD, Ikonnikov NS, Chesnokov AA, Larionov OV, Parmer VS, Kumar R, Kagan HB (1998) Tetrahedron: Asymmetry 9:851; Belokon YN, Kochetkov KA, Churkina TD, Ikonnikov NS, Chesnokov AA, Larionov OV, Singh I, Parmer VS, Vyskocil S, Kagan HB (2000) J Org Chem 65:7041
27. Belokon YN, North M, Churkina TD, Ikonnikov NS, Maleev VI (2001) Tetrahedron 57:2491; Belokon YN, Davies RG, North M (2000) Tetrahedron Lett 41:7245
28. Belokon YN, North M, Kublitski VS, Ikonnikov NS, Krasik PE, Maleev VI (1999) Tetrahedron Lett 40:6105
29. Trost BM, Radinov R, Grenzer EM (1997) J Am Chem Soc 119:7879
30. Kuwano R, Ito Y (1999) J Am Chem Soc 121:3236
31. Kuwano R, Nishio R, Ito Y (1999) Org Lett 1:837
32. Brunel JM, Tenaglia A, Buono G (2000) Tetrahedron: Asymmetry 11:3585
33. Hiroi K, Suzuki Y, Abe I, Hasegawa Y, Suzuki K (1998) Tetrahedron: Asymmetry 9:3797
34. Trost BM, Schroeder GM (1999) J Am Chem Soc 121:6757
35. Trost BM, Ariza X (1999) J Am Chem Soc 121:10727
36. Trost BM, Schroeder GM (2000) J Org Chem 65:1569
37. You SL, Hou X, Dai L, Zhu X (2001) Org Lett 3:149

38. Kaneko S, Yoshino T, Katoh T, Terashima S (1998) Tetrahedron 54:5471
39. Chen G, Deng Y, Gong L, Mi A, Cui X, Jiang Y, Choi MCK, Chan ASC (2001) Tetrahedron:
    Asymmetry 12:1567
40. Nakoji M, Kanayama T, Okino T, Takemoto Y (2001) Org Lett 3:3329
41. Hamada T, Chieffi A, Ahman J, Buchwald SL (2002) J Am Chem Soc 124:1261
42. Chieffi A, Kamikawa K, Ahman J, Fox JM, Buchwald SL (2001) 3:1897
43. Lee S, Hartwig JF (2001) J Org Chem 66:3402
44. Yamashita Y, Odashima K, Koga K (1999) Tetrahedron Lett 40:2803
45. O'Brien P (1998) J Chem Soc Perkin Trans 1 1439
46. Goldspink NJ, Simpkins NS, Beckman M (1999) Synlett 8:1292
47. Adams DJ, Simpkins NS, Smith TJN (1998) Chem Commun 1605
48. Henderson KW, Kerr WJ, Moir JH (2000) Chem Commun 479; Anderson JD, Garcia PG,
    Hayes D, Henderson KW, Kerr WJ, Moir JH, Fondekar KP (2001) Tetrahedron Lett 42:7111;
    Henderson KW, Kerr WJ, Moir JH (2001) Synlett 1253
49. Graf CD, Malan C, Harms K, Knochel P (1999) J Org Chem 64:5581

Supplement to Chapter 39
Combinatorial Approaches

Amir H. Hoveyda, Kerry E. Murphy

Department of Chemistry, Merkert Chemistry Center, Boston College, Chestnut Hill, Massachusetts 02467, USA
e-mail: amir.hoveyda@bc.edu

Keywords: Asymmetric synthesis, Chiral catalysis, Diversity-based approaches, Supported chiral catalysts, Solid-phase chemistry

1
Introduction

During the past four years, after the first edition of this series was published, several studies regarding the development of new and effective catalytic enantioselective reactions have emerged that utilize diversity-based approaches as a key component. Many of these investigations have delivered unique conditions that afford optically enriched or pure products that are not readily accessible by alternative methods. Moreover, a number of structural features are emerging that have proved critical for successful application of high-throughput screening approaches to chiral catalyst discovery. It must also be noted that many of the libraries constructed have been designed based on mechanistic knowledge and intuition; it is the synergistic relationship between mechanism and diversity-based protocols that is the primary driving force in this important area of research [1].

Below, various investigations involving the development of new metal-catalyzed enantioselective reactions that utilize the principles of diversity-based approaches are highlighted. Several key studies have also been disclosed recently that correspond to identification of non-metal-based catalysts; notable advances in this area are highlighted as well.

2
Diversity-Based Approaches to the Discovery of Catalysts for Enantioselective Synthesis

2.1
Catalytic Asymmetric Strecker Synthesis of Amino Acids

In 1998, Jacobsen and Sigman demonstrated that peptide-based ligands, such as the one shown in Eq. (1), can be used to access optically enriched amino nitriles. The identity of the optimal catalyst was determined through examination of parallel libraries of catalyst candidates [2]. Later, it was demonstrated that this protocol may be extended to additions to ketoimines, affording tertiary amino nitriles in high enantioselectivities [3].

$$(1)$$

As the follow up to our studies in connection to the development of Ti-catalyzed cyanide additions to *meso* epoxides [4], we developed the corresponding catalytic enantioselective additions to imines [5]. A representative example is shown in Scheme 1; chiral non-racemic products may be readily converted to the derived α-amino acids (not available through catalytic asymmetric hydrogenation methods). In these studies, we further developed and utilized the positional optimization approach effected by examination of parallel libraries of amino acid-based chiral ligands (e.g., 1 and 2). Thus, the facile modularity of these ligands and their ease of synthesis were further exploited towards the development of a new catalytic enantioselective method that delivers various ar-

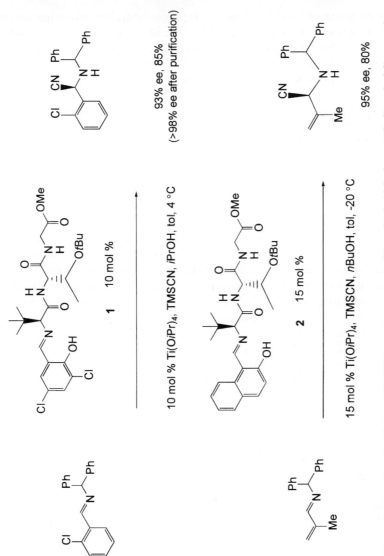

Scheme 1. Ti-catalyzed addition of TMSCN to imines promoted by salicyl aldehyde peptidic ligands delivers amino nitriles in high enantioselectivity

omatic and aliphatic amino acids. As before, in several cases, the identity of the optimal ligand varied depending on the substrate structure (see Scheme 1).

Optimal reaction conditions were established by the following systematic screening protocol: first, ten metal salts in five different solvents were screened in parallel in the presence of a salicyl aldehyde-derived peptidic chiral ligand (Val as AA1 Phe as AA2) and a representative imine substrate. The screening pointed to Ti(OiPr)$_4$ as the metal catalyst and toluene as the solvent that promote the most efficient and enantioselective C–C bond formations. Subsequent screening of seven cyanating agents highlighted TMSCN as optimal. Next, positional optimization of the chiral ligand was performed; these investigations indicated that *t*Leu and Thr(*t*Bu) as the AA1 and AA2 residues, respectively, afford the highest levels of asymmetric induction among the ligand structures examined. To enhance low catalyst turnovers, based on various mechanistic observations, we investigated the effect of a variety of protic additives (e.g., alcohols and amines) on reaction rates. This led us to establish that slow addition of *i*PrOH to the reaction mixtures results in significant increase in reaction efficiency without lowering of product optical purity [6].

2.2
Catalytic Asymmetric Alkylation of Imines

Detailed studies on the mechanism of the abovementioned Ti-catalyzed process (Scheme 1) led us to suggest a mechanistic model that underlines the significance of the peptidic moiety of this class of chiral ligands – not as passive providers of a chiral environment – but as active participants in the asymmetric bond-forming reactions. With such mechanistic paradigms as our guide, we set out to examine the possibility of developing a catalytic asymmetric alkylation of imines. Screening of parallel libraries was used to establish conditions for an effective set of protocols, including the optimal metal center, solvent, reaction temperature, and amine protecting group [7]. These screening studies indicated that Zr(O*i*Pr)$_4$ in combination with dipeptide amine ligand **3** (Scheme 2) (versus the corresponding Schiff base) delivers excellent reactivity and enantioselectivity. As was found with the Ti-catalyzed addition of cyanide to imines (Scheme 1), initial studies indicate that the AA2 moiety is crucial for achievement of both high reactivity and enantioselectivity.

The Zr-catalyzed asymmetric alkylation shown in Eq. (2) [8] illustrates two important principles: (1) The catalytic asymmetric protocol can be readily applied to the synthesis of non-aryl imines to generate homochiral amines that cannot be prepared by any of the alternative imine or enamine hydrogenation protocols. (2) The catalytic amine synthesis involves a three-component process that includes the in situ formation of the imine substrate, followed by its asymmetric alkylation. This strategy can also be readily applied to the preparation of arylamines. The three-component enantioselective amine synthesis suggests that such a procedure may be used to synthesize libraries of homochiral amines in a highly efficient and convenient fashion.

$$10 \text{ mol \% } \textbf{3},$$
$$10 \text{ mol \% Zr(O}i\text{Pr)}_4 \cdot \text{HO}i\text{Pr}$$
$$3 \text{ equiv Et}_2\text{Zn, toluene,}$$
$$0 \text{ to } 22 \,^\circ\text{C}$$

95% ee, 58%

(2)

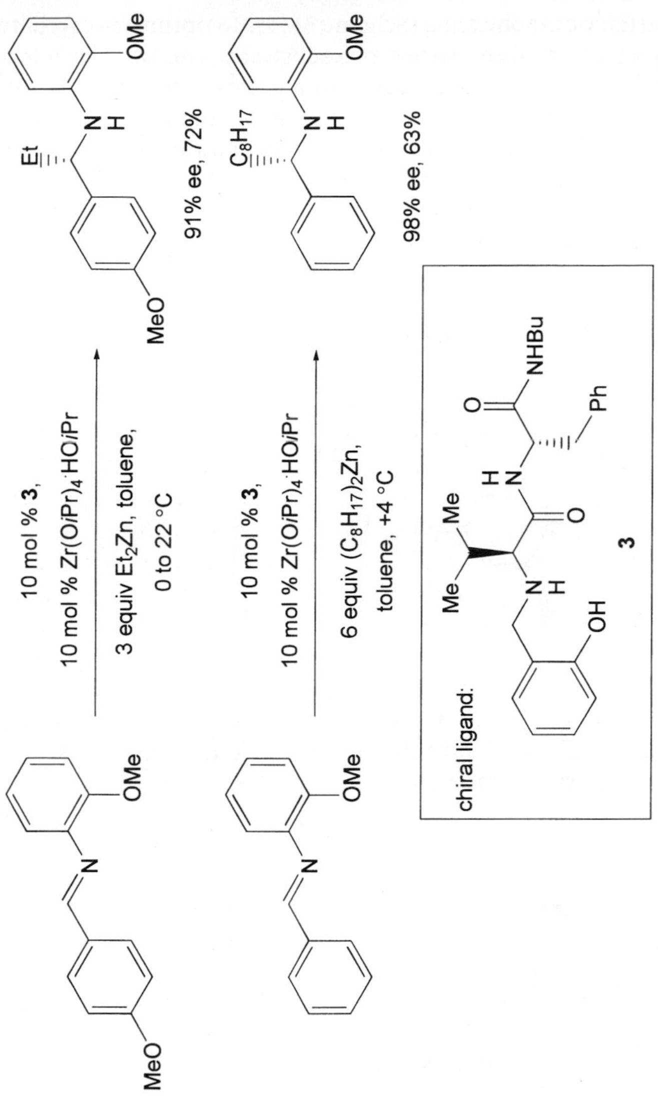

Scheme 2. With the peptide-based amine ligand a variety of alkylzincs can be added to imines in an enantioselective and catalytic manner

2.3
Catalytic Asymmetric Cyanide Additions to Ketones

The bifunctional mechanism proposed for the Ti-catalyzed Strecker suggests that electrophiles other than amines can be induced to undergo catalytic asymmetric additions with cyanide. The details of the optimal conditions (exactly what ligand structure, metal salt, solvent, and any needed additives) would however have to be identified through screening of parallel libraries as mentioned above. In this fashion, we have been able to develop an Al-catalyzed asymmetric synthesis of tertiary cyanohydrins (Scheme 3) [9]. To optimize reactivity, a parallel screening of twenty-three metals, two substrates, and two solvents was first performed using a representative ligand (salicyl aldehyde with Val as AA1 and Phe as AA2) attached to a solid support. Screening studies indicated Al(OiPr)$_3$ in toluene affords the most efficient C–C bond formations. Positional optimiza-

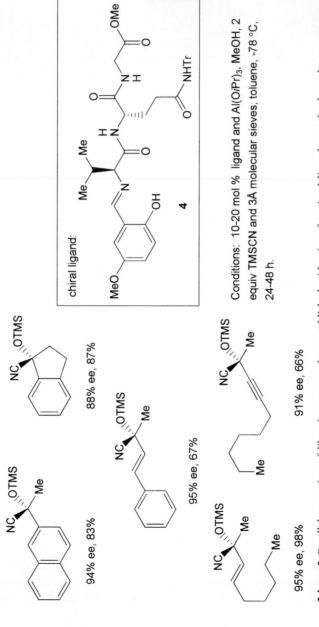

Scheme 3. Parallel screening of libraries was used to establish the identity of optimal ligand, metal salt, and any necessary additives in the development of Al-catalyzed asymmetric synthesis of tertiary cyanohydrins

tion of the ligand on solid support provided 5-methoxy-salicylaldehyde as the Schiff base, Val as AA1 and Gln(Trt) as AA2, giving 51% ee. Synthesis of the ligand in solution capped with a glycine methyl ester resulted in a 15% increase in enantioselectivity. As was the case with the abovementioned Strecker process, a range of additives were screened which led us to establish that the addition of two equivalents of 3 Å molecular sieves and 10 mol % MeOH enhances reaction efficiency and enantioselectivity. Three additional important points should be noted: (1) The chiral ligand **4** is prepared in six easy steps from commercially available materials in 75% overall yield. (2) Peptide **4** can be recovered for use in subsequent reactions to afford similar levels of reactivity and selectivity. (3) High enantioselectivities are observed even with ketones where there is not a significant steric difference between the two substituents (see Scheme 3).

2.4
Catalytic Asymmetric Conjugate Additions

Another important class of transformations that has benefited from diversity-based approaches are catalytic asymmetric conjugate additions to unsaturated carbonyls. Gennari and co-workers have reported on the utility of various chiral sulfonamides as ligands for Cu-catalyzed additions of Et_2Zn to cyclic enones [10]. A representative example is shown in Eq. (3). Although the levels of enantioselectivity attained are not competitive with other available methods involving six- and seven-membered ring enones [11], this disclosure indicated the influence of cooperative effects that can exist among various modules of the chiral ligand candidate.

$$(3)$$

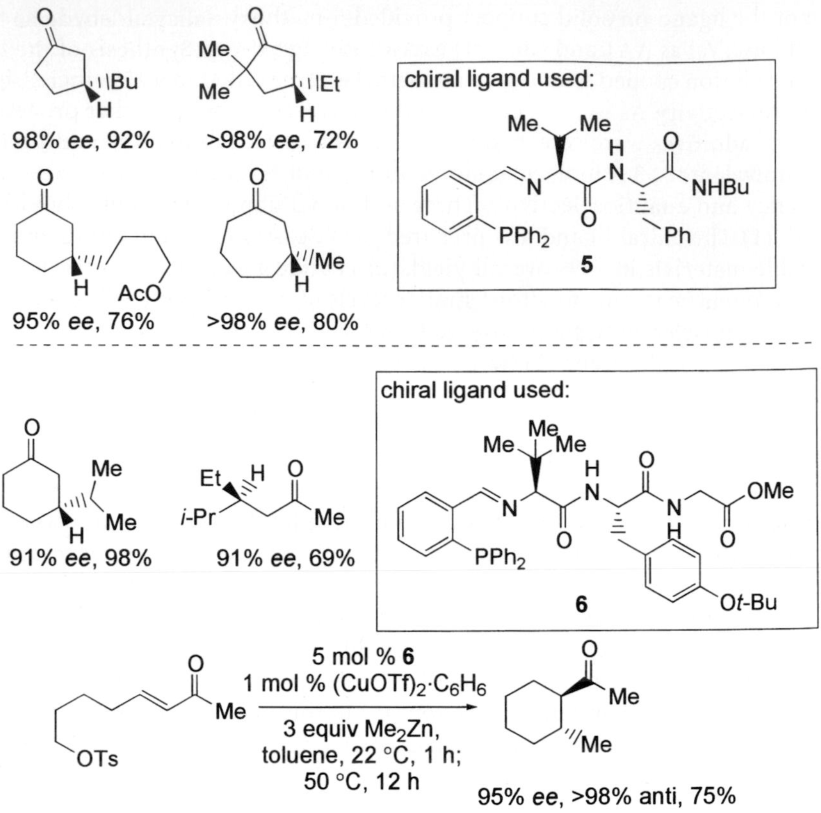

Scheme 4. Cu-catalyzed asymmetric conjugate additions to various cyclic and acyclic enones with peptidic phosphines as the chiral ligand

The possibility of using the modular Schiff base polypeptides as ligands that promote asymmetric conjugate additions to enones has been explored in our laboratories [12]. Screening of ligand, Cu salt, and solvent libraries indicated that phosphine-based dipeptides, such as 5 and 6 shown in Scheme 4, promote conjugate additions with excellent reactivity and efficiency in the presence of CuOTf. Use of a hydroxyl-containing ligand (e.g., 3 in Scheme 2) leads to the formation of racemic products. The Cu-catalyzed C–C bond-forming reactions can be effected with a wide range of dialkylzinc reagents (see Scheme 4). Most importantly, in contrast to the previously reported protocols, the present method provides an efficient and highly enantioselective conjugate addition protocol for cyclopentenones.

The ease of modularity of these peptide ligands proved critical in these studies as well. As an example, use of *i*Pr₂Zn as the alkylating agent leads to moderate levels of asymmetric induction when 5 is used as the chiral ligand. To address this selectivity problem, we set out to identify an improved chiral ligand through the positional optimization strategy to establish that reactions with *i*Pr$_2$Zn (such as the cyclohexenone adduct in Scheme 4) when promoted by **6** deliver products in higher enantioselectivity (91% ee versus 72% ee with **5**). Chiral peptide ligand **6** allows us to offer a solution to another important and long-standing problem in asymmetric catalysis [13]: effective enantioselective conjugate additions to acyclic substrates (see Scheme 4).

As shown in Eq. (4), parallel screening of ligand libraries has allowed us to establish that a closely related peptide-based phosphine ligand promotes the catalytic asymmetric conjugate addition of alkylzincs to nitroalkenes [14]. Not only are the corresponding alkyl nitrones obtained efficiently and in high diastereo- and enantioselectivity, appropriate acid workup can deliver the derived ketone directly.

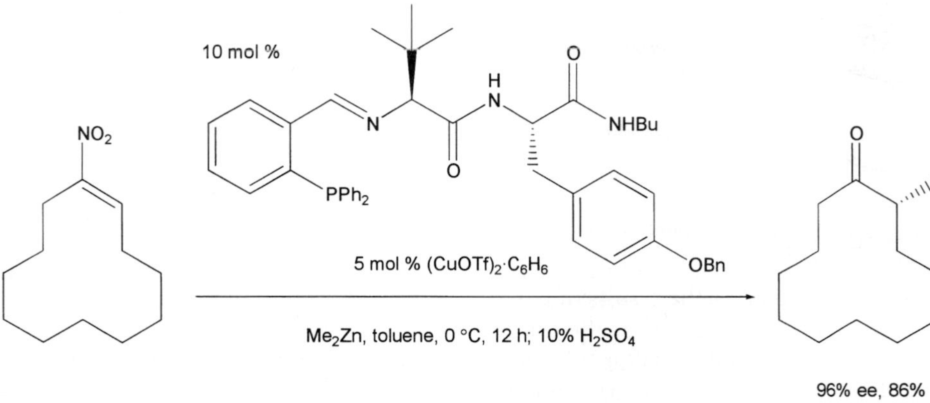

10 mol %

NO$_2$

PPh$_2$

5 mol % (CuOTf)$_2$·C$_6$H$_6$

Me$_2$Zn, toluene, 0 °C, 12 h; 10% H$_2$SO$_4$

NHBu

OBn

O

Me

96% ee, 86%

(4)

Miller and co-workers, as part of their program in connection to the development of polypeptides as practical and readily modular chiral catalysts and new screening protocols for combinatorial synthesis, have been able to achieve efficient catalytic enantioselective conjugate additions of azides to acyclic enones [15]. As the example in Eq. (5) illustrates, the resulting β-azidocarbonyls can be easily modified to afford N-containing heterocycles.

(5)

2.5
Catalytic Asymmetric Allylic Substitution Reactions

In the late nineties, the results of two studies were disclosed in which diversity-based approaches served as the centerpiece of the search performed to identify effective chiral ligands for Pd-catalyzed allylic substitutions (Scheme 5) [16]. Although these studies focused on one of the most extensively examined reactions in asymmetric catalysis and in the end do not offer a better alternative to the existing protocols, they helped establish several critical parameters regarding screening of parallel libraries of chiral ligand candidates. It is noteworthy that, in comparison to chiral ligands such as 5 and 6 (Scheme 4), coordinating phosphines are incorporated within the peptide framework in 8 in an entirely different fashion.

With the success of the peptide-based ligands in promoting the addition of alkylzinc reagents to imines, we set out to examine whether a similar process may be effected with olefinic starting materials. One class of reactions of principal interest are the allylic substitution reactions. We were particularly interested in transformations that utilize the less explored "hard" alkylating agents which can enantioselectively deliver the problematic quaternary carbon centers. In 2001, through systematic screening of parallel libraries designed based on metal coordination chemistry, we were able to develop a new class of pyridyl Schiff base peptides (see 9 and 10, Scheme 6) that accomplish part of the above objective [17].

The results shown in Scheme 6 are representative and point out which ligand provides the highest selectivity for a particular substrate (<2% S_N2 product in all cases). The catalytic enantioselective synthesis of quaternary carbons represents a unique and effective method for the regio- and enantioselective prepa-

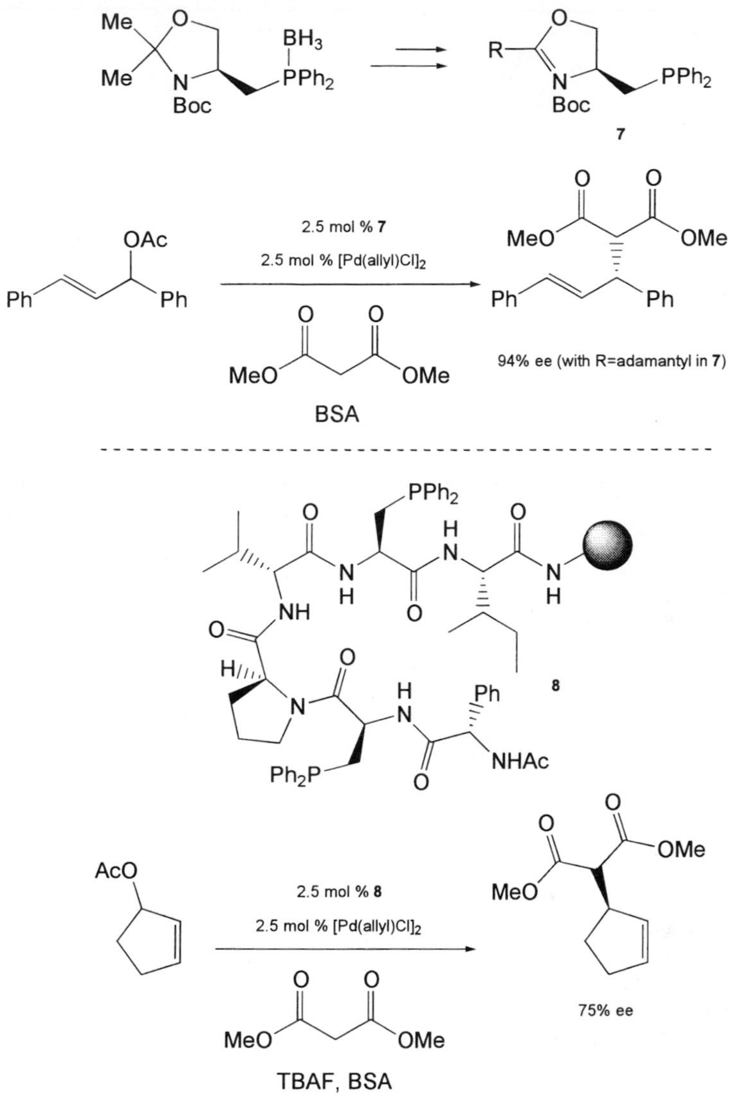

Scheme 5. Diversity-based approaches used in the late nineties in connection to the development of catalytic asymmetric allylic substitution reactions

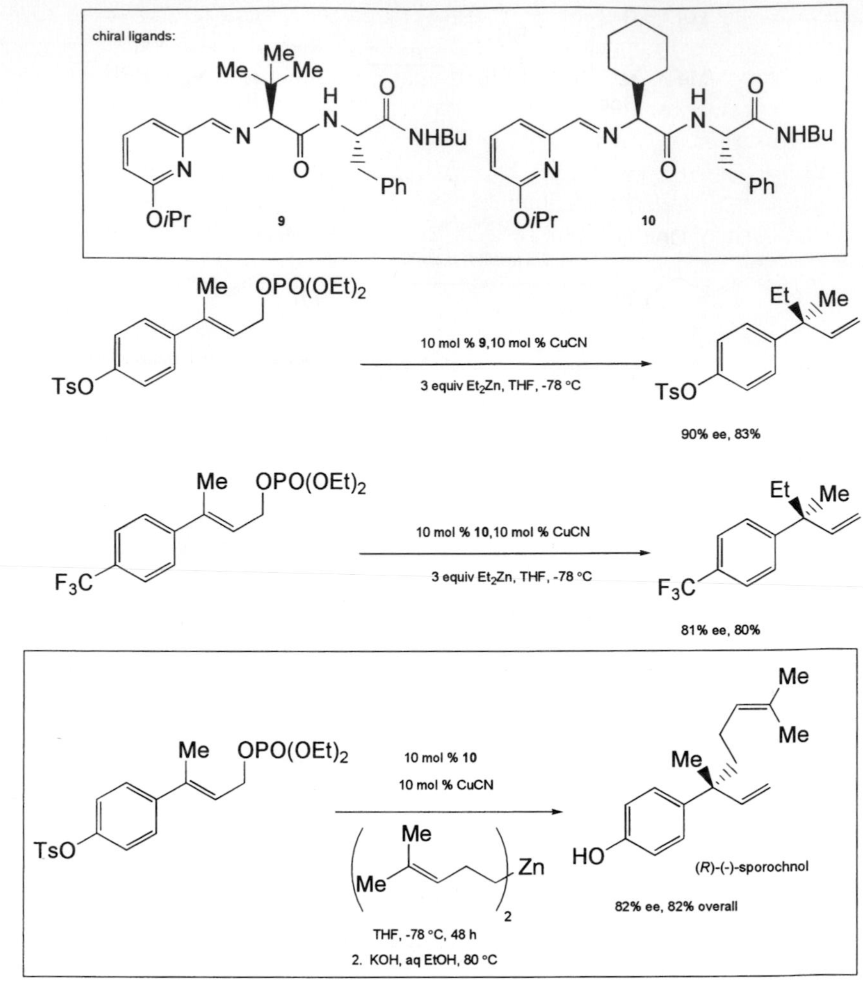

Scheme 6. Parallel screening of ligand libraries indicates that pyridyl dipeptides can serve as effective ligands for Cu-catalyzed allylic substitution reactions that afford quaternary carbon centers enantioselectively

ration of this important class of homochiral compounds. The significance of the new technology was demonstrated in the context of a brief enantioselective total synthesis of fish deterrent sporochnol (Scheme 6).

2.6
Catalytic Asymmetric Hetero-Diels–Alder Reactions

A more recent application of diversity-based approaches to chiral catalyst identification is reported by Ding in connection to cycloadditions of Danishefsky dienes to a variety of aryl or unsaturated aldehydes [18]. The activity of various chiral complexes formed in the presence of Ti(OiPr)$_4$ (104-member library) was screened leading to the optimal conditions shown in Eq. (6). A noteworthy aspect of this work is that the effect of combining two different ligands was carefully examined; this perhaps is based on the reported success of various tetraalkoxide derivatives derived from group 4 transition metals [19]. The Ti-diolate catalysts that were identified deliver exceptional levels of enantioselectivity, particularly when reactions are carried out neat.

99% *ee*, 82%

(6)

2.7
Catalytic Asymmetric Epoxidation of Olefins

Jacobsen and co-workers have used combinatorial metal-ligand libraries to identify a new alkene epoxidation catalyst [20]. Libraries of modular chiral ligands bearing peptidic and metal-binding segments were examined in this study. Five different amino acids with potential metal-binding sites were joined to a solid support through three different linkers. Another class of ligands was included based on the fact that salicyl imines have been shown to be effective epoxidation catalysts. Twelve different capping agents were also used. The combination of all the components resulted in 192 different ligand structures which were pooled with 30 different metal salts. The screening was effected in three steps. First, the whole catalyst library was screened for reaction conditions, indicating H$_2$O$_2$ to be the optimal oxidant. Next, catalyst sub-libraries were prepared, which involved combining all 192 ligands with one metal and subjecting the substrate to the catalyst mixture under the reaction conditions; these studies pointed to FeCl$_2$ as the optimal metal salt. In the last segment, various ligand components were divided into twelve batches, each with its own end cap: a pyridine end cap was shown to be the most desirable. To test the validity of the screen, individual screening of every ligand with FeCl$_2$ was performed and the screen pointed to the same end cap class as the best, which underlines the validity of screening ligand combinations. In the end, these researchers identified a metal catalyst that promotes the enantioselective oxidation of β-methylstyrene in 20% ee (see Eq. (7)). It is pointed out that, although the enantioselectivity is less than ideal, a lead catalyst has been discovered with a modular structure which could be optimized by parallel screening.

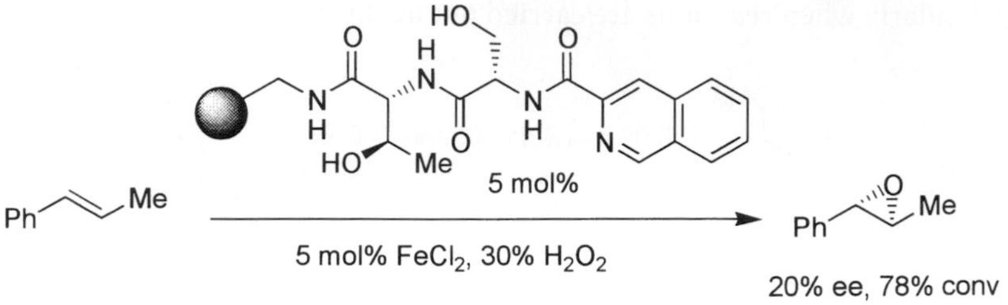

(7)

Yamamoto has used the modularity of another type of α-amino acid-based chiral ligand to promote enantioselective epoxidations of allylic alcohols [21]. Thus, as illustrated in Eq. (8), parallel libraries of various ligand candidates were prepared and the identity of the optimal ligand 13 was established through positional optimization.

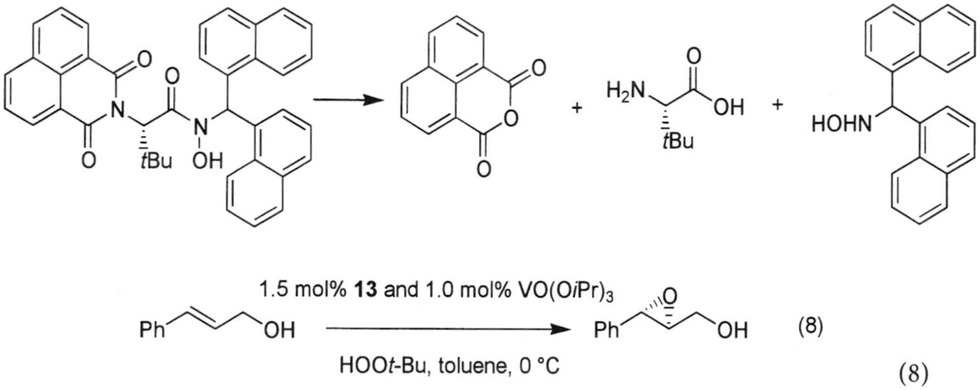

(8)

2.8
Catalytic Asymmetric Acylations and Phosphorylations

Miller and co-workers have exploited the highly modular nature of peptide structures, along with mechanistic knowledge regarding peptide conformational control and the principles of nucleophilic catalysis to develop an assortment of synthetically unique, useful, and practical asymmetric transformations [22]. Representative examples are illustrated in Scheme 7. Particularly noteworthy is the asymmetric phosphorylation; the resultant phosphate ester can be readily

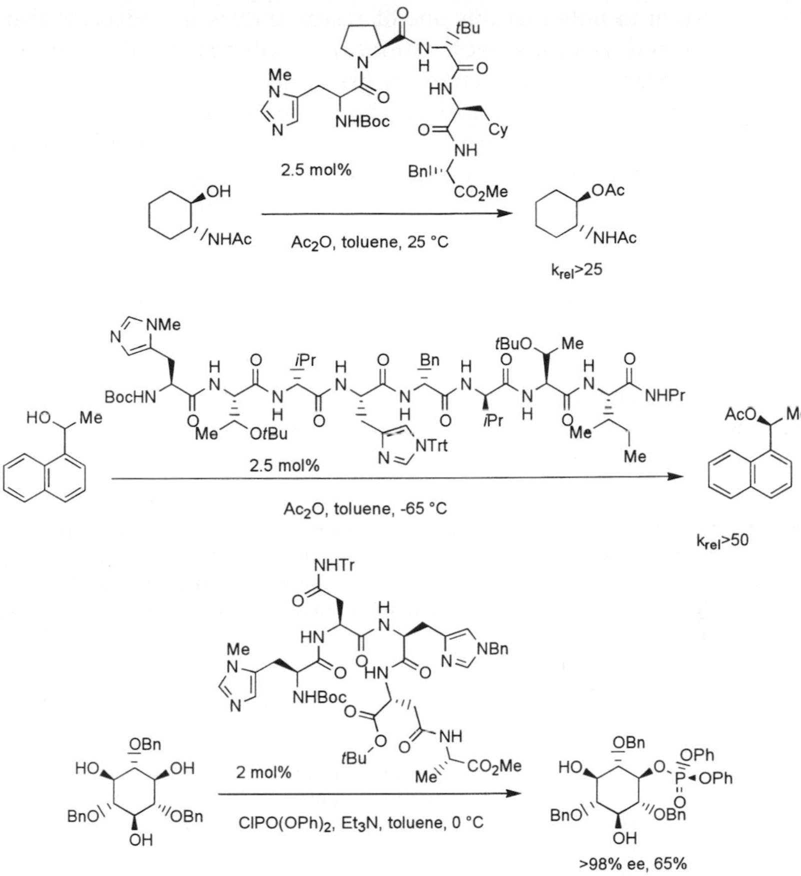

Scheme 7. Screening of libraries of various polypeptides has led to the identification of numerous synthetically useful catalytic enantioselective acylation and phosphorylation reactions

deprotected to afford the natural product D-*myo*-inositol-1-phosphate with exceptional efficiency in two steps from *myo*-inositol.

3
Conclusions and Future Outlook

The research described herein bears testimony to the fact that use of combinatorial protocols is becoming increasingly popular with researchers in the important field of chiral catalyst discovery. The majority of the above examples represent cases in which combinatorial approaches have led to the development of a new method in organic synthesis. There are numerous other disclosures involving optimization of established protocols that benefited from diversity-based approaches as well [23].

It is important to note that this line of research does not advocate that we abandon rational or rigorous investigations of detailed mechanisms of important processes. Elements of design and a priori decisions are still required in determining what collection of catalysts should be prepared; the framework is broader and initial bias that may be based on a few initial observations has less of a chance to point us in the wrong direction. Although optimal ligands, metal centers, protecting groups, and solvents have been typically determined through systematic screening of parallel libraries, mechanistic insight along with basic chemical intuition are also critical ingredients along the way.

It should be noted that, in spite of the above important advances, many fundamental issues regarding the facility of library syntheses and analysis of the resulting data need to be addressed [24]. Research along such lines and application of high-throughput screening to the discovery of new and effective chiral catalysts will undoubtedly constitute an exciting and critical branch of organic chemistry for many years to come.

Acknowledgments: We are grateful to the National Institutes of Health for supporting our programs in this area (GM-47480 and GM-57212). We thank Professor Marc Snapper who generously collaborated with us on some of the studies mentioned above. We warmly acknowledge the experimental and intellectual contributions made by the co-workers mentioned in the reference section.

References

1. For a more in depth discussion, see: Hoveyda AH (2002) Diversity-based identification of efficient homochiral organometallic catalysts for enantioselective synthesis. In: Nicolaou KC, Hanko R, Hartwig W (eds) Handbook of combinatorial chemistry. Wiley-VCH, New York, Chap 33, p 991
2. Sigman MS, Jacobsen EN (1998) J Am Chem Soc 120:4901; Sigman MS, Vachal P, Jacobsen EN (2000) Angew Chem Int Ed Engl 39:1279
3. Vachal P, Jacobsen EN (2000) Org Lett 2:867
4. Cole BM, Shimizu KD, Krueger CA, Harrity JP, Snapper ML, Hoveyda, AH (1996) Angew Chem Int Ed Engl 35:1668; Shimizu KD, Cole BM, Krueger CA, Kuntz KW, Snapper ML, Hoveyda AH (1997) Angew Chem Int Ed Engl 36:1704
5. Krueger CA, Kuntz KW, Dzierba CD, Wirschun WG, Gleason JD, Snapper ML, Hoveyda AH (1999) J Am Chem Soc 121:4284; Porter JR, Wirschun WG, Kuntz KW, Snapper ML, Hoveyda AH (2000) J Am Chem Soc 122:2657
6. For detailed studies regarding the mechanism of the Ti-catalyzed cyanide additions to imines promoted by amino acid-based ligands, see: Josephsohn NS, Kuntz KW, Snapper ML, Hoveyda AH (2001) J Am Chem Soc 123:11594
7. Porter JR, Traverse JF, Hoveyda AH, Snapper ML (2001) J Am Chem Soc 123:984
8. Porter JR, Traverse JF, Hoveyda AH, Snapper ML (2001) J Am Chem Soc 123:10409
9. Deng H, Isler MP, Snapper ML, Hoveyda AH (2002) Angew Chem Int Ed Engl 41:1009
10. Chataigner I, Gennari C, Piarulli U, Ceccarelli S (2000) Angew Chem Int Ed 39:916; Chataigner I, Gennari C, Ongeri S, Piarulli U, Ceccarelli S (2001) Chem Eur J 7:2628; Ongeri S, Piarulli U, Jackson RFW, Gennari C (2001) Eur J Org Chem 803
11. For a review on asymmetric conjugate addition reactions see: Krause N, Hoffmann-Roder A (2001) Synthesis 171
12. Degrado SJ, Mizutani H, Hoveyda AH (2001) J Am Chem Soc 123:755

13. Mizutani H, Degrado SJ, Hoveyda AH (2002) J Am Chem Soc 124:779
14. Luchaco-Cullis CA, Hoveyda AH (2002) J Am Chem Soc 124:8192
15. Guerin DJ, Miller SJ (2002) J Am Chem Soc 124:2134; Horstmann TE, Guerin DJ, Miller SJ
    (2000) Angew Chem Int Ed Engl 39:3635
16. Porte AM, Reibenspies J, Burgess K (1998) J Am Chem Soc 120:9180; Gilbertson SR, Col-
    libee SE, Agarkov A (2000) J Am Chem Soc 122:6522
17. Luchaco-Cullis CA, Mizutani H, Murphy KE, Hoveyda AH (2001) Angew Chem Int Ed Engl
    40:1456
18. Long J, Hu J, Shen X, Ji B, Ding K (2002) J Am Chem Soc 124:10
19. For example, see: Ishitani H, Ueno M, Kobayashi S (1997) J Am Chem Soc 119:7153
20. Francis MB, Jacobsen EN (1999) Angew Chem Int Ed Engl 38:937
21. Hoshino Y, Yamamoto H (2000) J Am Chem Soc 122:10452
22. Copeland GT, Miller SJ (1999) J Am Chem Soc 121:4306; Harris RF, Nation AJ, Copeland
    GT, Miller SJ (2000) J Am Chem Soc 122:11270; Copeland GT, Miller SJ (2001) J Am Chem
    Soc 123:6496; Jarvo ER, Evans CA, Copeland GT, Miller SJ (2001) J Org Chem 66:5522;
    Papaioannou N, Evans CA, Blank JT, Miller SJ (2001) 3:2879; Sculimbrene BR, Miller SJ
    (2001) J Am Chem Soc 123:10125
23. Ding K, Ishii A, Mikami K (1999) Angew Chem, Int Ed Engl 38:497; Brouwer AJ, van der
    Linden HJ, Liskamp RMJ (2000) J Org Chem 65:1750; Moreau C, Frost CG, Murrer B (1999)
    Tetrahedron Lett 40:5617; Gao X, Kagan HB (1998) Chirality 10:120; Kobayashi S, Kusak-
    abe K-I, Haruro I (2000) Org Lett 2:1225
24. Reetz MT (2001) Angew Chem Int Ed Engl 40:284; Reetz MT (2002) Angew Chem Int Ed
    Engl 41:1335

Chapter 43
Acylation Reactions

Elizabeth R. Jarvo, Scott J. Miller

Department of Chemistry, Merkert Chemistry Center, Boston College, Chestnut Hill, MA
02467–3860, USA
e-mail: Scott.Miller.1@bc.edu

Keywords: Acylation, Kinetic resolution, Desymmetrization, Nucleophilic catalysis

1
Introduction

The asymmetric acylation reaction has proven utility in the synthesis of biologically relevant targets. This is demonstrated by the plethora of applications of li-
pases and esterases in total syntheses [1]. While these enzymes often display su-
perb selectivities, their application to a broad class of substrates may be difficult and unpredictable [2]. To increase access to these materials in optically pure
form, over the past decade several groups have developed small molecule catalysts for the asymmetric acylation reaction [3, 4]. In addition, these catalysts

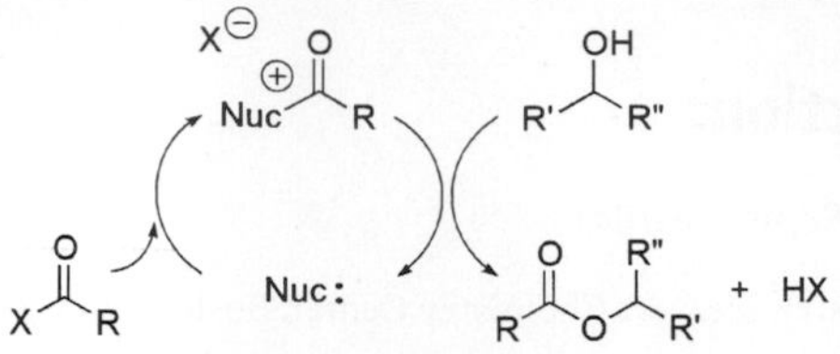

Scheme 1. Nucleophilic catalysis of acyl transfer

have proven to be a useful testing ground for new hypotheses in the general field of asymmetric catalyst development.

The most commonly used catalyst substructures include organocatalysts such as *N*-methyl imidazole (NMI), *N,N*-dimethylaminopyridine (DMAP), 4-pyrrolidinopyridine (PPY), and phosphines [5, 6]. These compounds are thought to catalyze the acyl transfer reaction by a nucleophilic mechanism (Scheme 1). Most efforts towards asymmetric catalysts for this reaction have focused on localizing one of these Lewis basic moieties in the midst of a chiral scaffold. Catalysts based on the DMAP/PPY framework have relied on innovative approaches to chirality transfer from remote asymmetric centers, since substitution on the 2-position of the pyridine ring drastically reduces catalyst activity [7]. These catalysts have been utilized to effect the kinetic resolution of racemic alcohols and for the desymmetrization of *meso* diols. Lewis acids are also known to catalyze these transformations. However, there have been fewer reports of such asymmetric catalysts in the literature. This Chapter will focus only on catalytic reactions; reactions mediated by chiral promoters and the use of optically pure acylating agents and alcohols as resolving agents will not be discussed [8]. Other pathways for the synthesis of these optically pure functional groups exist (e.g., ketone reduction, organometallic additions to carbonyls and imines, epoxidation, dihydroxylation, ring opening of epoxides, enamine hydrogenation, and oxidation of alcohols) [9]; however, these methods will not be discussed here.

2
Catalysts for Kinetic Resolution and Desymmetrization of Alcohols

2.1
Phosphines

Vedejs and co-workers have explored the use of chiral phosphines as acyl transfer catalysts. The viability of this approach was proven when phosphine **1** was shown to catalyze the resolution of secondary alcohols with promising selectivities (Scheme 2) [10, 11].

After further optimization of catalyst structure, phosphine catalyst **2** was found to be very effective in the kinetic resolution of aryl alkyl carbinols (k_{rel}=31–369, Scheme 3) [12]. Reactions exhibit high selectivity factors when

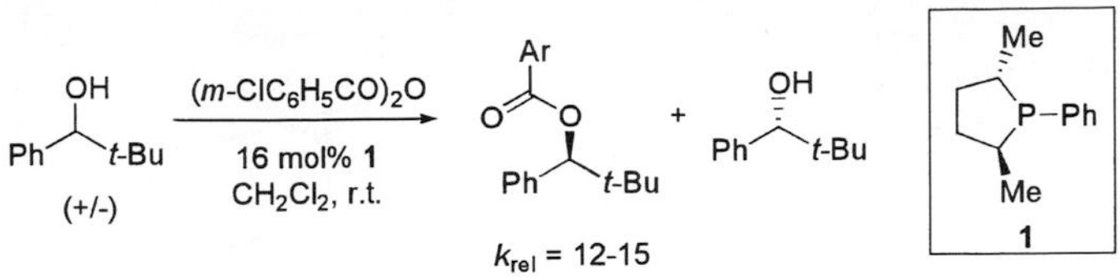

Scheme 2. Kinetic resolution using chiral phosphine catalyst 1

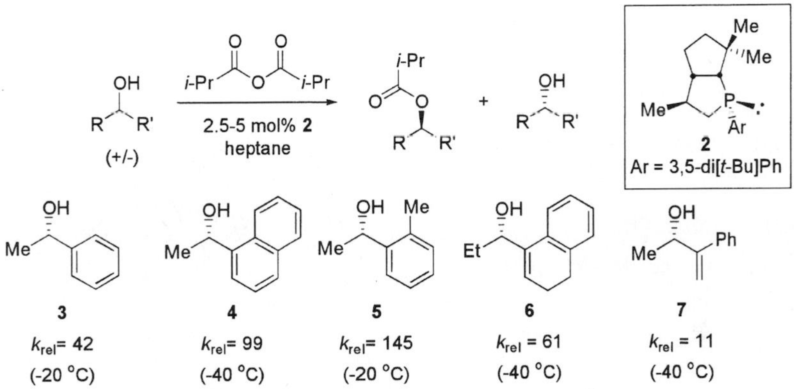

Scheme 3. Chiral phosphine catalyst for kinetic resolution of secondary alcohols

performed in heptane or toluene as solvent, with isobutyric anhydride as the acylating agent, and at low temperatures (-20 °C to -40 °C). Catalyst loadings vary from 0.6–12.1 mol% (generally 3–4 mol% is used). Inactivation of the catalyst by oxidation serves to slow reactivity when oxygen is not rigorously excluded from the reaction. Allylic alcohols are resolved with moderate to high selectivities (k_{rel}=4–82) [13]. Cyclic substrates wherein the orientation of the alcohol with respect to the olefin is restricted by factors other than allylic strain exhibit drastically reduced selectivities. A parallel kinetic resolution of alcohol 4 has been achieved using phosphine 2 and a lipase, in a 3-phase system that ensures selective reagent activation by each catalyst [14].

2.2
Metallocenes

Fu and co-workers have detailed the use of planar chiral DMAP and PPY analogs as catalysts for the resolution of secondary unsaturated alcohols (Fig. 1) [15]. Both ferrocene and ruthenocene-based catalysts have been examined, with the iron-based catalysts generally proving less reactive but more selective [16]. Catalysts are prepared in racemic form and are subsequently resolved by preparative chiral HPLC.

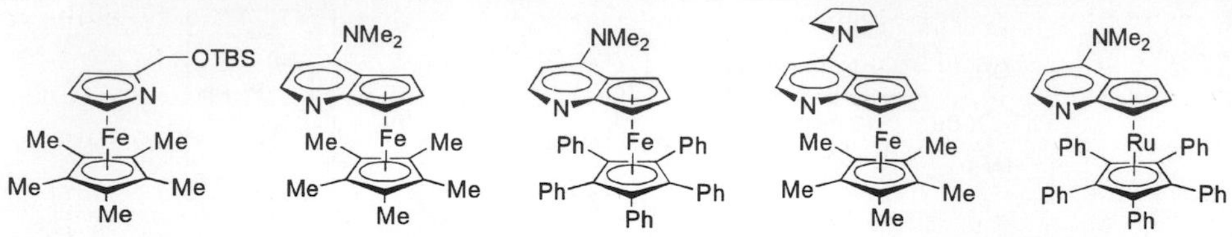

Fig. 1. Planar chiral nucleophilic catalysts

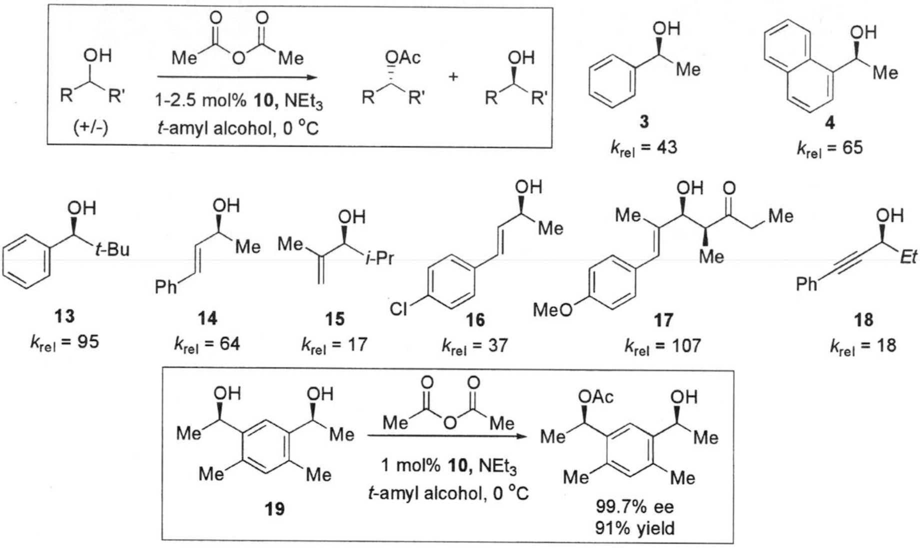

Scheme 4. Planar chiral DMAP analog **10** as a catalyst for kinetic resolution of secondary alcohols

Of the catalysts examined, DMAP analog **10** yields the highest selectivities in the kinetic resolution of a wide variety of aryl alkyl carbinols (k_{rel}=32–95, see Scheme 4) [17]. Low catalyst loadings (1 mol%) are used and high selectivities are obtained using *t*-amyl alcohol as solvent at low temperatures (0 °C to -20 °C), with a tertiary amine base as a stoichiometric additive. A broad substrate scope may be resolved, and in general, substrates with increased steric bulk are resolved with higher selectivities. *Meso* diol **19** may be subjected to desymmetrization with excellent enantioselectivity. Allylic alcohols may also be resolved effectively with catalyst **10** (k_{rel}=4.7–80) [18]. *Trans*-phenyl-substituted olefins such as **14** are resolved with the highest selectivities. With this more reactive class of substrates, reactions performed without exogenous tertiary amine base give higher selectivities and catalyst loadings remain low (1–2.5 mol%). The utility of this reaction was demonstrated in the preparation of optically pure allylic alcohols **16** and **17**, intermediates in the syntheses of (-)-baclofen and epothilone A, respectively. Propargylic alcohols (e.g., **18**) are also resolved with moderate to good levels of selectivity (k_{rel}=3.8–20) [19].

The acyl pyridinium salt of catalyst **10** has been isolated and characterized by X-ray and NMR spectroscopy [20]. The salt was prepared by acylation with acetyl chloride and exchange of chloride counterion with SbF_6^-. In the solid state and in solution, there is evidence that the oxygen atom of the acetyl group is disposed towards the fused cyclopentadienyl ring. Furthermore, the two cyclopentadiene rings are tilted away from each other by approximately 8°. This distortion may minimize steric interactions, supporting the idea that one face of the acyl pyridinium is effectively blocked by the pentaphenylcyclopentadienyl ring.

2.3
PPY Analogs

Fuji and co-workers have demonstrated the use of a PPY derivative that utilizes remote stereochemistry and an interesting "induced fit" process to control selectivity [21]. Upon acylation of catalyst **20**, a conformational change occurs, stabilizing the intermediate N-acyliminium ion **21** (Fig. 2a,b). Chemical shifts in the ^{1}H NMR and nOes observed support a π–π interaction between the electron-rich naphthyl ring and the electron-deficient pyridinium ring. This blocks the top face of the catalyst and directs attack of the alcohol from the bottom face. Catalyst **20** effects resolutions of diol-monoesters and amino alcohol derivatives such as **22** and **23** with moderate to good selectivity factors (k_{rel}=4.7–21, see Fig. 2c) [22].

2.4
Peptide-Based Catalysts

Miller et al. have shown that short peptides containing alkylated histidine residues can be used as catalysts for the kinetic resolutions of secondary and some tertiary alcohols. These catalysts are also postulated to effect catalysis by a nucleophilic mechanism. The backbone amides and ancillary functionality are proposed to govern selectivity through catalyst–substrate contacts (e.g., hy-

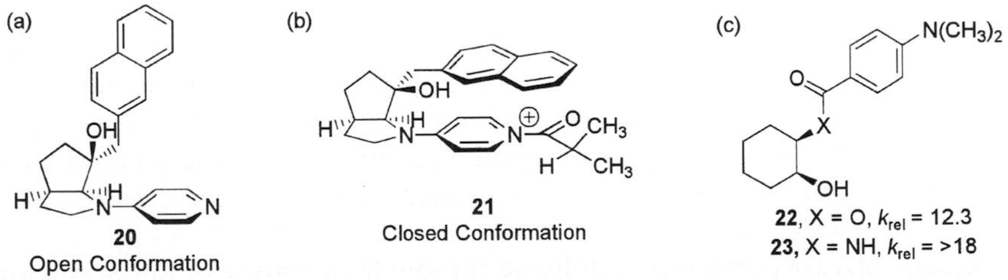

Fig. 2. Fuji's chiral PPY catalyst **20**. (a) Preferred conformation of **20**. (b) Preferred conformation of N-acyliminium ion **21**. (c) Substrates **22** and **23**

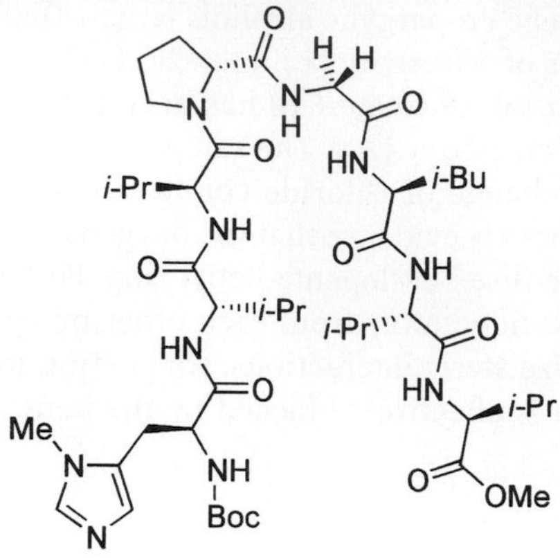

24

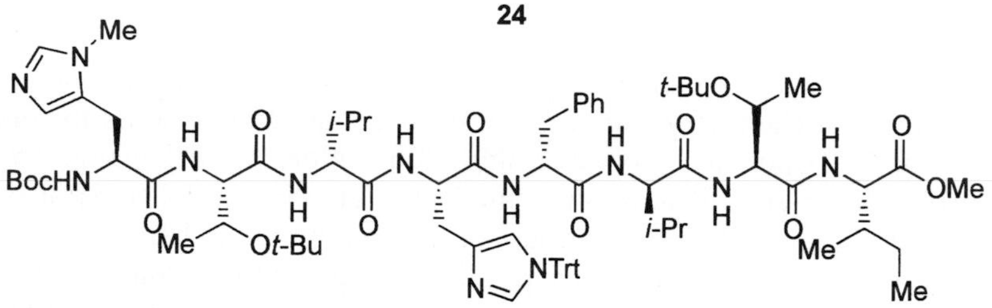

25

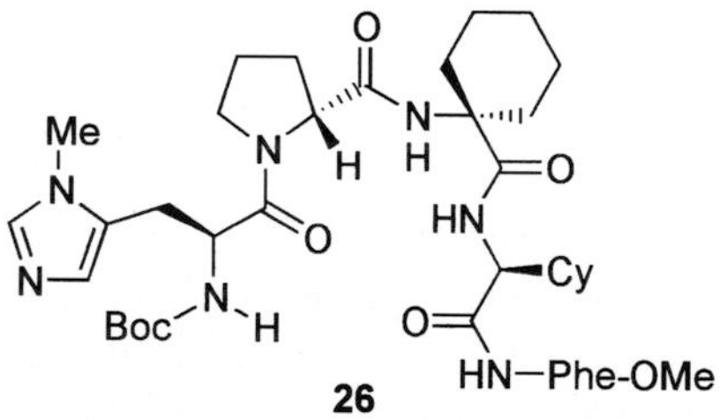

26

Fig. 3. Peptide-based catalysts for kinetic resolution of alcohols

drogen bonding, π-stacking, ion pairing, etc.) that could stabilize the transition state for reaction of one enantiomer. Preliminary studies drew heavily from the peptide design literature [23], focusing on catalysts that were biased towards β-turn conformations in solution (e.g., peptides **24** and **26**), thus decreasing catalyst flexibility (Fig. 3).

Resolutions are performed in toluene at room temperature or lower, and low catalyst loadings (0.5–2.5 mol%) are required. Acetamide-functionalized substrates such as **27** are subject to enantioselective acylation with catalyst **24** [24].

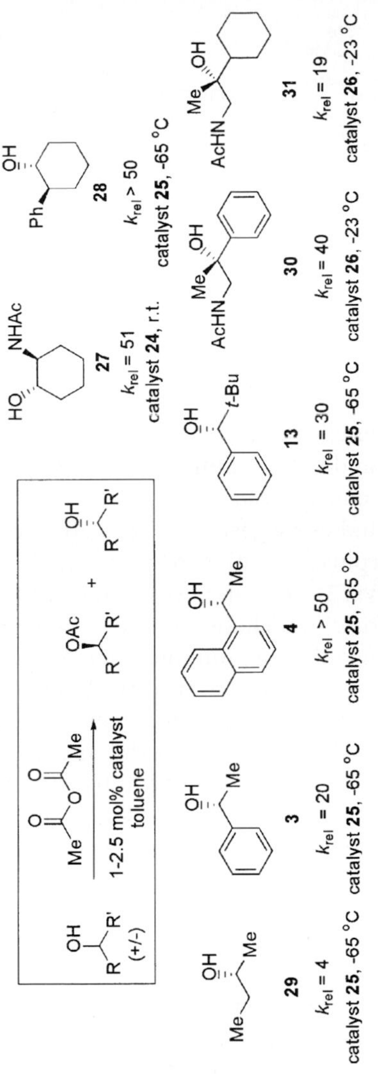

Scheme 5. Kinetic resolution of alcohols with peptide-based catalysts

A variety of unfunctionalized secondary alcohols, including saturated and unsaturated carbinols, are resolved by catalyst **25** with moderate to high selectivities (k_{rel}=4 to >50, see Scheme 5) [25]. Octapeptide **25** was discovered by screening a split-pool library of peptide catalyst candidates for acylation of 1-phe-
nylethanol (**3**), using a reactivity-based fluorescence screen [26], followed by
structure optimization with directed libraries. While substrates with increased
steric bulk about the alcohol are resolved with highest selectivities, even 2-buta-
nol is resolved with modest selectivity (k_{rel}=4). Peptide-based catalysts have also
been applied to the resolution of tertiary alcohols, a relatively unexplored area
of nonenzymatic asymmetric acylation catalysis [27–29]. By using a fluores-

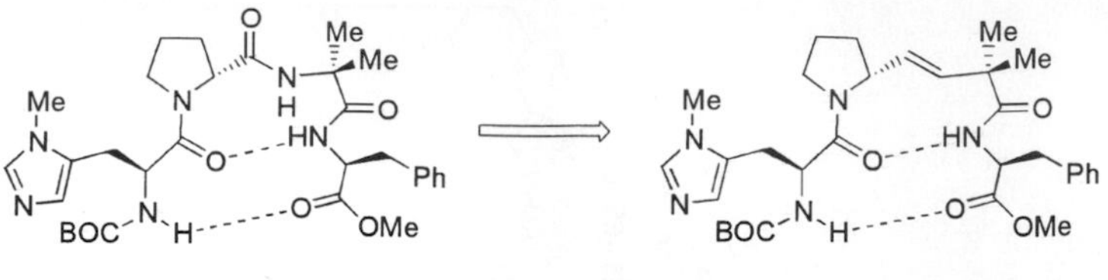

Fig. 4. Isosteric alkene substitution eliminates catalyst selectivity

cence-based reactivity assay in a spatially segregated enantiomer screen, peptide **26** was found to acylate acetamide-functionalized tertiary alcohols such as **30** and **31** with good to excellent selectivities (k_{rel}=19 to >50) [30]. Higher catalyst loadings (10 mol%) and tertiary amine base are required to obtain acceptable conversions (approximately 50%).

Solution-state NMR studies suggest that the catalysts containing L- and D-Pro adopt β-turns and β-hairpins in solution, respectively. Reactions exhibit first-order dependence on catalyst **24**, consistent with a monomeric catalyst in the rate-determining step of the reaction. These catalysts exhibit enantiospecific rate acceleration, in comparison to the reaction rate when NMI is employed as catalyst. An isosteric replacement of an alkene for a backbone amide in a tetrapeptide catalyst (catalysts **32** and **33**, Fig. 4) has lent credence to a proposed mechanism of rate acceleration [31]. While catalyst **32** exhibits a k_{rel}=28 with substrate **27**, alkene-containing catalyst **33** is not selective in this kinetic resolution and also affords a reduced reaction rate. This suggests that the prolyl amide is kinetically significant in the stereochemistry-determining step of the reaction.

2.5
Tertiary Diamines

Oriyama and co-workers have shown that diamines derived from proline, **37** and **38**, catalyze the desymmetrization of several *meso* diols [32] and the kinetic resolution of certain secondary alcohols [33]. Acylation is achieved by using benzoyl chloride at -78 °C and an achiral tertiary amine (NEt$_3$ or Hunig's base) as a stoichiometric additive (see Scheme 6). Very low catalyst loadings (0.3 mol%) are tolerated and products are obtained with high levels of enantioselectivity. Various cyclic and acyclic *meso* diols may be subjected to desymmetrization with good yields and enantioselectivities. Particularly interesting is the ability of catalyst **38** to selectively acylate *meso*-1,3-propanediols with very good to excellent enantiomeric excesses, albeit in modest yields [34]. Kinetic resolution of racemic alcohols proceeds with highest selectivities for cyclic substrates. β-Halohydrins are also resolved efficiently [35]. Under the basic reaction conditions, some racemization of starting materials is observed, raising the possibility of a dynamic kinetic resolution [36]. Diamine **37** may be attached to sol-

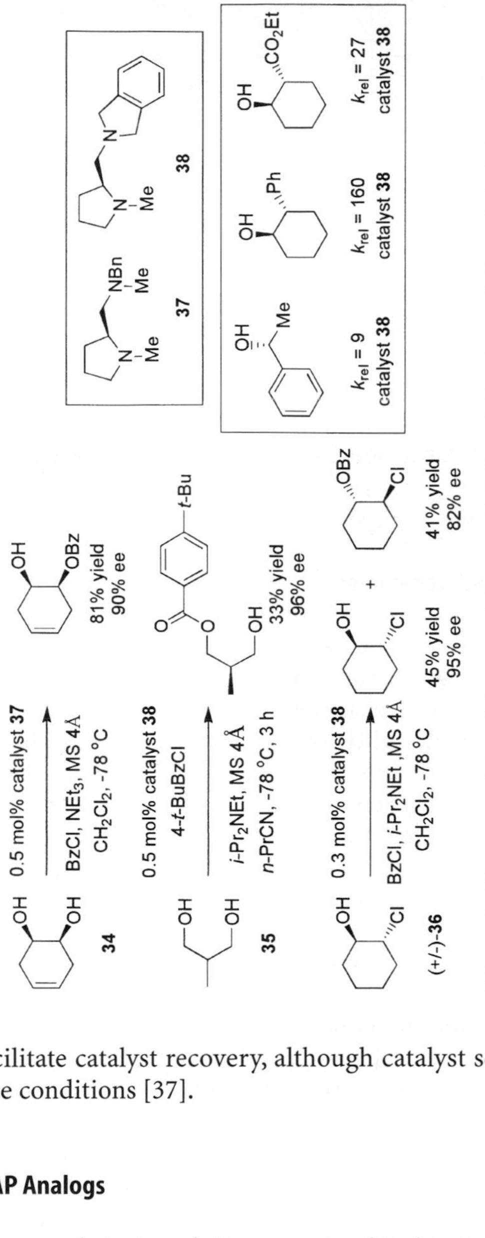

Scheme 6. Diamine-catalyzed desymmetrization of diols and kinetic resolution of alcohols

id support to facilitate catalyst recovery, although catalyst selectivity is diminished under these conditions [37].

2.6
Axially Chiral DMAP Analogs

An interesting approach to translating remote chirality into enantiomer discrimination relies upon blocking one face of the heterocycle using axially chiral DMAP analogs. To this end, Spivey and co-workers have examined catalysts **39** and **40**, obtained in optically pure form by preparative HPLC [38]. Catalyst **39** is effective for the kinetic resolution of aryl alkyl carbinols with good selectivities (k_{rel}=8.4–27, see Scheme 7). The substitution on the 4-position nitrogen

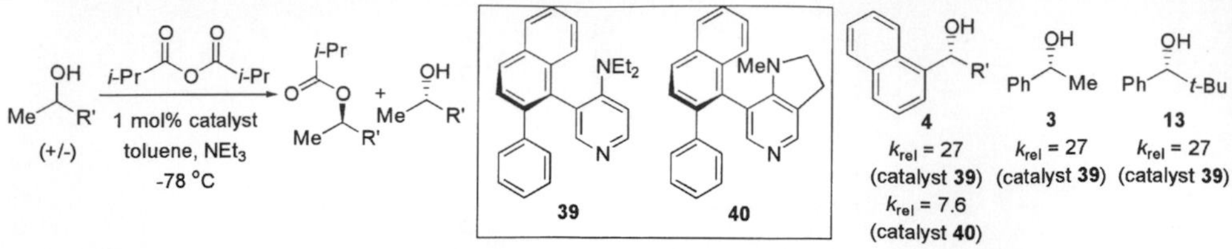

Scheme 7. Axially chiral catalysts for kinetic resolution of secondary alcohols

proved to be critical to the catalyst's success (cf. resolution of **4** with **39** and **40**, Scheme 7). This substituent proved to have a large effect on the energy barrier to rotation about the biaryl bond [39].

3
Catalysts for the Kinetic Resolution and Desymmetrization of Acylating Agents

3.1
Cinchona Alkaloids and Derivatives

Asymmetric acyl transfer reactions may be performed to convert a *meso*, prochiral or racemic acylating agent to an optically pure product in high yield. Such desymmetrizations have indeed been accomplished using enzymatic [40] and nonenzymatic methods [41, 42]. Early examples of small molecule catalysts for these processes utilized the cinchona alkaloids as catalysts with promising results. Oda and co-workers employed cinchonine as a catalyst to desymmetrize mono- and bicyclic anhydrides such as **41** with moderate to good enantiose-lectivities (23–70% ee, see Scheme 8) [43]. Aitken et al. employed quinine to effect desymmetrization of anhydride **43** (76% ee) [44]. Bohm combined the use of quinidine with an achiral base, pempidine (1,2,2,5,5-pentamethylpiperidine), to accomplish catalytic desymmetrizations of *meso* bicyclic and tricyclic anhydrides (e.g., **45**) [45]. Enantioselectivities are only slightly lower than with stoichiometric use of quinidine, however, reaction rates suffer.

More recently, Deng et al. have applied cinchona alkaloid derivatives, namely the Sharpless AD ligands, as excellent catalysts for both desymmetrizations and kinetic resolutions. As observed in the dihydroxylation reaction, pseudoenantiomeric catalysts provide comparable and opposite absolute stereospecificities in most cases. Desymmetrization of anhydrides proceeds with high enantioselectivities and good to excellent yields (70–99% yield, 90–98% ee, see Scheme 9) [46]. Highest enantioselectivities are obtained with bicyclic anhydrides with α-stereogenic centers. Racemic monosubstituted succinic anhydrides undergo a parallel kinetic resolution, in which catalyst **47** mediates a different reaction for each enantiomer of starting material [(*S*)-**48** goes to (*S*)-**49** while (*R*)-**48** goes to (*R*)-**50**, see Scheme 9] [47]. Succinate monoesters are obtained with very good enantiomeric excesses and yields that correlate to a simple kinetic resolution

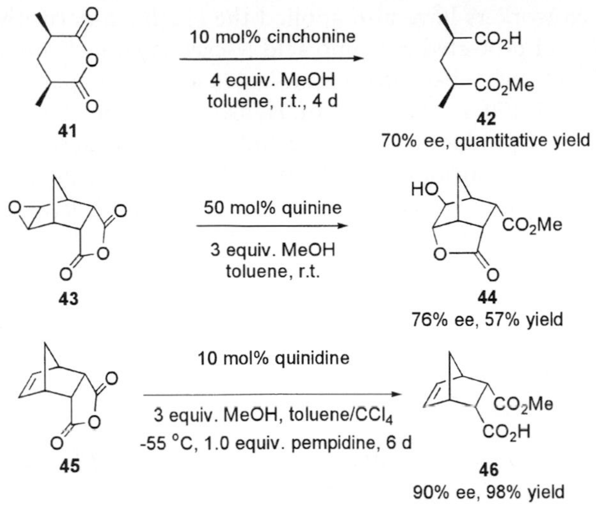

Scheme 8. Catalytic enantioselective desymmetrizations of anhydrides using cinchona alkaloid catalysts

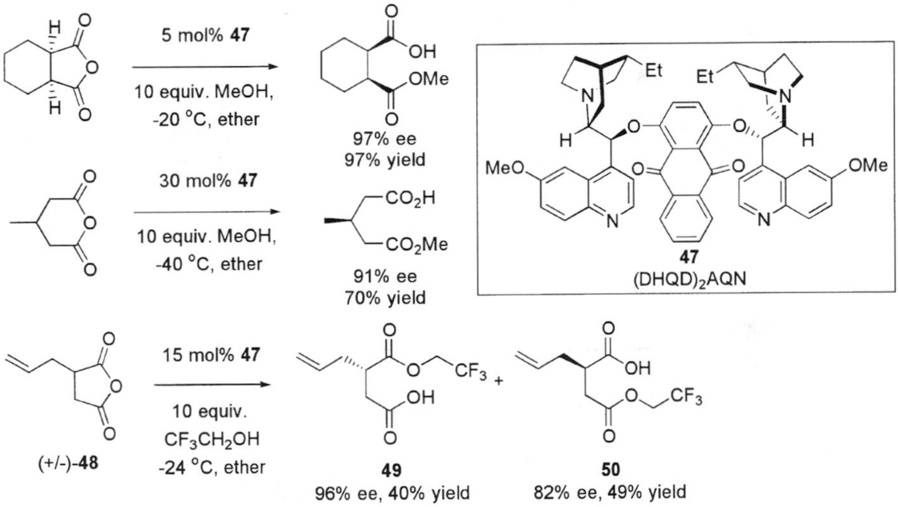

Scheme 9. Desymmetrization and parallel kinetic resolution of cyclic anhydrides by (DHQD)$_2$AQN

with $k_{rel}>100$ for alkyl and aryl substituted anhydrides. 3-Substituted succinates (e.g., **49**) are provided in higher optical purity (91–98% ee, 36–40% yield) than 2-substituted succinates (e.g., **50**, 66–87% ee, 41–50% yield), perhaps as a consequence of the proximity of the stereocenter to the reacting center. Mixtures of aryl succinate monesters may be reduced directly to the butyrolactones to facilitate separation.

Deng and co-workers have also applied the cinchona derivatives to the kinetic resolution of protected α-amino acid *N*-carboxyanhydrides **51** [48]. A variety of alkyl and aryl-substituted amino acids may be prepared with high selectivities (k_{rel}=23–170, see Scheme 10). Hydrolysis of the starting material, in the presence of the product and catalyst, followed by extractive workup allows for recovery of ester, carboxylic acid, and catalyst. The catalyst may be recycled with little effect on selectivity (run 1, k_{rel}=114; run 2, k_{rel}=104). The reaction exhibits first-order dependence on methanol and catalyst and a kinetic isotope effect (k_{MeOH}/k_{MeOD}=1.3). The authors postulate that this is most consistent with a mechanism wherein rate-determining attack of alcohol is facilitated by (DHQD)$_2$AQN acting as a general base. 5-Alkyl 1,3-dioxolanes **52** may also

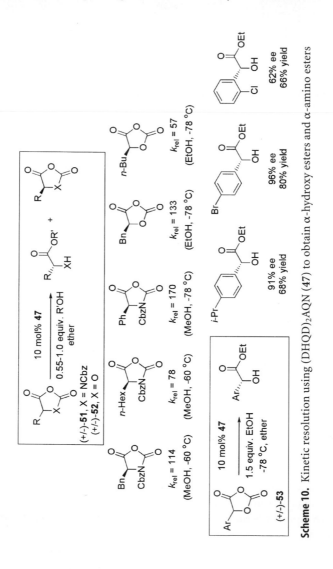

Scheme 10. Kinetic resolution using (DHQD)$_2$AQN (**47**) to obtain α-hydroxy esters and α-amino esters

be resolved efficiently, yielding α-hydroxy esters and acids in high enantiomeric excesses [49]. Aryl substitution reduces the pK_a of the α-proton significantly so that cinchona catalyst **47** may racemize the starting material at a rate faster than ethanolysis. A dynamic kinetic resolution results, in which both enantiomers of starting material **53** are transformed to (*R*)-hydroxy ester. A variety of substrates with electron-rich and electron-poor aryl substitutions yield products with high enantiomeric excesses and in good yields (91–96% ee, 65–80% yield). *Ortho*-substituted aryl groups are the exception: these substrates racemize more slowly and so enantioselectivity suffers at high conversion.

3.2
Metallocenes

Fu and co-workers have also applied their planar chiral catalyst **9** to dynamic kinetic resolution of racemic azalactones [50]. Azalactones **54** racemize under the reaction conditions, allowing all material to be funneled to optically pure product. Protected (*S*)-amino acids **55** are formed in excellent yields with moderate enantioselectivities (83–98% yield, 44–61% ee, see Scheme 11). Use of more sterically encumbered alcohols as nucleophiles increases enantioselectivities but reaction rates become slower.

In addition, azaferrocene catalyst **8** has been utilized by Fu and co-workers to perform enantioselective additions of alcohols to prochiral ketenes [51]. Aryl alkyl ketenes are substituted with MeOH to give α-aryl ester products in good enantioselectivities and very good yields, with higher enantiomeric excesses obtained for products with larger alkyl groups (Scheme 12). Use of 2,6-di-*t*-butylpyridinium triflate as a proton shuttle substantially enhances the enanti-

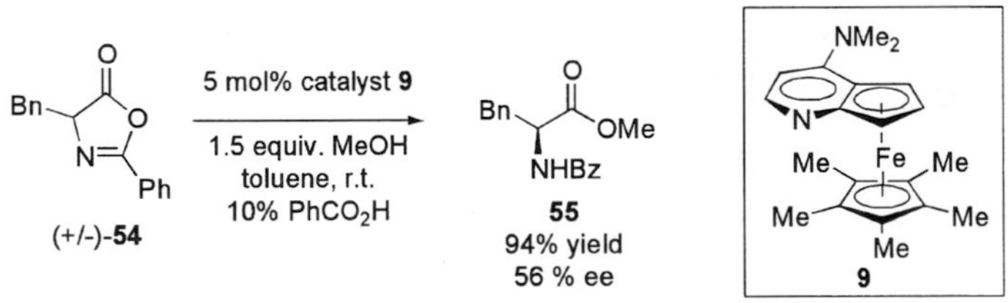

Scheme 11. Dynamic kinetic resolution of azalactones

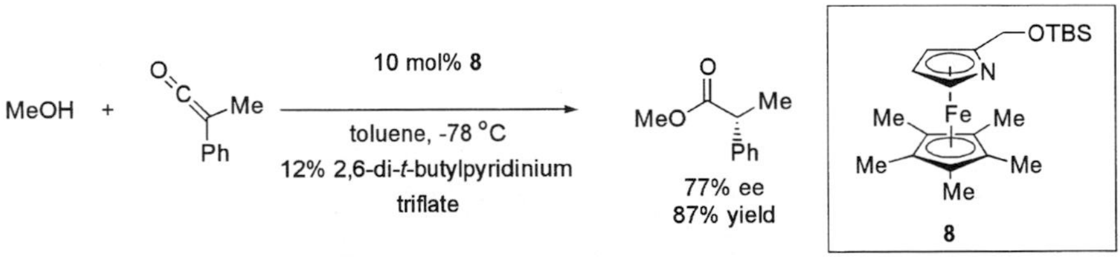

Scheme 12. Enantioselective addition of alcohols to ketenes

oselectivity of the reaction. Furthermore, the reaction exhibits a primary kinetic isotope effect ($k_H/k_D=3.2$).

3.3
Lewis Acids

Narasaka et al. demonstrated the utility of titanium-ligand complexes in the resolution of chiral α-aryl esters [52]. Ti(Oi-Pr)$_4$–ligand **56** complex resolves 2-pyridine thioesters with high selectivities (k_{rel}=26–42, see Scheme 13). Seebach and co-workers have examined titanium-TADDOLate complexes as reagents for the ring opening of *meso* anhydrides, dioxolanones, and azalactones [53]. Addition of an achiral isopropoxide source renders the desymmetrization of *meso*

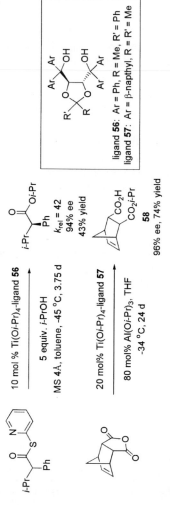

Scheme 13. Kinetic resolution of thiopyridyl esters and desymmetrization of *meso* anhydrides

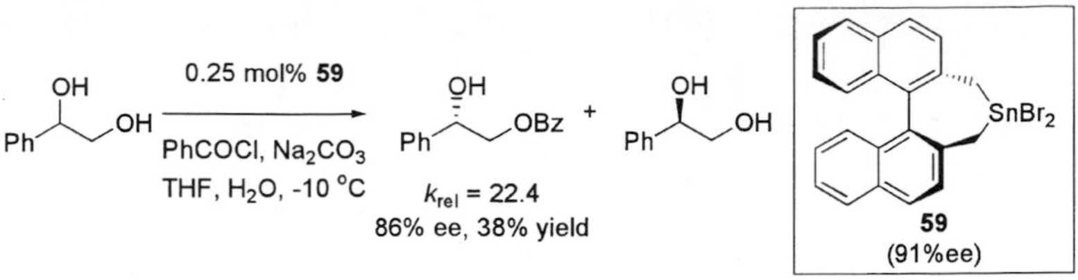

Scheme 14. Kinetic resolution of 1,2-diols using organotin catalyst **59**

diols catalytic in Ti-ligand complex, although with lower selectivities and longer reaction times than had been reported for the stoichiometric process (cf. 98% ee and 88% yield of **58** in 7 d for stoichiometric process) [54].

Organotin complex **59** has been used in the kinetic resolution of 1,2-diols [55]. This catalyst is postulated to react with the diol to form a stannylene acetal that is subsequently benzoylated. Utilization of Na_2CO_3 as a solid base and addition of controlled amounts of water allow for resolution of certain diols with modest to good selectivities (k_{rel}=3.2–22.4, when enantiomeric excesses are corrected for the ee of catalyst, see Scheme 14) [56].

4
Kinetic Resolution of Amines

While several efficient catalysts for the kinetic resolution of alcohols have been reported, the nonenzymatic kinetic resolution of amines is far less precedented. The challenge in this case is heightened since the high nucleophilicity of amines allows for uncatalyzed attack on standard acylating agents (e.g., Ac_2O) [57]. Among the approaches to this problem, chiral acylating agents have been used for diastereoselective acylation of amines, obviating the need for intervention by a chiral catalyst [58]. In the course of studies of enantioselective rearrangement processes, Fu and co-workers recognized the potential for use of O-azalactone **60** in amine resolution [59]. This acylating agent reacts more quickly with metallocene catalyst **11** than with primary amines. Kinetic studies are consistent with a fast reaction between PPY analog **11** and O-azalactone **60**, yielding an ion pair which is the resting state of the catalyst. Attack of the amine on the ion pair follows. Resolutions of benzylic primary amines proceed with very good levels of selectivity (k_{rel}=11–27, see Scheme 15). Increased steric bulk on the substrate appears to lead to increased selectivities. The applicability of this method to the resolution of functionalized amines is demonstrated with substrate **63**.

5
Conclusions

The past decade has seen a large increase in the number of small molecule catalysts for asymmetric alcohol acylation. Representative members of virtually all

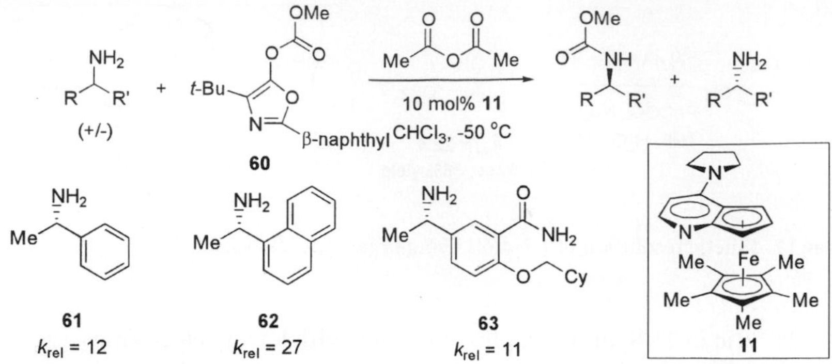

Scheme 15. Planar chiral PPY analog **11** as a catalyst for kinetic resolution of primary amines

substrate classes (unsaturated, saturated, and tertiary carbinols) have been resolved with moderate to high selectivities. The future undoubtedly holds further improvements and broader applicability. Extension of these reactions into dynamic kinetic resolutions (by coupling to starting material racemization) and parallel kinetic resolutions [60] (by coupling to a second selective reaction of opposite absolute stereospecificity) will further increase their synthetic utility. Excellent results have been obtained in the desymmetrization and resolution of acylating agents using cinchona-, metallocene- and titanium-based catalysts. While significant advances have been made, the application of a wider variety of catalysts to these transformations should yield interesting results [61]. The kinetic resolution of amines still stands as a significant challenge to synthetic chemists and creative solutions to this problem are anticipated.

References

1. Wong C-H, Whitesides GM (1994) Enzymes in synthetic organic chemistry. Elsevier, Oxford; Schmid RD, Verger R (1998) Angew Chem Int Ed 37:1608–1633
2. Griffith DA, Danishefsky SJ (1996) J Am Chem Soc 118:9526–9538
3. Spivey AC, Maddaford A, Redgrave AJ (2000) Org Prep Proceed Int 32:331–365; Somfai P (1997) Angew Chem Int Ed Engl 36:2731–2733
4. As a contemporary alternative to small molecule catalysts, directed evolution of a lipase to increase the enzyme's selectivity for a desired substrate has been demonstrated. Liebeton K, Zonta A, Schimossek K, Nardini M, Lang K, Dijkstra BW, Reetz MT, Jaeger K-E (2000) Chem Biol 7:709–718
5. Höfle G, Steglich W, Vorbruggen H (1978) Angew Chem Int Ed Engl 17:569–583
6. Vedejs E, Diver ST (1993) J Am Chem Soc 115:3358–3359
7. Vedejs E, Chen X (1996) J Am Chem Soc 118:1809–1810
8. For examples of kinetic resolution of alcohols using chiral acylating agents, see: Evans DA, Anderson JC, Taylor MK (1993) Tetrahedron Lett 34:5563–5566; Vedejs E, Chen X (1997) J Am Chem Soc 119:2584–2585
9. Jacobsen EN, Pfaltz A, Yamamoto H (eds) (1999) Comprehensive asymmetric catalysis. Springer, Berlin Heidelberg New York

10. Vedejs E, Daugulis O, Diver ST (1996) J Org Chem 61:430–31
11. $k_{rel}=k_{fast\ enantiomer}/k_{slow\ enantiomer}$=selectivity factor, see: Kagan HB, Fiaud JC (1988) Top Stereochem 18:249–330
12. Vedejs E, Daugulis O (1999) J Am Chem Soc 121:5813–5814
13. Vedejs E, MacKay JA (2001) Org Lett 3:535–536
14. Vedejs E, Rozners E (2001) J Am Chem Soc 123:2428–2429
15. Fu GC (2000) Acc Chem Res 33:412–420
16. Garrett CE, Lo MM-C, Fu GC (1998) J Am Chem Soc 120:7479–7483
17. Ruble JC, Fu GC (1996) J Org Chem 61:7230–7231; Ruble JC, Latham HA, Fu GC (1997) J Am Chem Soc 119:1492–1493; Ruble JC, Tweddell J, Fu GC (1998) J Org Chem 63:2794–2795
18. Bellemin-Laponnaz S, Tweddell J, Ruble JC, Breitling FM, Fu GC (2000) J Chem Soc Chem Commun 1009–1010
19. Tao B, Ruble JC, Hoic DA, Fu GC (1999) J Am Chem Soc 121:5091–5092
20. See: ref. [19]
21. Kawabata T, Nagata M, Takasu K, Fuji K (1997) J Am Chem Soc 119:3169–3170
22. Kawabata T, Yamamoto K, Momose Y, Yoshida H, Nagaoka Y, Fuji K (2001) J Chem Soc Chem Commun 2700–2701
23. Gellman SH (1998) Curr Opin Chem Biol 2:717–725; Venkatraman J, Shankaramma SC, Balaram P (2001) Chem Rev 101:3131–3152
24. Jarvo ER, Copeland GT, Papaioannou N, Bonitatebus PJ Jr, Miller SJ (1999) J Am Chem Soc 121:11638–11643; Copeland GT, Jarvo ER, Miller SJ (1998) J Org Chem 63:6784–6785; Miller SJ, Copeland GT, Papaioannou N, Horstmann TE, Ruel EM (1998) J Am Chem Soc 120: 1629–1630
25. Copeland GT, Miller SJ (2001) J Am Chem Soc 123:6496–6502
26. Copeland GT, Miller SJ (1999) J Am Chem Soc 121:4306–4307; Harris RF, Nation AJ, Copeland GT, Miller SJ (2000) J Am Chem Soc 122:11270–11271
27. Biocatalytic resolutions of tertiary carbinol centers are traditionally difficult (k_{rel} values typically <10), see: Franssen MCR, Goetheer ELV, Jongenjan H, deGroot A (1998) Tetrahedron Lett 39:8345–8348, and references cited therein
28. A notable exception is the application of retroaldolase catalytic antibodies: List B, Shabat D, Zhong G, Turner J M, Li A, Bui T, Anderson J, Lerner RA, Barbas CF (1999) J Am Chem Soc 121:7283–7291
29. DMAP is well known to catalyze the acylation of tertiary alcohols, while NMI is notably less effective. Guibe-Jampel E, Bram G, Vilkas M (1973) Bull Soc Chim Fr 1021–1027; Ref. [5]
30. Jarvo ER, Evans CA, Copeland GT, Miller SJ (2001) J Org Chem 66:5522–5527
31. Vasbinder MM, Jarvo ER, Miller SJ (2001) Angew Chem Int Ed 40:2824–2827
32. Oriyama T, Imai K, Sano T, Hosoya T (1998) Tetrahedron Lett 39:3529–3532; Oriyama T, Hosoya T, Sano T (2000) Heterocycles 52:1065–1069
33. Sano T, Imai K, Ohashi K, Oriyama T (1999) Chem Lett 265–266
34. Oriyama T, Taguchi H, Terakado D, Sano T (2002) Chem Lett 26–27
35. Sano T, Miyata H, Oriyama T (2000) Enantiomer 5:119–123
36. For examples of dynamic kinetic resolution, see: Ohkuma T, Noyori R (1999) Hydrogenation of carbonyl groups. In: Jacobsen EN, Pfaltz A, Yamamoto H (eds) Comprehensive asymmetric catalysis. Springer, Berlin Heidelberg New York, chap 6.1; Huerta FF, Minidis ABE, Bäckvall J-E (2001) Chem Soc Rev 30:321–331
37. Clapham B, Cho C-W, Janda KD (2001) J Org Chem 66:868–873; Cordova A, Tremblay MR, Clapham B, Janda KD (2001) J Org Chem 66:5645–5648
38. Spivey AC, Fekner T, Spey SE (2000) J Org Chem 65:3154–3159
39. Spivey AC, Charbonneau P, Fekner T, Hochmuth DH, Maddaford A, Malardier-Jugroot C, Redgrave A, Whitehead MA (2001) J Org Chem 66:7394–7401
40. Schoffers E, Golebiowski A, Johnson CR (1996) Tetrahedron 52:3769–3826

41. For reviews of desymmetrization reactions, see: Spivey AC, Andrews BI (2001) Angew Chem Int Ed 40:3131–3134; Willis MC (1999) J Chem Soc Perkin Trans 1 1765–1784; Poss CS, Schreiber SL (1994) Acc Chem Res 27:9–17
42. For examples of desymmetrization and kinetic resolution of lactides by polymerization, see: Ovitt TM, Coates GW (2002) J Am Chem Soc 124:1316–1326
43. Hirataka J, Inagaki M, Yamamoto Y, Oda J (1987) J Chem Soc Perkin Trans 1 1053–1058
44. Aitken RA, Gopal J, Hirst JA (1998) J Chem Soc Chem Commun 632–634
45. Bolm C, Schiffers I, Dinter CL, Gerlach A (2000) J Org Chem 65:6984–6991
46. Chen Y, Tian S-K, Deng L (2000) J Am Chem Soc 122:9542–9543
47. Chen Y, Deng L (2001) J Am Chem Soc 123:11302–11303
48. Hang J, Tian S-K, Tang L, Deng L (2001) J Am Chem Soc 123:12696–12697
49. Tang L, Deng L (2002) J Am Chem Soc 124:2870–2871
50. Liang J, Ruble JC, Fu GC (1998) J Org Chem 63:3154–3155
51. Hodous BL, Ruble JC, Fu GC (1999) J Am Chem Soc 121:2637–2638
52. Narasaka K, Kanai F, Okudo M, Miyoshi N (1989) Chem Lett 1187–1190
53. Seebach D, Jaeschke G, Wang YM (1995) Angew Chem Int Ed 34:2395–2396; Seebach D, Jaeschke G, Gottwald K, Matsuda K, Formisano R, Chaplin DA (1997) Tetrahedron 53: 7539–7556
54. Jaeschke G, Seebach D (1998) J Org Chem 63:1190–1197
55. Iwasaki F, Maki T, Nakashima W, Onomura O, Matsumura Y (1999) Org Lett 1:969–972; Iwasaki F, Maki T, Onomura O, Nakashima W, Matsumura Y (2000) J Org Chem 65:996–1002
56. For a discussion of kinetic resolutions with enantiomerically impure catalysts, see: Blackmond DG (2001) J Am Chem Soc 123:545–553; Ismagilov RF (1998) J Org Chem 63:3772–3774
57. Enzymatic kinetic resolution of amines: Faber K (2000) Biotransformations in organic chemistry, 4th edn. Springer, Berlin Heidelberg New York, Sect. 3.1.3
58. For examples, see: Ie Y, Fu GC (2000) J Chem Soc Chem Commun 119–120; Al-Sehemi AG, Atkinson RS, Fawcett J, Russell DR (2000) Tetrahedron Lett 41:2239–2242
59. Arai S, Bellemin-Laponnaz S, Fu GC (2001) Angew Chem Int Ed 40:234–236
60. Eames J (2000) Angew Chem Int Ed 39:885–888
61. Note added in proof: The benzoylation of diols using zinc- and copper-based catalysts has been reported, see Trost BM, Mino T (2003) J Am Chem Soc 125:2410-2411; Matasumura Y, Maki T, Murakami S, Onomura O (2003) J Am Chem Soc 125:2052-2053

Chapter 44
Metathesis Reactions

Amir H. Hoveyda[1], Richard R. Schrock[2]

[1] Department of Chemistry, Merkert Chemistry Center, Boston College, Chestnut Hill,
Massachusetts 02467, USA
e-mail: amir.hoveyda@bc.edu

[2] Department of Chemistry Massachusetts Institute of Technology, Cambridge,
Massachusetts 02319, USA
e-mail: rrs@mit.edu

Keywords: Asymmetric synthesis, Chiral catalysis, Mo-based catalysts, Natural product synthesis, Olefin metathesis, Recyclable catalysts, Ru-based catalysts, Supported chiral catalysts

1
Introduction

The field of organic synthesis has been strongly influenced by catalytic olefin metathesis during the past decade [1]. Ru- and Mo-catalyzed olefin metathesis are now used regularly to prepare a wide range of compounds, including small, medium, and large rings [2]. Metathesis-based approaches are being employed with such increasing frequency that they are now considered relatively routine.

With regard to the synthesis of optically pure materials, however, olefin metathesis has largely served an auxiliary role. In cases where ring-closing metathesis (RCM) is called for [2a–c, 2e–f, 2h], an already-optically pure diene substrate is treated with an achiral metal catalyst to deliver a non-racemic cyclic unsaturated product. Alternatively, a racemic product obtained by metathesis may be catalytically resolved [2b]. Optically enriched cyclic alkenes are similarly employed in instances where ring-opening metathesis (ROM) is needed [2d, 2g]. Although such strategies have led to a number of notable and impressive accomplishments in asymmetric synthesis, some of the unique attributes of catalytic olefin metathesis can only be realized if chiral optically pure catalysts for olefin metathesis are available. This claim is tied to the fact that one of the most useful characteristics of metathetic processes is their ability to promote efficient skeletal rearrangements: simple achiral or racemic substrates can be transformed into complex non-racemic organic molecules with an effective chiral metathesis catalyst. In numerous instances, products that are rendered readily available by a chiral metathesis catalyst would only be accessible, and often less selectively, by a much longer route if alternative synthetic methods were to be used.

2
Mo-Based Chiral Olefin Metathesis Catalysts

2.1
The Catalyst Construct

The makeup of Mo-based complexes, represented by 1 [3], offers an attractive opportunity for the design, synthesis, and development of effective chiral metathesis catalysts. This claim is based on several factors: 1) Mo-based catalysts such as 1 possess a modular structure [4] involving imido and alkoxide moieties that do not disassociate from the metal center in the course of the catalytic cycle. Any structural alteration of these ligands may thus lead to a notable effect on the reaction outcome and could be employed to control selectivity and reactivity. 2) Alkoxide moieties offer an excellent opportunity for incorporation of chirality within the catalyst structure through installment of non-racemic tethered chiral bis(hydroxy) ligands. 3) Mo-based complexes provide appreciable levels of activity and may be utilized to prepare highly substituted alkenes.

With the above considerations in mind, we prepared and examined a myriad of chiral Mo-based catalysts for both asymmetric RCM (ARCM) and ROM (AROM) transformations [5]. In this article, several efficient and enantioselective reactions that are catalyzed by these chiral complexes are discussed [6]. The structural modularity inherent to the Mo-based systems allows screening of catalyst pools, so that optimal reactivity and selectivity levels are identified expeditiously.

2.2
Mo-Catalyzed Kinetic Resolution with Hexafluoro-Mo Catalysts [7]

The preparation and catalytic activity of chiral complex 2, based on the original Mo-alkylidene 1, has been reported by Grubbs and Fujimura [8]. These workers report on the kinetic resolution of various dienes [9]. As illustrated by the resolution of 3, however, levels of enantiodifferentiation were typically low ($k_{rel}<3$).

2.3
Chiral Biphen-Mo Catalysts

To examine the possibility of a more efficient catalytic olefin metathesis, we prepared chiral Mo-based catalysts, **4a** and **4b** [10]. This approach was not without precedence: related chiral Mo complexes were initially synthesized in 1993 and were used to promote polymer synthesis [6]. We judged that these biphen-based complexes would be able to initiate olefin metathesis with high levels of asymmetric induction due to their rigidity and steric attributes. Chiral complexes **4a** and **4b** are orange solids, stable indefinitely when kept under inert atmosphere.

4a R = *i* Pr

4b R = Me

X-ray structure of **4a**

2.4
Catalytic Kinetic Resolution Through Mo-Catalyzed ARCM

The catalytic kinetic resolution of various dienes through ARCM can be carried out in an efficient manner at 22 °C in the presence of 5 mol % **4a** [10]. As the data in Scheme 1 illustrate, 1,6-dienes **5–7** are resolved with excellent levels of enantiocontrol (k_{rel}>20) [11]. Chiral complex **4a** readily promotes the resolution of allylic ethers **8–10** as well [12].

The higher levels of enantioselectivity attained through the use of **4a** (versus **2**) is likely due to a preference for ARCM reactions to proceed through intermediates such as **I** (Scheme 1). The intermediacy (higher reactivity) of the *anti* Mo-alkylidene (alkylidene C–C *anti* to Mo=N) is based on previous mechanistic studies [13]. The stereochemistry of olefin–transition metal association is according to the position of the LUMO of the chiral complex [13b, 14]. The 1,1-disubstituted olefin interacts with Mo away from the protruding *t*-Bu group of the diolate and *i*-Pr groups of the imido ligands (see X-ray of **4a**).

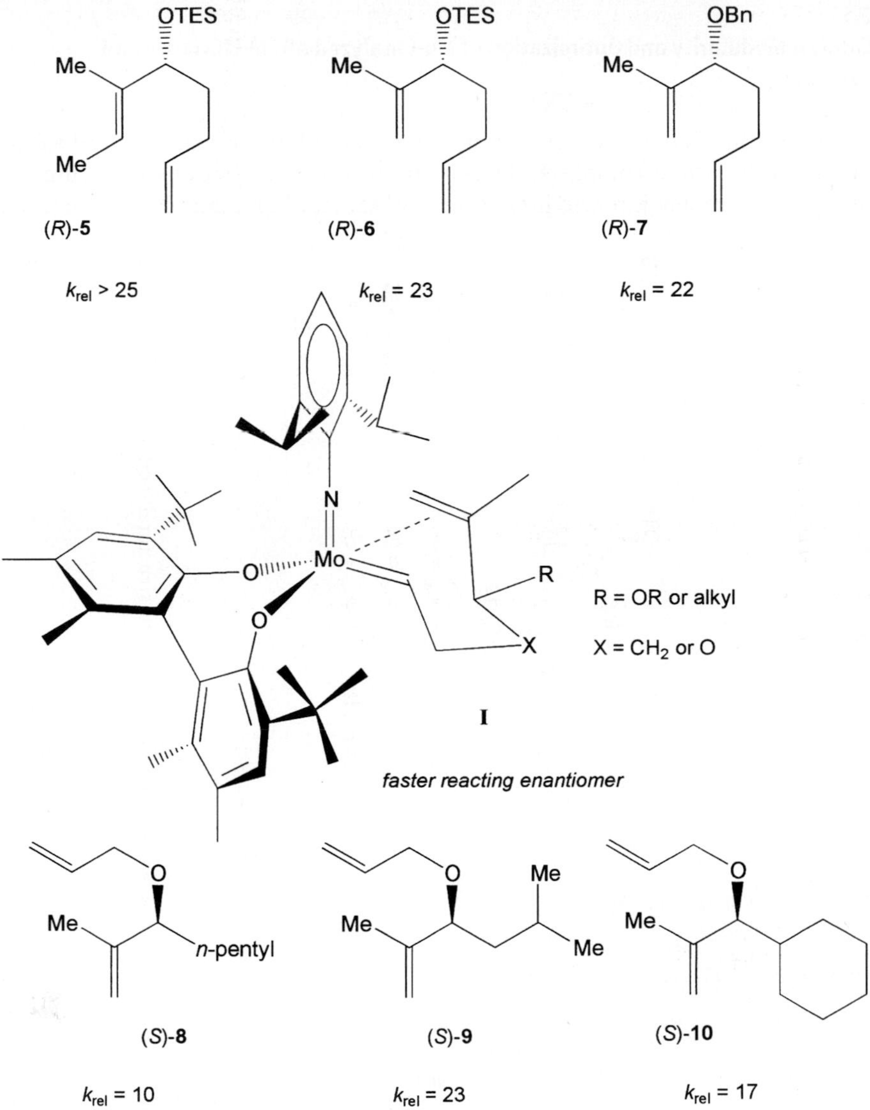

Scheme 1. Mo-catalyzed kinetic resolution of 1,6-dienes through ARCM

2.5
Catalyst Modularity and Optimization of Mo-Catalyzed ARCM Efficiency and Selectivity

In spite of the high asymmetric induction observed in the Mo-catalyzed ARCM of 1,6-dienes, when complexes **4a** and **4b** are used in reactions involving 1,7-dienes, inferior asymmetric induction is obtained. For example, as illustrated

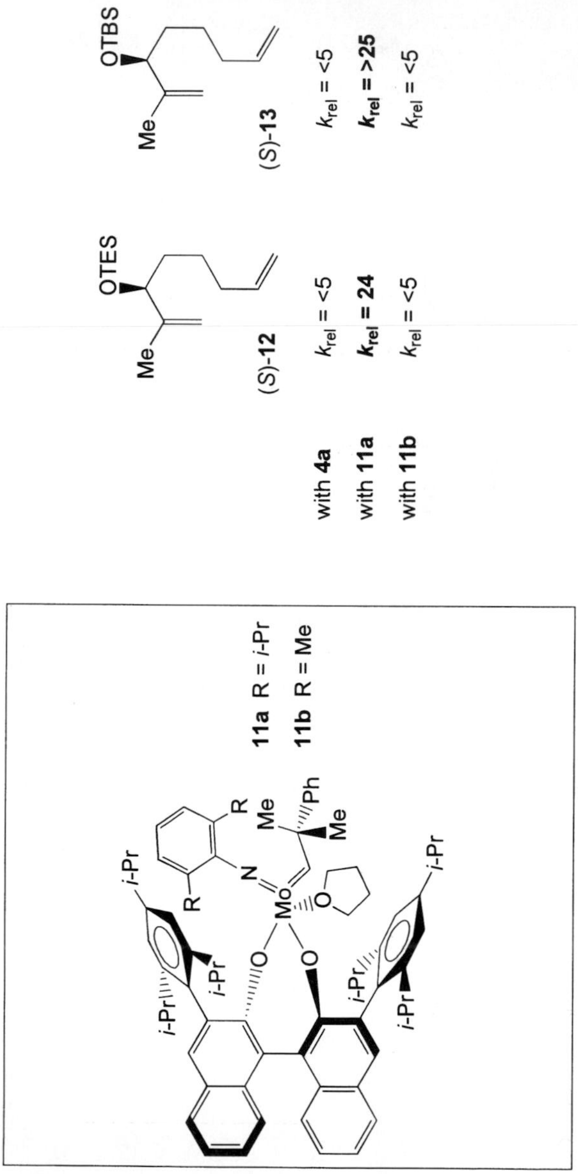

Scheme 2. Mo-catalyzed kinetic resolution of 1,7-dienes and the importance of subtle structural modification of the chiral catalysts

in Scheme 2, dienes **12** and **13** are not resolved with useful selectivity (k_{rel}<5) when **4a** is employed as the catalyst. To address this shortcoming, we took advantage of the modular character of the Mo complexes and prepared other chiral complexes. As depicted in Scheme 2, we discovered that BINOL-based catalyst **11a** promotes the RCM of dienes **12** and **13** with outstanding levels of selectivity (k_{rel}=24 and >25, respectively) [15]. BINOL-based complex **11b**, bearing the (dimethyl)phenylimido ligand (versus (diisopropyl)phenylimido of **11a**), is not an efficient catalyst for kinetic resolutions of the dienes **12** and **13**.

The data in Scheme 3 illustrate that a wide range of 1,7-dienes can be resolved with high selectivity and efficiency. These findings provide further evidence regarding the importance of the availability of a diverse set of chiral catalysts. Although BINOL-based complexes (e.g., **11a**) typically promote ARCM

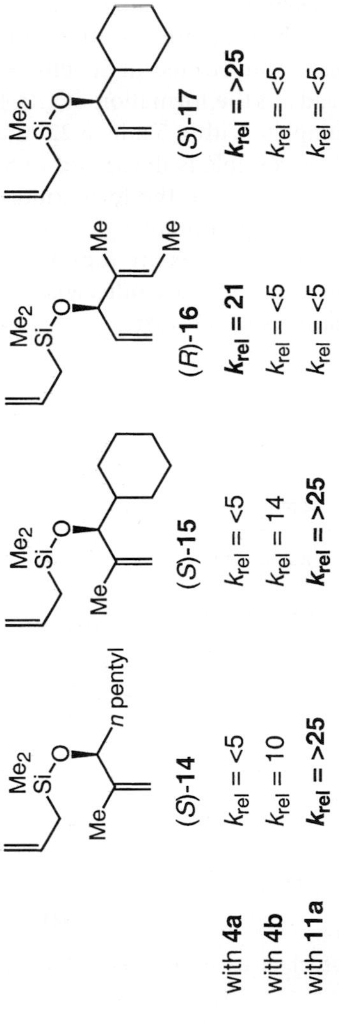

Scheme 3. Small structural changes within the substrate structure can alter the identity of the optimum chiral metathesis catalyst

	(S)-14	(S)-15	(R)-16	(S)-17
with **4a**	k_{rel} = <5	k_{rel} = <5	k_{rel} = 21	k_{rel} = >25
with **4b**	k_{rel} = 10	k_{rel} = 14	k_{rel} = <5	k_{rel} = <5
with **11a**	k_{rel} = >25	k_{rel} = >25	k_{rel} = <5	k_{rel} = <5

of 1,7-dienes with higher selectivity than the biphen-based catalysts (e.g., **4a**), such a generalization is not always justified. As expected, **11a** catalyzes the kinetic resolution of 1,7-dienes **14** and **15** with $k_{rel} > 25$. Unlike biphen complex **4a**, however, the closely related catalyst **4b** also provides appreciable enantioselection, albeit less effectively than **11a**. With substrates **16** and **17**, in which two terminal alkenes are involved, the situation is reversed: now, it is the biphen-based complex **4a** that is the efficient catalyst. Although each catalyst is not optimal in every instance, efficient kinetic resolution of a wide range of chiral oxygenated 1,6- and 1,7-dienes can be achieved by various chiral Mo complexes.

2.6
Catalytic Asymmetric Synthesis Through Mo-Catalyzed ARCM

The arena in which catalytic asymmetric olefin metathesis can have the largest impact on organic synthesis is the desymmetrization of readily accessible achiral molecules. Two examples are illustrated in Scheme 4. Treatment of achiral triene **18** with 2 mol % **4a** leads to the formation of (*R*)-**19** in 99% *ee* and 93% yield [12]. The reaction is complete within 5 min at 22 °C and, importantly, does not require a solvent. Another example is illustrated in Scheme 4 as well; here, BINOL complex **11a** is used to promote the formation of optically pure (*R*)-**21** from siloxy triene **20** in nearly quantitative yield. Once again, solvent is not needed [15]. Readily accessible substrates are rapidly transformed to non-racemic optically enriched molecules that are otherwise significantly more difficult to access without generating solvent waste.

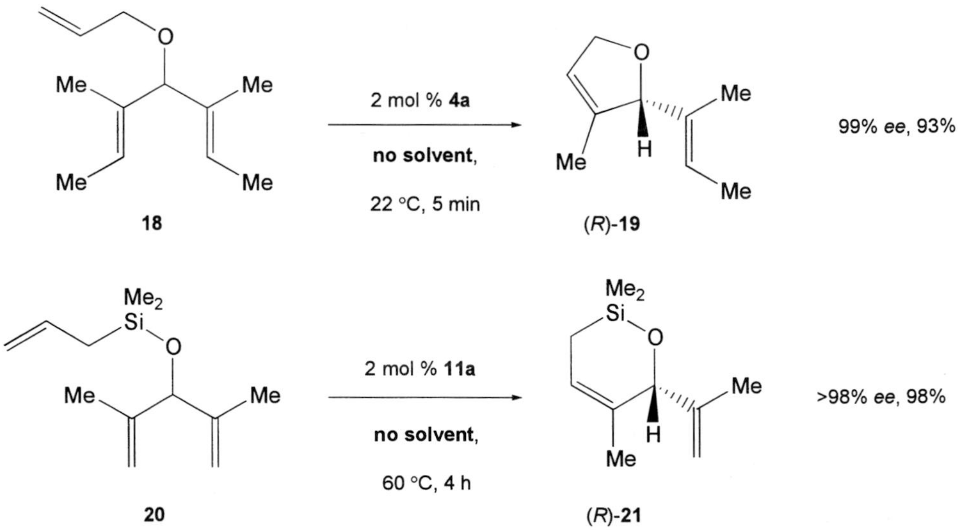

Scheme 4. Mo-catalyzed ARCM of achiral trienes can be effected efficiently, enantioselectively, and in the absence of solvent

In connection with reactions where a solvent is required, it must be noted that all transformations promoted by chiral Mo catalysts may be carried out in toluene (in addition to benzene) or various alkanes (e.g., *n*-pentane) with equal efficiency (see below for specific examples). Moreover, although 5 mol % catalyst is typically used in our studies, 1–2 mol % loading often delivers equally efficient and selective transformations.

As the above studies predicate, reaction of **18** is significantly less efficient with **11a** (<5% conversion in 18 h) and that of **20** proceeds only to 50% conversion in 24 h in the presence of **4a** (65% ee). Remarkably, in the latter transformation, even in a 0.1 M solution, the major product is the homodimer formed through metathesis of the terminal alkenes. The absence of homodimer generation when **11a** is used as the catalyst, particularly in the absence of any solvent, bears testimony to the high degree of catalyst–substrate specificity in these catalytic C–C bond-forming reactions.

The catalytic desymmetrization shown in Scheme 5 involves a *meso*-tetraene substrate: optically pure unsaturated siloxane **23** is obtained in >99% ee and 76% yield [16]. The unreacted siloxy ether moiety is removed to afford optically pure **24**. Mo-alkylidenes derived from both enantiotopic terminal alkenes in **22** are likely involved. Since the initial metal-alkylidene generation is rapidly reversible, the major product arises from the rapid RCM of the "matched" segment of the tetraene. If any of the "mismatched" RCM takes place, a subsequent and more facile matched RCM leads to the formation of *meso*-bicycle. Such a by-product is absent from the unpurified mixture containing **23**, indicating the exceptionally high degree of stereodifferentiation induced by the chiral Mo com-

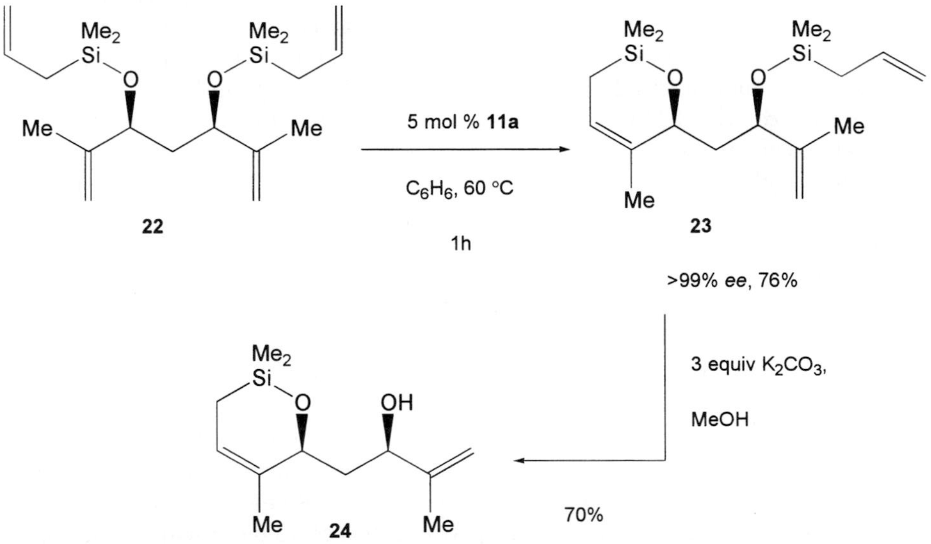

Scheme 5. Mo-catalyzed desymmetrization of *meso* tetraenes proceeds to afford optically pure heterocycles

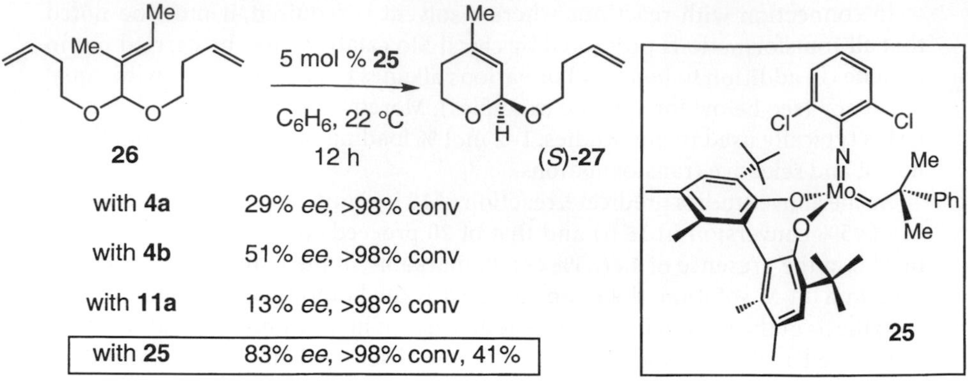

Scheme 6. Chiral complex 25, bearing a 2,6-dichloro imido ligand is the catalyst of choice for asymmetric synthesis of acetals

plex in this transformation. As before, catalyst **4a** is not effective in promoting ARCM of **22**.

Incorporation of electron-withdrawing groups within either the imido or diolate segments of Mo complexes might result in higher levels of catalytic activity, since the Lewis acidity of the transition metal center is enhanced. As the representative examples in Scheme 6 depict, such structural modifications have a profound effect on the levels of enantioselectivity as well. In the desymmetrization of acetal **26**, dichlorophenylimido complex **25** provides substantially higher levels of asymmetric induction than biphen- or BINOL-based catalysts that carry 2,6-dialkylphenylimido moieties (e.g., **11a**). Acetals of the type represented by **27** in Scheme 6 retain their stereochemical integrity through various routine operations such as silica gel chromatography and can be readily functionalized to deliver a range of chiral non-racemic heterocyclic compounds [14].

The Mo-catalyzed ARCM technology summarized above has been utilized in a brief and enantioselective total synthesis of *exo*-brevicomin (**30**) by Burke. The key step, as illustrated in Scheme 7, is the desymmetrization of achiral triene **28** [17].

Mo-catalyzed ARCM may be used in the enantioselective synthesis of medium ring carbo- and heterocycles [18]. As shown in Scheme 8, medium ring tertiary siloxanes (e.g., **35**), are prepared with high levels of enantioselectivity. These processes can be effected efficiently in preparative scale and at low catalyst loading (e.g., **33**→**35**). Such attributes render the catalytic enantioselective method attractive from the practical point of view.

The representative transformation in Scheme 9 illustrates that the optically enriched siloxanes obtained by Mo-catalyzed ARCM can be further functionalized to afford tertiary alcohols (e.g., **39**) with excellent enantio- and diastereomeric purity. Conversion of **38** to **39** in Scheme 9 is carried out without solvent, at 1 mol % catalyst loading and on 1-g scale (only 30 mg of catalyst **4a** is needed) [18].

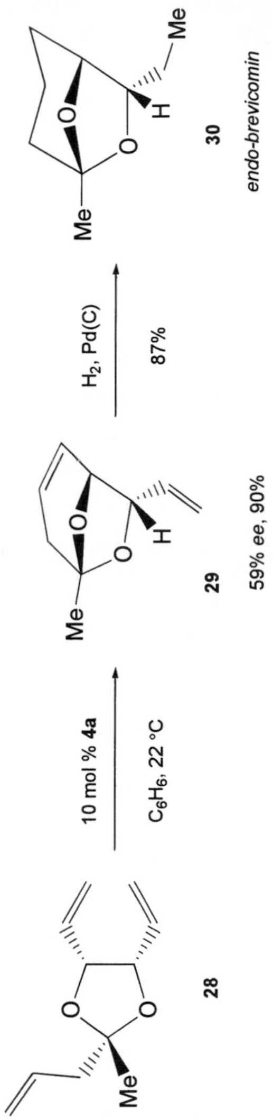

Scheme 7. Application of Mo-catalyzed ARCM to the synthesis of brevicomin

Most recent studies indicate that ARCM can be used to synthesize small and medium ring N-containing unsaturated heterocycles in high yield and with excellent ee through catalytic kinetic resolution and asymmetric synthesis [19]. As the representative data in Scheme 10 illustrate, levels of enantioselectivity can vary depending on the nature of the arylamine (compare **44** to **46**). As the synthesis of **48** indicates (cf. Scheme 10), the facility and selectivity with which medium ring unsaturated amines are obtained by the Mo-catalyzed protocol is particularly noteworthy.

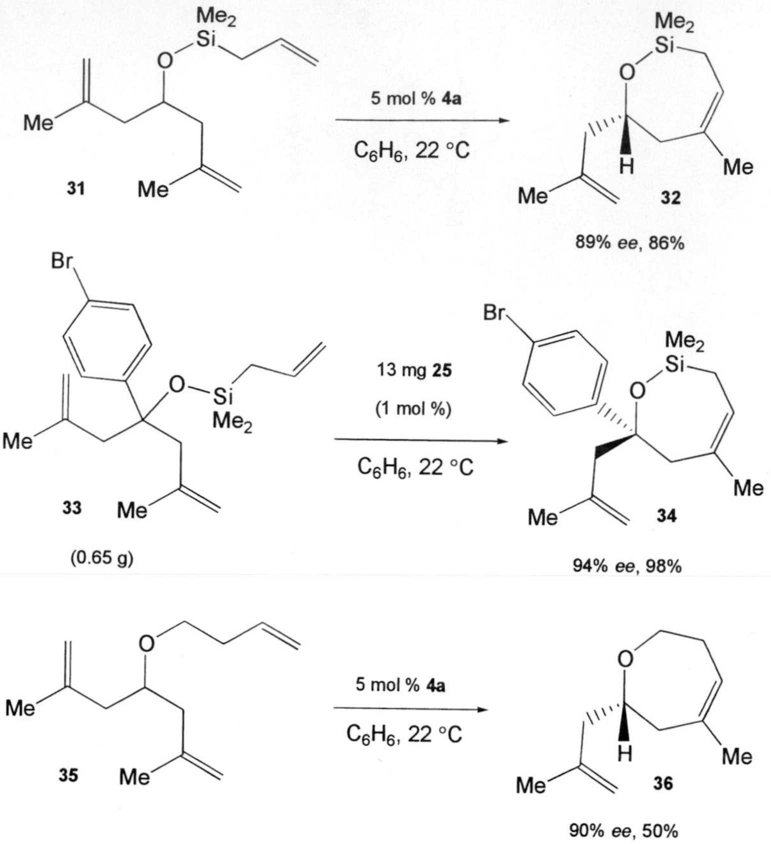

Scheme 8. Mo-catalyzed tandem ARCM can be used to synthesize seven-membered carbo- and heterocyclic structures efficiently and in optically enriched form

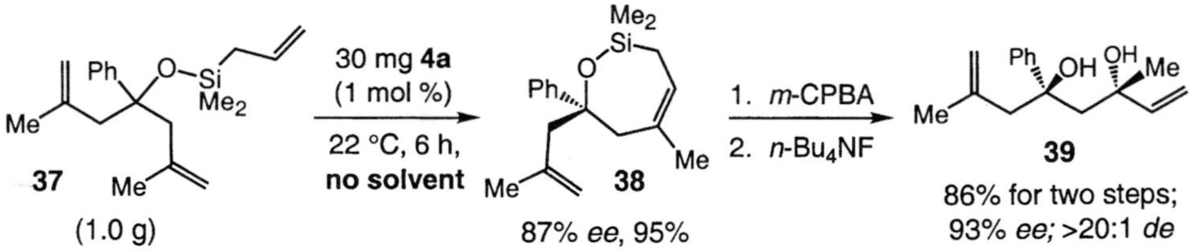

Scheme 9. Mo-catalyzed tandem ARCM can be used to synthesize synthetically versatile intermediates such as 1,3-tertiary diols in high enantio- and diastereopurity

Unlike carbocyclic and oxygen-containing heterocyclic systems, catalytic enantioselective synthesis of eight-membered ring amines proceeds efficiently and with excellent enantioselectivity. These catalytic ARCM reactions can be carried out in the absence of solvent as well. Representative data regarding cat-

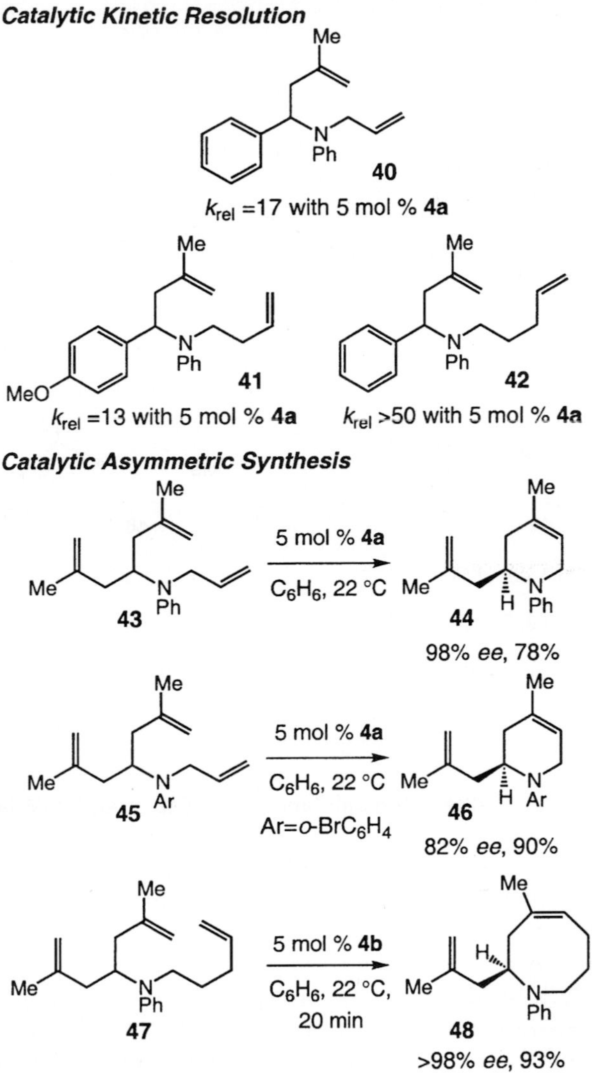

Catalytic Kinetic Resolution

40

k_{rel} =17 with 5 mol % **4a**

41

k_{rel} =13 with 5 mol % **4a**

42

k_{rel} >50 with 5 mol % **4a**

Catalytic Asymmetric Synthesis

5 mol % **4a**

C$_6$H$_6$, 22 °C

43   **44**

98% *ee*, 78%

5 mol % **4a**

C$_6$H$_6$, 22 °C

Ar=*o*-BrC$_6$H$_4$

45   **46**

82% *ee*, 90%

5 mol % **4b**

C$_6$H$_6$, 22 °C,

20 min

47   **48**

>98% *ee*, 93%

Scheme 10. Enantioselective synthesis of amines through Mo-catalyzed ARCM

alytic enantioselective synthesis of various N-containg heterocycles without the use of solvent is depicted in Scheme 11. This efficient enantioselective method again highlights the ability of asymmetric metathesis to deliver synthetically versatile materials that are otherwise difficult to prepare.

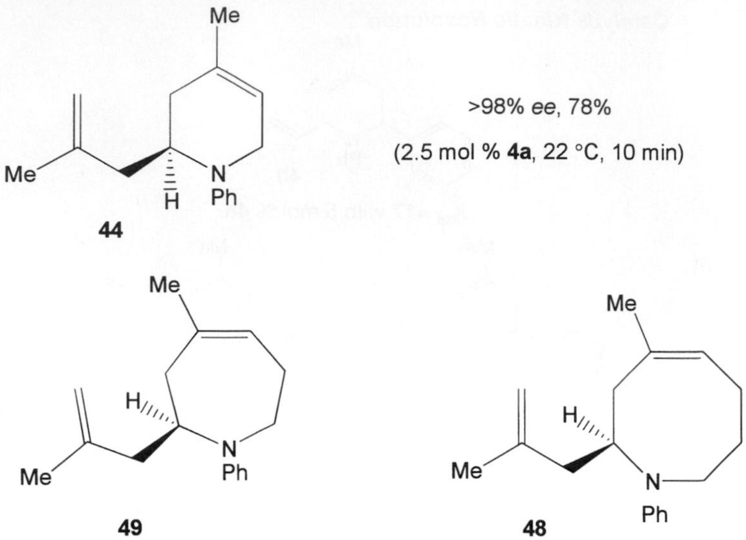

>98% ee, 78%

(2.5 mol % **4a**, 22 °C, 10 min)

44

49   **48**

95% ee, 95%   97% ee, >98%

(3 mol % **4a**, 22 °C, 3.5 h)   (4 mol % **25**, 22 °C, 7 h)

Scheme 11. Catalytic enantioselective synthesis of amines in the absence of solvent through Mo-catalyzed ARCM

2.7
Catalytic Asymmetric Synthesis Through Tandem Mo-Catalyzed AROM/RCM

The appreciable levels of asymmetric induction observed in the catalytic ARCM reactions mentioned above suggest a high degree of enantiodifferentiation in the association of olefinic substrates and chiral complexes. This stereochemical induction may also be exploited in asymmetric ring-opening metathesis (AROM). Catalytic ROM transformations [20] offer unique and powerful methods for the preparation of complex molecules [2d, 2g]. The chiral Mo-alkylidenes that are products of AROM reactions can be trapped either intramolecularly (RCM) or intermolecularly (cross metathesis, CM) to afford a range of optically enriched adducts.

Transformations shown in Schemes 12–14 constitute the first examples of catalytic AROM reactions ever reported. *Meso*-triene **50** is converted to chiral heterocyclic triene **32** in 92% ee and 68% yield in the presence of 5 mol % **4a** (Scheme 12) [21]. Presumably, stereoselective approach of the more reactive cyclobutenyl alkene in the manner shown in Scheme 12 (**II**) leads to the enantioselective formation of Mo-alkylidene **III**, which in turns reacts with an adjacent terminal olefin to deliver **51**. Another example in Scheme 12 involves the net rearrangement of *meso*-bicycle **52** to bicyclic structure **54** in 92% ee and 85% yield. The reaction is promoted by 5 mol % **4a** and requires the presence of di-

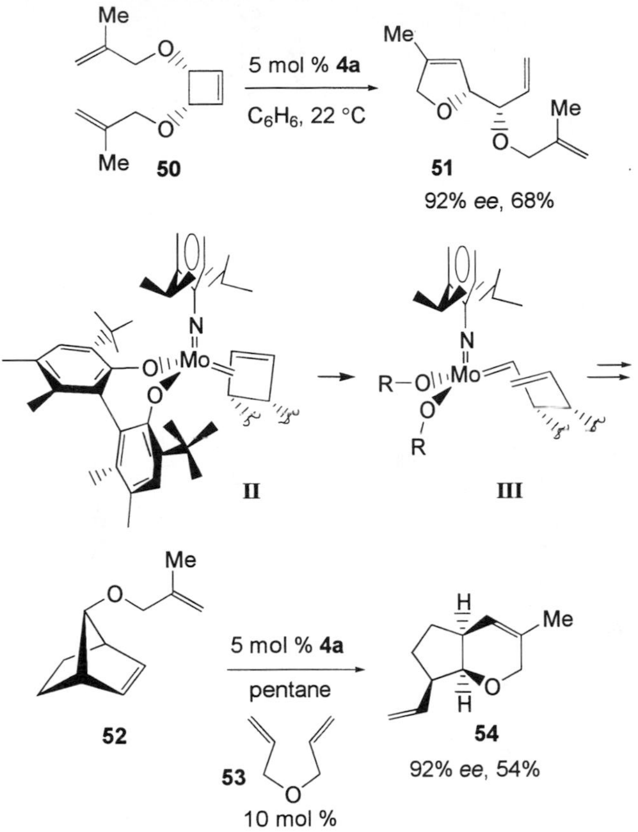

Scheme 12. Mo-catalyzed tandem AROM / RCM allows access to complex heterocyclic structures efficiently and in optically enriched form

allyl ether **53** [22]. Mechanistic studies suggest that initial reaction of **53** with **4a** leads to the formation of the substantially more reactive chiral Mo=CH$_2$ complex (versus neophylidene **4a**) which then can react with the sterically hindered norbornyl alkene to initiate the catalytic cycle.

In contrast to **52** (Scheme 12), diastereomer **55** (Scheme 13), because of its more exposed and highly reactive strained olefin, undergoes rapid polymerization in the presence of **4a**. The less reactive Ru complex **56** [23] can however be used under an atmosphere of ethylene to effect a tandem ROM/CM to generate **57**. The resulting triene can be induced to undergo Mo-catalyzed ARCM (5 mol % **4a**) to afford optically pure **58**, the AROM/RCM product that would be obtained from **55**.

The Mo-catalyzed transformations shown in Scheme 14 may also be viewed as AROM/RCM processes [24]. Furthermore, it is possible that initiation occurs at the terminal olefin, followed by an ARCM involving the cyclic alkene. Regardless of these mechanistic possibilities, the enantioselective rearrangements

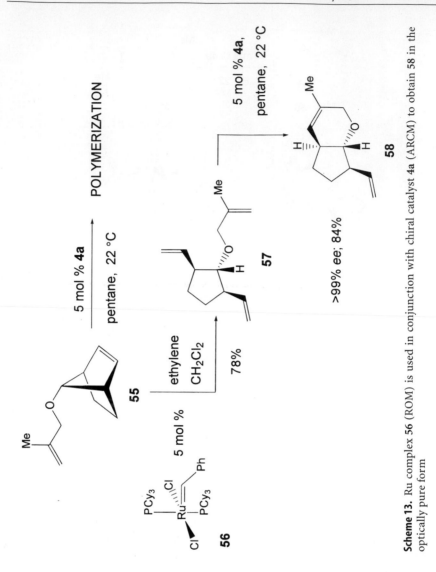

Scheme 13. Ru complex **56** (ROM) is used in conjunction with chiral catalyst **4a** (ARCM) to obtain **58** in the optically pure form

shown in Scheme 14, catalyzed by binaphtholate-based catalyst **11a**, deliver unsaturated pyrans bearing a tertiary ether site with excellent efficiency and enantioselectivity. This class of heterocycles would not be readily accessible by an enantioselective synthesis of the precursor diene, followed by RCM promoted by an achiral catalyst. The requisite optically enriched pure tertiary ether or alcohol cannot be easily accessed by any available methods. It should be noted that in this class of asymmetric reactions, biphenolate-based complexes provide significantly lower levels of enantioselectivity (e.g., **4a** affords **60** in 15% ee). The asymmetric synthesis of the pyran portion of the anti-HIV agent tiprana-

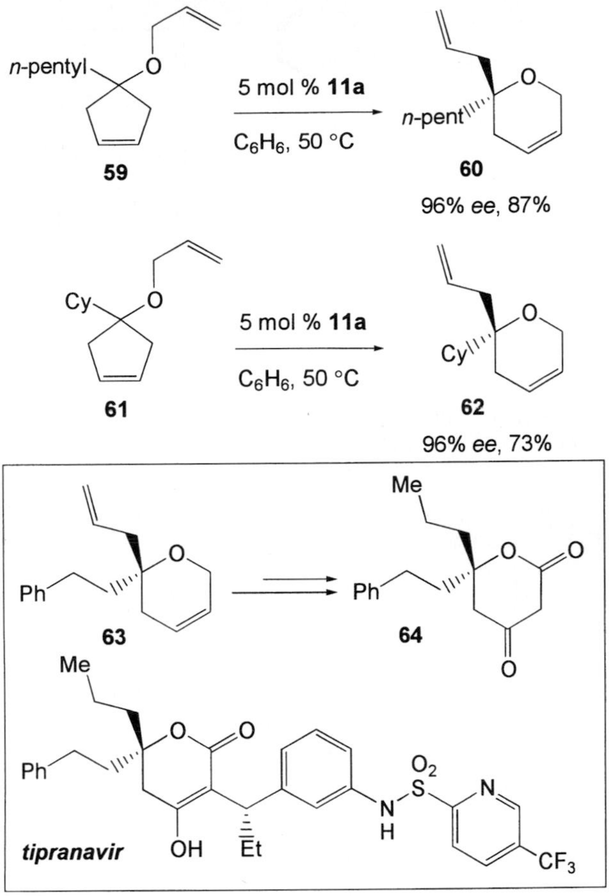

Scheme 14. Mo-catalyzed enantioselective rearrangement of *meso*-cyclopentenes to chiral unsaturated pyrans

vir (Scheme 14) serves to demonstrate the significant potential of the method in asymmetric synthesis of biomedically important agents.

The non-racemic pyrans shown in Scheme 14 can be accessed by Mo-cata-lyzed ARCM of the corresponding trienes. The example shown in Eq. (1) is illus-trative. Interestingly, elevated temperatures are required for high levels of enan-tioselectivity; under conditions shown in Scheme 15, trienes such as **65** afford the desired pyrans in significantly lower ee (e.g., **66** is obtained in 30% ee at 50 °C). Clearly, detailed mechanistic studies must be carried out before the ori-gin of such variations in selectivity is understood.

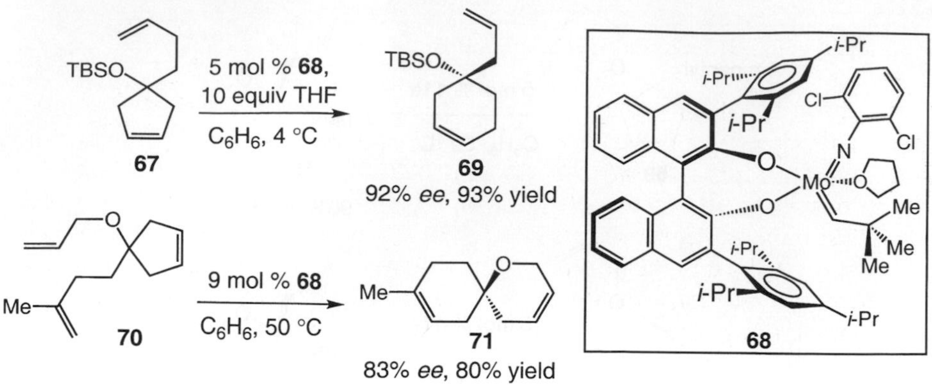

Scheme 15. Enantioselective synthesis of carbocyclic tertiary ethers and spirocycles through Mo-catalyzed asymmetric olefin metathesis

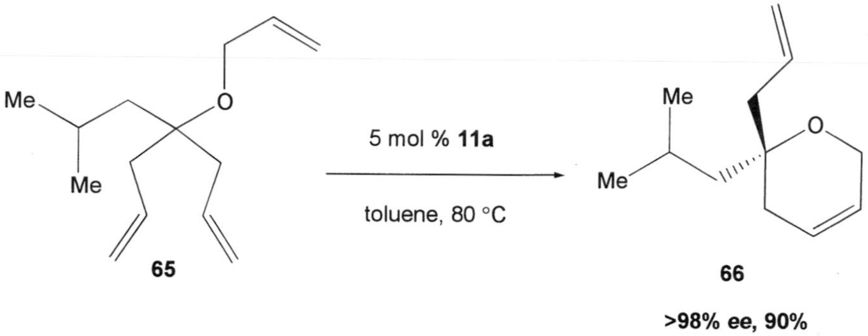

$$(1)$$

A process related to those shown in Scheme 14 involves the asymmetric Mo-catalyzed conversion of tertiary carbocyclic cyclopentenyl ethers to the corresponding cyclohexenyl ethers with enantioselectivity (e.g., **67**→**69**, Scheme 15) [25]. A remarkable and unusual attribute of this class of transformations is that significantly higher levels of enantioselectivity are observed when ten substrate equivalents of THF are used as an additive. As an example, **69** (Scheme 15) is formed in only 58% ee in the absence of THF (<5% conversion is observed when THF is used as solvent). As also shown in Scheme 15 (**70**→**71**), enantioenriched spirocycles may be accessed easily by a similar approach; in this case, no additive effect is observed.

2.8
Catalytic Asymmetric Synthesis Through Tandem Mo-Catalyzed AROM/CM

The chiral Mo-alkylidene complex derived from AROM of a cyclic olefin may also participate in an intermolecular cross metathesis reaction. As depicted in Scheme 16, treatment of *meso*-72a with a solution of 5 mol % 4a and 2 equivalents of styrene leads to the formation of optically pure 73 in 57% isolated yield and >98% *trans* olefin selectivity [26]. The Mo-catalyzed AROM/CM reaction can be carried out in the presence of vinylsilanes: the derived optically pure 74 (Scheme 16) may subsequently be subjected to Pd-catalyzed cross-coupling reactions, allowing access to a wider range of optically pure cyclopentanes.

The Mo-catalyzed AROM/CM can be effected on highly functionalized norbornyl substrates (e.g., 76 and 77 in Scheme 17) and those that bear tertiary ether sites (e.g., 79–81). Although initial studies indicate that the relative orientation of the heteroatom substituent versus the reacting olefin can have a significant influence on reaction efficiency, the products shown in Scheme 17 repre-

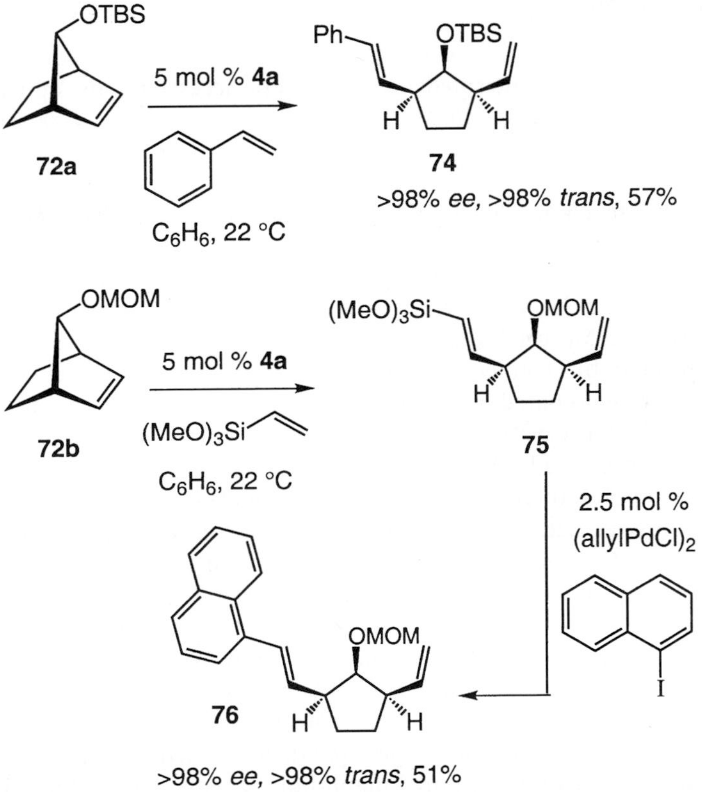

Scheme 16. Mo-catalyzed tandem AROM/CM proceeds with high enantioselectivity and olefin stereocontrol

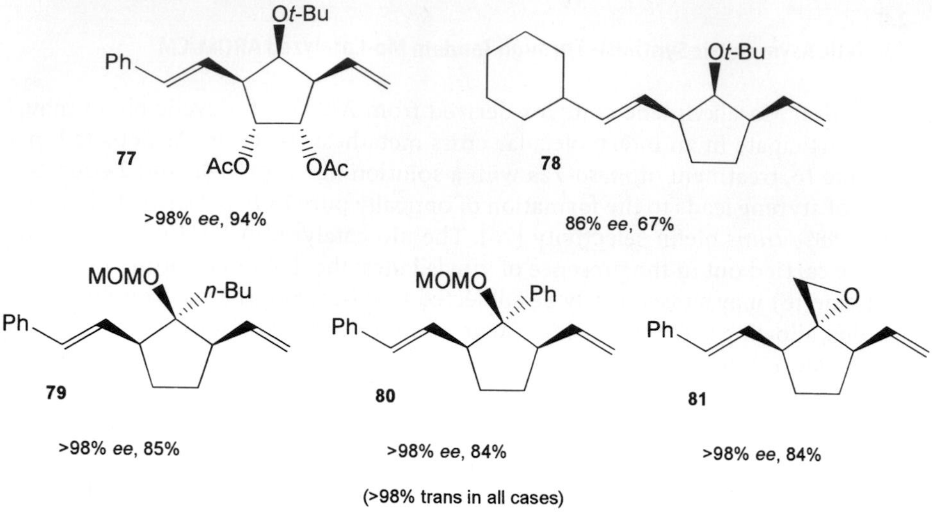

Scheme 17. Mo-catalyzed tandem AROM/CM delivers highly functionalized cyclopentanes in the optically pure form

sent versatile synthetic intermediates that can be accessed in the optically pure form by Mo-catalyzed AROM/CM.

2.9
Towards User-Friendly and Practical Chiral Mo-Based Catalysts for Olefin Metathesis

Although the main focus of our programs have been on issues of reactivity and enantioselectivity, we have recently begun to address the important issue of practicality in Mo-catalyzed asymmetric metathesis. Two key advances have been reported in this connection: (1) The availability of a general chiral Mo catalyst that can be prepared in situ from commercially available compounds. (2) The synthesis of a recyclable polymer-supported chiral Mo catalyst. These advances are summarized below.

Up to this point, there have been two general classes of chiral Mo catalysts discussed: Biphenolate-based complexes such as **4** and binaphtholate systems represented by **11** (Scheme 2). From a practical point of view, binaphthol-based systems have an advantage: the synthesis of the optically pure diolate begins from the inexpensive and commercially available (R) or (S)-binaphthol. In contrast, access to the optically pure biphenol ligand in **4** and its derivatives requires resolution of the racemic samples by fractional crystallization of the derived phosphorus(V) mentholates [2]. Accordingly, we prepared chiral Mo complex **82** [27], bearing a "biphenol-type" ligand; however, this is synthesized from the readily available optically pure binaphthol (Scheme 18). Complex **82** shares structural features with both the biphen- (**4**) and BINOL-based (**11**) systems and represents an intriguing possibility regarding the range of starting ma-

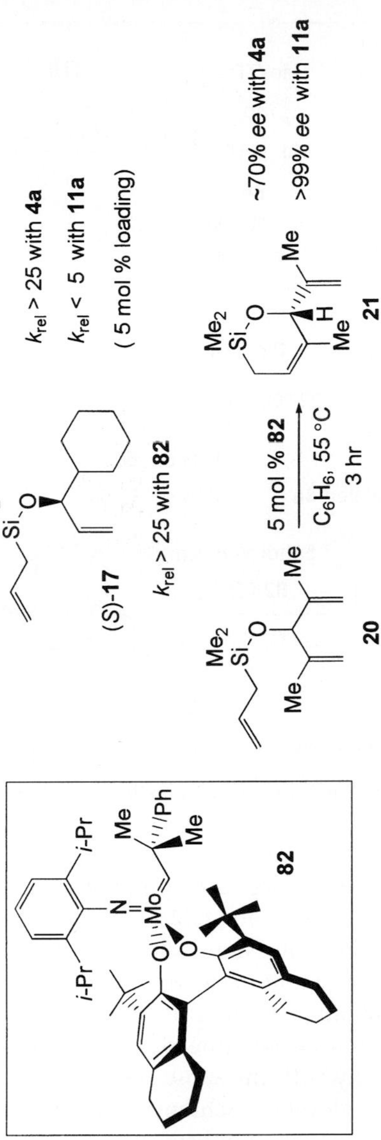

Scheme 18. Chiral complex **82** represents a hybrid between biphen- and BINOL-based catalysts and provides a unique selectivity profile that is often not seen with the latter two classes individually

terials for which it may be a suitable catalyst. Two examples are depicted in Scheme 18: catalyst **82** delivers compounds of high optical purity where either biphen- or BINOL-based complexes are ineffective. It is not in all instances that **82** operates as well as **4a** and **11a**. As an example, in the presence of 5 mol % **82**, triene **18** (Scheme 4) is converted to furan **19** in 77% ee and 73% yield (versus 99% ee and 93% yield with **4a**). It should be noted that catalyst **82** serves as an additional example where modification of the chiral alkoxide ligand can lead to substantial (and not easily predictable) variation in selectivity.

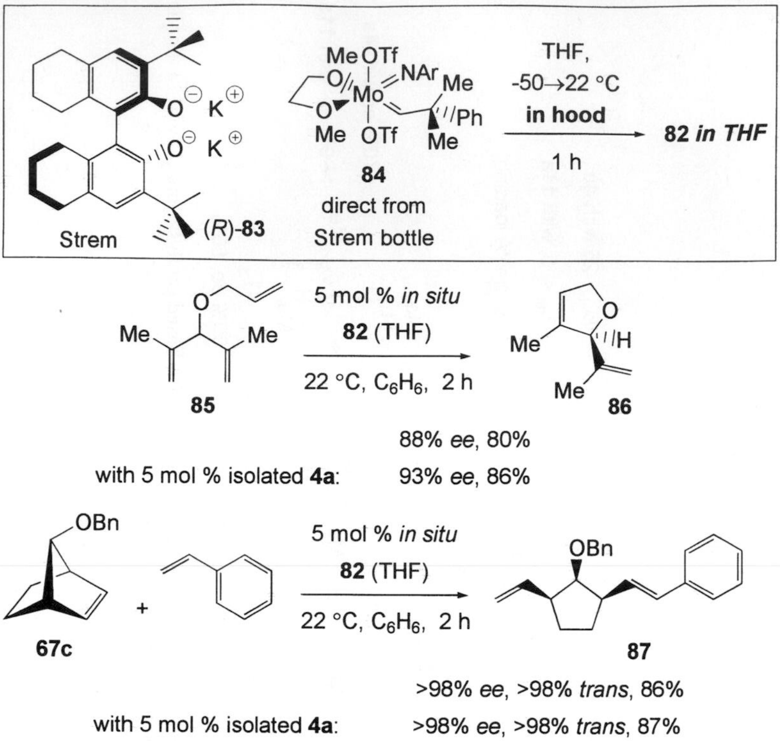

Scheme 19. In situ preparation and utility of chiral metathesis catalyst **82**

It is not only that catalyst **82** is more easily prepared than **4**; more important-ly, as illustrated in Scheme 19, a solution of **82**, obtained by the reaction of com-mercially available reagents bis(potassium salt) **83** and Mo triflate **84** (Strem), can be directly used to promote enantioselective metathesis. Similar levels of re-activity and selectivity are obtained with in situ **82** as with isolated and purified **4a** or **82** (cf. Scheme 19). Moreover, asymmetric olefin metatheses proceed with equal efficiency and selectivity with the same stock solutions of (R)-**83** and **84** after two weeks. The use of a glovebox, Schlenck equipment, or vacuum lines is not necessary (even with the two-week old solutions).

More recently, we have synthesized and studied the activity of **89**, the first supported chiral catalyst for olefin metathesis (Scheme 20) [28]. Catalyst **89** ef-ficiently promotes a range of ARCM and AROM processes; a representative ex-ample is shown in Scheme 20. Rates of reaction are lower than observed with the corresponding monomeric complex, but similar levels of enantioselectivity are observed. Although **89** must be kept under rigorously dry and oxygen-free conditions, it can be recycled. Catalyst activity, however, is notably diminished by the third cycle. As the data and the figure in Scheme 20 show, the product so-lution obtained by filtration contains significantly lower levels of metal impu-

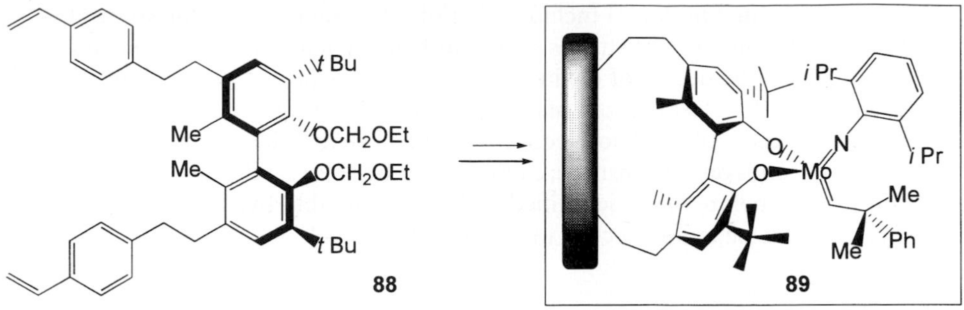

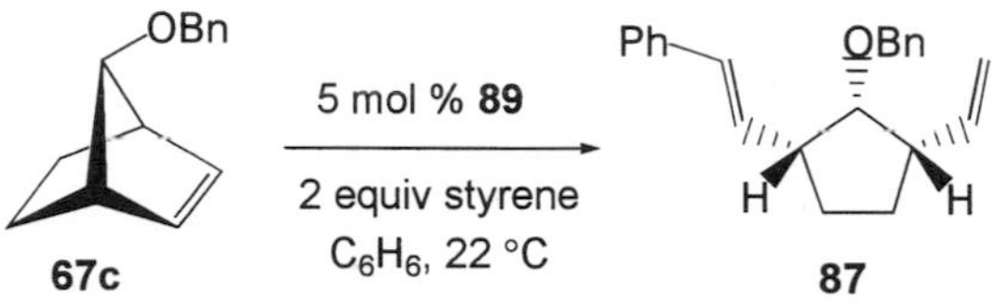

Cycle 1: >98% conv, 30 min; 97% *ee*
product contains 3% of total Mo initially used

Cycle 2: 98% conv, 30 min; 98% *ee*
product contains 10% of total Mo initially used

Cycle 3: 55% conv, 16 h; 89% *ee*
product contains 16% of total Mo initially used

Scheme 20. The first recyclable and supported chiral catalyst for olefin metathesis, **89** delivers reaction products that contain significantly less metal impurity. The two dram vials show unpurified **87** from a reaction catalyzed by **4a** (*left*) and **89** (*right*)

rity than observed with the monomeric catalysts, where >90% of the Mo used is found in the unpurified product (ICP–MS). This first generation of supported chiral Mo catalysts is, as should perhaps be expected, less active than the parent system (**4a**). The lower levels of activity exhibited by **89** may be due to inefficient diffusion of substrate molecules into the polymer. The supported catalyst is expected to be less susceptible to bimolecular decomposition of highly reactive methylidene intermediates [29]. Synthesis of more rigid polymer supports or those that represent lower Mo loading should further minimize bimolecular decomposition pathways and lead to a more robust class of catalysts.

3
Towards Chiral Ru-Based Olefin Metathesis Catalysts

In 2001, Grubbs and co-workers reported a class of Ru catalysts [(cf. **90**, Eq. (2)] [30] that bear a chiral monodentate *N*-heterocyclic carbene ligand [31]. The reactions illustrated in Eq. (2) include the highest ee reported in this study (13–90% ee). These important studies clearly indicate that asymmetric induction is dependent on the degree of olefin substitution (cf. Schemes 19 and 4 for comparison with the Mo-catalyzed reactions of the same substrates). As is the case with nearly catalytic enantioselective reactions [4], the identity of the optimal catalyst depends on the substrate; a number of chiral Ru catalysts were prepared and screened before **90** was identified as the most suitable. In addition, reactions were shown to be more selective in the presence of NaI.

R = H   **85→86**   39% *ee*, 22% conv

R =Me **18→19**   90% *ee*, 82% conv

$$(2)$$

Most recently, we have reported the synthesis and structure of a new chiral bidentate imidazolinium ligand and a derived chiral Ru-based carbene **91** (Scheme 21) [32]. The chiral Ru complex **91** is stereogenic at the metal center, can be prepared in >98% diastereoselectivity and purified by silica gel chromatography with undistilled solvents. This air-stable Ru complex efficiently catalyzes ring-closing and ring-opening metathesis and is recyclable. As the representative cases in Scheme 21 illustrate, the chiral complex is highly effective (0.5–10 mol % loading) in promoting enantioselective ring-opening/cross metathesis reactions (up to >98% ee). These enantioselective transformations can be effected in air (cf. Scheme 21), with unpurified solvent, and with substrates that would only polymerize with Mo-based catalysts.

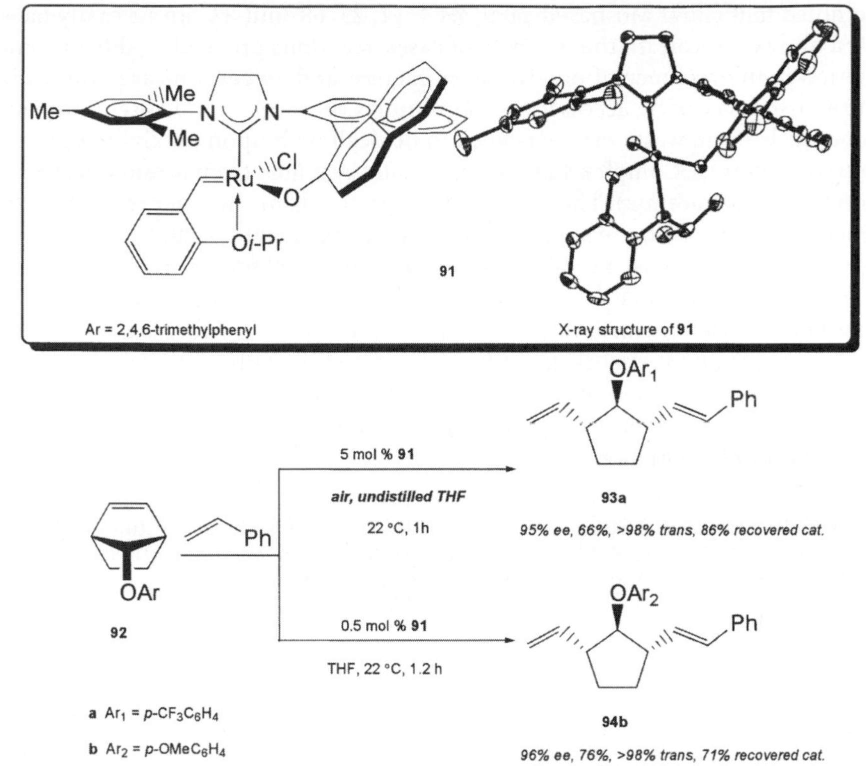

Me
Me
Me
N
N
Ru
Cl
O
Oi-Pr
91
Ar = 2,4,6-trimethylphenyl
X-ray structure of 91
OAr
Ph
92
5 mol % 91
air, undistilled THF
22 °C, 1h
OAr1
Ph
93a
95% ee, 66%, >98% trans, 86% recovered cat.
0.5 mol % 91
THF, 22 °C, 1.2 h
OAr2
Ph
94b
96% ee, 76%, >98% trans, 71% recovered cat.
a Ar1 = p-CF3C6H4
b Ar2 = p-OMeC6H4

Scheme 21. Air-stable chiral Ru-based catalyst for olefin metathesis can be used for highly effective and selective AROM/CM reactions

4 Conclusions and Outlook

The exciting results of the above investigations clearly indicate that the modular Mo-based construct initially reported for catalyst **1** can be exploited to generate a range of highly efficient and selective chiral catalysts for olefin metathesis. Both ARCM and AROM reactions can be promoted by these chiral catalysts to obtain optically enriched or pure products that are typically unavailable by other methods or can only be accessed by significantly longer routes (for example, see tertiary ethers formed in Scheme 15). Substantial variations in reactivity and selectivity arising from subtle changes in catalyst structures support the notion that synthetic generality is more likely if a range of catalysts are available [4].

The chiral Mo-based catalysts discussed herein are more sensitive to moisture and air than the related Ru-based catalysts [1]. However, these complexes, remain the most effective and general asymmetric metathesis catalysts and are significantly more robust than the original hexafluoro-Mo complex **1**. It should be noted that chiral Mo-based catalysts **4, 11, 25, 68,** and **82** can be easily handled on a large scale. In the majority of cases, reactions proceed readily to completion in the presence of only 1 mol % catalyst and, in certain cases, optically pure materials can be accessed within minutes or hours in the absence of solvents; little or no waste products need to be dealt with upon obtaining optically pure materials. Complex **4a** is commercially available from Strem, Inc. (both antipodes and racemic). The advent of the protocols for in situ preparation of chiral Mo catalyst **82,** the supported and recyclable complex **89,** and the emergence of chiral Ru-based catalysts **90** and **91** augur well for future development of practical chiral metathesis catalysis. The above attributes collectively render these new classes of chiral catalysts extremely attractive for future applications in efficient, catalytic, enantioselective, and environmentally conscious protocols in organic synthesis. Finally, the arrival of effective chiral Ru catalysts is likely the harbinger of upcoming highly effective and truly practical chiral Ru-based catalysts for olefin metathesis.

Acknowledgements: Financial support was provided from the National Institutes of Health (GM-59426 to A.H.H. and R.R.S.) and the National Science Foundation (CHE-9905806 and CHE-0213009 to A.H.H. and CHE-9700736 to R.R.S.). We are grateful to all our co-workers whose names appear in the references for invaluable intellectual and experimental contributions.

References

1. For select reviews on catalytic olefin metathesis, see: Grubbs RH, Chang S (1998) Tetrahedron 54:4413;Furstner A (2000) Angew Chem Int Ed 39:3012
2. For example, see: Xu Z, Johannes CW, Houri AF, La DS, Cogan DA, Hofilena GE, Hoveyda AH (1997) J Am Chem Soc 119:10302; Meng D, Su D-S, Balog A, Bertinato P, Sorensen EJ, Danishefsky SJ, Zheng Y-H, Chou T-C, He L, Horwitz SB (1997) J Am Chem Soc 119:2733; Nicolaou KC, Winssinger N, Pastor J, Ninkovic S, Sarabia F, He Y, Vourloumis D, Yang Z, Li T, Giannakakou P, Hamel E (1997) Nature 387:268; Johannes C.W., Visser MS, Weatherhead GS, Hoveyda AH (1998) J Am Chem Soc 120:8340; Delgado M, Martin JD (1999) J Org Chem 64:4798; Fuerstner A, Thiel OR (2000) J Org Chem 65:1738; Limanto J, Snapper ML (2000) J Am Chem Soc 122:8071–8072; Smith AB, Kozmin SA, Adams CM, Paone DV (2000) J Am Chem Soc 122:4984
3. Schrock RR, Murdzek JS, Bazan GC, Robbins J, DiMare M, O'Regan M (1990) J Am Chem Soc 112:3875; Bazan GC, Oskam JH, Cho HN, Park LY, Schrock RR (1991) J Am Chem Soc 113:6899
4. Kuntz KW, Snapper ML, Hoveyda AH (1999) Curr Opin Chem Biol 3:313; Shimizu KD, Snapper ML, Hoveyda AH (1999) In: Jacobsen EN, Pfaltz A, Yamamoto H (eds) Comprehensive asymmetric catalysis, vol 3. Springer, Berlin Heidelberg New York, pp 1389–1399; For a report regarding synthesis of various Mo complexes, see: Oskam, JH, Fox HH, Yap KB, McConville DH, O'Dell R, Lichtenstein BJ, Schrock RR (1993) J Organomet Chem 459:185
5. For a previous brief overview of this Mo-catalyzed asymmetric olefin metathesis, see: Hoveyda, AH, Schrock RR (2001) Chem Eur J 7:945
6. For early reports regarding the preparation of chiral Mo-based catalysts used for ROMP, see: McConville DH, Wolf JR, Schrock RR (1993) J Am Chem Soc 115:4413; Totland KM, Boyd TJ, Lavoie GG, Davis WM, Schrock RR (1996) Macromolecules 29:6114

7. Throughout this article, the identity of the recovered enantiomer shown is that which is obtained by the catalyst antipode illustrated. Moreover, transformations with BINOL-based complexes (e.g., **11**) were carried out with the opposite antipode of the catalyst versus that illustrated. Because (S)-biphen and (R)-BINOL complexes were employed in our studies, this adjustment has been made to facilitate comparison between biphen- and BINOL-based catalysts
8. Fujimura O., Grubbs RH (1996) J Am Chem Soc 118:2499; Fujimura O., Grubbs RH (1998) J Org Chem 63:824
9. For a review on metal-catalyzed kinetic resolution, see: Hoveyda AH, Didiuk MT (1998) Curr Org Chem 2:537
10. Alexander JB, La DS, Cefalo DR, Hoveyda AH, Schrock RR (1998) J Am Chem Soc 120:4041
11. The value for k_{rel} is calculated by the equation reported by Kagan: Kagan HB, Fiaud JC (1988) Top Stereochem 53:708
12. La DS, Alexander JB, Cefalo DR, Graf DD, Hoveyda AH, Schrock RR J Am Chem Soc 120:9720
13. Oskam, JH, Schrock RR (1993) J Am Chem Soc 115:11831; Fox HH, Schofield MH, Schrock RR (1994) Organometallics 13:2804
14. Schrock RR (1995) Polyhedron 14:3177; Wu YD, Peng ZH (1997) J Am Chem Soc 119:8043
15. Zhu SS, Cefalo DR, La DS, Jamieson JY, Davis WM, Hoveyda AH, Schrock RR (1999) J Am Chem Soc 121:8251
16. Weatherhead GS, Houser JH, Ford GJ, Jamieson JY, Schrock RR, Hoveyda AH (2000) Tetrahedron Lett 41:9553
17. Burke SD, Muller N, Beudry CM (1999) Org Lett 1:1827
18. Kiely AF, Jernelius JA, Schrock RR, Hoveyda AH (2002) J Am Chem Soc 124:2868
19. Dolman SJ, Sattely ES, Hoveyda AH, Schrock RR (2002) J Am Chem Soc 124:6991
20. For representative studies regarding non-asymmetric ROM reactions, see: Randall ML, Tallarico JA, Snapper ML (1995) J Am Chem Soc 117:9610; Zuercher WJ, Hashimoto M, Grubbs RH (1996) J Am Chem Soc 118:6634; Harrity JPA, Visser MS, Gleason JD, Hoveyda AH (1997) J Am Chem Soc 119:1488; Schneider, MF; Lucas N, Velder J, Blechert S (1997) Angew Chem Int Ed 36:257; Cuny FD, Cao J, Hauske JR (1997) Tetrahedron Lett 38:5237
21. Weatherhead GS, Ford JG, Alexanian EJ, Schrock RR, Hoveyda AH (2000) J Am Chem Soc 122:8071
22. Harrity JPA, La DS, Cefalo DR, Visser MS, Hoveyda AH (1998) J Am Chem Soc 120:2343
23. Schwab P, France MB, Ziller JW, Grubbs RH (1995) Angew Chem Int Ed 34:2039
24. Cefalo DR, Kiely AF, Wuchrer M, Jamieson JY, Schrock RR, Hoveyda AH (2001) J Am Chem Soc 123:3139
25. Teng X, Cefalo D, Schrock RR, Hoveyda AH (2002) J Am Chem Soc 122:10779
26. La DS, Ford GJ, Sattely ES, Bonitatbus PJ, Schrock RR, Hoveyda AH (1999) J Am Chem Soc 121:11603
27. Aeilts SL, Cefalo DR, Bonitatebus PJ Jr, Houser JH, Hoveyda AH, Schrock RR (2001) Angew Chem Int Ed 40:1452
28. Hultsizch KC, Jernelius JA, Hoveyda AH, Schrock RR Angew Chem Int Ed 41:589
29. Robbins J, Bazan GC, Murdzek JS, O'Regan MB, Schrock RR (1991) Organometallics 10:2902
30. Seiders TJ, Ward DW, Grubbs RH (2001) Org Lett 3:3225
31. Herrmann WA, Goossen LJ, Kocher C, Artus GRJ (1996) Angew Chem Int Ed 35:2806
32. VanVeldhuizen JJ, Garber, SB, Kingsbury JS, Hoveyda AH (2002) J Am Chem Soc 124:4954

Subject Index